T0201630

# ELECTROMAGNETIC OPTICS OF THIN-FILM COATINGS

Three experts in the field of thin-film optics present a detailed and self-contained theoretical study of planar multilayers and how they can be effectively exploited in both traditional and modern applications.

Starting with a discussion of the relevant electromagnetic optics, the fundamental optical properties of multilayers are introduced using an electromagnetic approach based on a direct solving of Maxwell's equations by Fourier transforms. This powerful approach is illustrated through the comprehensive description of two of the most important phenomena in multilayers, i.e., giant field enhancement in dielectric stacks and light scattering from thin-film optical filters. The same approach is extended to the description of the operation of planar microcavities and the balance of energy between radiated and trapped light.

This book will be valuable to researchers, engineers, and graduate students with interests in nanophotonics, optical telecommunications, observational astronomy, and gravitational wave detection.

CLAUDE AMRA is Director of Research at the Centre National de la Recherche Scientifique (CNRS), Institut Fresnel, which he co-founded and led. He was previously Director of Laboratoire d'Optique des Surfaces et des Couches Minces and then Deputy Scientific Director of the Institute of Systems and Engineering Sciences within CNRS.

MICHEL LEQUIME is an emeritus professor at Ecole Centrale Marseille, Institut Fresnel. He was Head of Division at BERTIN, led the Optical Thin Films Group at the Institut Fresnel, and served as Secretary to the French Optical Society.

MYRIAM ZERRAD is a research engineer at Aix-Marseille University, Institut Fresnel. She heads the Scattering Group and leads the Laser Interferometer Space Antenna (LISA) team at the Institut Fresnel.

# ELECTROMAGNETIC OPTICS OF THIN-FILM COATINGS

## Light Scattering, Giant Field Enhancement, and Planar Microcavities

**CLAUDE AMRA**

*Institut Fresnel, CNRS*

**MICHEL LEQUIME**

*Institut Fresnel, Centrale Marseille*

**MYRIAM ZERRAD**

*Institut Fresnel, Aix-Marseille University*

CAMBRIDGE
UNIVERSITY PRESS

# CAMBRIDGE
## UNIVERSITY PRESS

University Printing House, Cambridge CB2 8BS, United Kingdom

One Liberty Plaza, 20th Floor, New York, NY 10006, USA

477 Williamstown Road, Port Melbourne, VIC 3207, Australia

314–321, 3rd Floor, Plot 3, Splendor Forum, Jasola District Centre, New Delhi – 110025, India

79 Anson Road, #06–04/06, Singapore 079906

Cambridge University Press is part of the University of Cambridge.

It furthers the University's mission by disseminating knowledge in the pursuit of education, learning, and research at the highest international levels of excellence.

www.cambridge.org
Information on this title: www.cambridge.org/9781108488877
DOI: 10.1017/9781108772372

First published 2021

Printed in the United Kingdom by TJ Books Limited, Padstow Cornwall

*A catalogue record for this publication is available from the British Library.*

ISBN 978-1-108-48887-7 Hardback

# Contents

# About the Authors

**Claude Amra** joined the Centre National de la Recherche Scientifique (CNRS) in 1986 after defending a PhD on the scattering of light by roughness in interference filters; he then worked at the laboratory for optical surfaces and thin films (Laboratoire d'Optique des Surfaces et des Couches Minces [LOSCM], CNRS, École Nationale Supérieure de Physique in Marseille). His work expanded electromagnetic modeling to heterogeneous volumes and then on to luminescent microcavities for free space and modal optics. In parallel, he studied optical metrology and addressed the inverse problems inherent in characterizing disordered media, before becoming involved with the technology of optical thin films.

From 1991, he initiated and actively contributed to the introduction of new topics at LOSCM: photo-induced thermal and laser damage, trapped light, near-field microscopy, and multiscale roughness; ellipsometry of speckle patterns, achromatic thin-film light absorbers, mid infrared spectral splitters, dense wavelength multiplexing, deposition technologies using ion beam sputtering, etc.

In 1996, Claude Amra was appointed director of LOSCM, which he then reconstructed thematically. In 2000, he co-founded the Institut Fresnel, a new mixed research unit incorporating CNRS, Aix-Marseille University, and the École Centrale Marseille; this unit draws together sciences and technologies of optics, electromagnetism, and signal and image processing on the campus of Marseille Nord. Over the course of his two terms as director from 2000 to 2008, the Institut Fresnel emerged and cemented its position on the European stage with one dedicated building housing all the staff, and a second building funded to house a technology center (the Espace Photonique).

In 2009, he was named Deputy Scientific Director of the Institute of Systems and Engineering Sciences within CNRS (INSIS, Paris); since 2016, he has chaired Section 8 of the National Committee for Scientific Research, covering photonics, electronics, micro- and nanotechnologies and micro- and nanosystems, electrical energy, electromagnetism, and antennas. Prior to that, he chaired or participated in numerous committees from the national funding or research assessment agencies and the national universities' council, while also organizing numerous international conferences.

Within the context of the CONCEPT team that he has been running since 2012 at the Institut Fresnel, Claude Amra's research interests cover light scattering in interference filters, giant optical field enhancements, polarization and coherence in disordered media, terahertz probing and mimetic thermal radiation for applications in the space and military fields, energy and the environment, biomedical optics, and precision agronomy.

**Michel Lequime** received his master's in engineering from the Institut d'Optique Graduate School in 1974 and his PhD degree from Paris-Sud University in 1977. His doctoral research was devoted to the study of third-order nonlinear optical processes at the picosecond scale, especially in one-dimensional conjugated polymers (École Polytechnique, Laboratoire d'Optique Quantique).

In 1979, he joined BERTIN, a French contract research organization, to participate in the creation of an optical group at the company: he was initially appointed a study engineer, then head of department, and finally head of division. The division's activities dealt mainly with space optics (multicolor camera for the observation of Halley's Comet – European Space Agency [ESA] Giotto probe, optical beacon for the ESA Semi-Conductor Inter Satellite Link EXperiment [SILEX]) and optical fiber sensors (inline color measurement of refined fuels, pressure and temperature measurements for down-hole applications).

In 2000, he joined the École Centrale Marseille as Professor of Optics and Project Management and the Institut Fresnel as Senior Scientist. From 2002 to 2014, he was in charge of the Optical Thin Films Group at the Institut Fresnel. He is currently an emeritus professor at Centrale Marseille and continues to pursue his research activities at the Institut Fresnel on the use of metamaterials in optical thin-film coatings and on the development of instrumentation for the comprehensive characterization of the scattering properties of multilayer components.

Michel Lequime served as Secretary to the French Optical Society from 2009 to 2012 and is the author of more than 280 publications and communications in the field of optics.

**Myriam Zerrad** is an engineer who graduated in 2003 from the École Centrale Marseille and the École Nationale Supérieure de Physique in Marseille, obtaining a PhD in physics from Aix-Marseille University in 2007.

As a research engineer at Aix-Marseille University and a member of the CONCEPT team at the Institut Fresnel, she is a recognized expert on the phenomena of light scattering in optical interference coatings, involving modeling, instrumentation, and reconstruction. Since 2010, she has widened her area of expertise to include the optical probing of disordered media using statistical analysis of speckle patterns and has moved on to the synthesis of resonant planar components and the numerical modeling of thermal metamaterials.

She now heads the Scattering Group within the Institut Fresnel, with a dozen people working on her different areas of expertise. As part of her research activities, she developed the DIFFUSIF instrumentation platform optimized for an accurate metrology of scattered light and/or using scattered light in metrological applications. Today, this platform is a reference for various space agencies and for workers in the field of space optics, both academic and industrial.

In the context of developing the new generation of gravitational wave detector (Laser Interferometer Space Antenna [LISA]), Myriam Zerrad is a member of the international consortium and leads the LISA team at the Institut Fresnel; the institute has overall responsibility for the metrology and modeling of light scattered by coatings at the component level. She has authored more than 300 publications and communications in the fields of optics and photonics.

# Preface

David Lynch's *Dune* starts with the striking statement that "*A beginning is a very delicate time,*" and this introduction is no different. However, this *fragility* clearly must not present an obstacle to any rational and concise answer to the two questions that inevitably arise from the publication of a new piece of work, namely *Why?* and *How?*

Let us begin by explaining the *Why?*

Interference optical filters represent a rather special scientific domain – in more ways than one. To start with, there are as many scientists and engineers involved in the industry as there are in the universities (indeed, the commercial market is highly lucrative, with the global optical coatings market projected to exceed USD 12 billion by 2023 at a compound annual growth rate of 7%[1]), and for decades these two communities have been accustomed to collaborating freely and with the greatest mutual respect. This is also a case in which perfection is required, and the impossible a necessity: we must not forget that optical filtering involves finding the structure of a stack consisting of several hundred layers at nanometer thicknesses that, once created, gives rise to myriad elementary waves whose interference effects will shape with unequaled precision the spectrum of light transmitted and/or reflected by the planar component thus created. Finally, this specialist area does not, in our view, receive the academic recognition it deserves, not least for the complexity of approach required to control, via the solutions to a multitude of inverse problems, the phenomena of spatial and spectral dispersion that are consequent to solutions of Maxwell's equations in a **nonperiodic stratified medium**.

Therefore, a book that might shed some light on these profound concepts and this elegance of treatment seemed to us to be highly desirable.

---

[1] Kenneth Research, *Global Optical Coating Market Research Report* (2017).

Clearly, much of the work dedicated to this topic is dominated by Angus Macleod's book *Thin Film Optical Filters*,[2] already in its fifth edition, the first being published in 1986. This book, intended for fabricators and users, aims to give as complete a description as possible of thin-film optical coatings (design, fabrication, characterization, and applications). Our own purpose is very different, since the book you have before you is *purely theoretical*; we do not mean by this that subjects resulting in applications are not considered, but that the book is restricted to providing an explanatory mathematical description. Here *mathematical* does not mean divorced from physical meaning; we shall see that one of the fundamental notions on which this book is based is that of **complex admittance**, an idea that generalizes to stationary waves the concept of effective refractive index, a concept usually associated with propagating waves. Over the course of the chapters of this book, the reader will gradually discover that the matrix formalism often prominent in the description of stacks of optical thin films is barely touched on, being replaced (much more usefully, in our opinion) by that of admittance, since this is indeed imbued with a physical meaning on which the designer's intuition can draw.

Reading these last sentences, one could be forgiven for thinking that we have already unconsciously started on the *How?* We are far from it, however, since the illumination of this idea of admittance and the engineering with which it is associated clearly constitutes one of the aims of this book, and to finally convince the reader of this, we refer him or her to the chapter on the scattering of light by multilayer stacks (for example), or to the sections dealing with microcavities and giant enhancements of the electromagnetic field.

In publishing this book, another of our objectives was to show how the massive use of the **Fourier transformation** and the concomitant panoply of properties enabled the uniform solution of Maxwell's equations in a medium invariant in translation in a plane. All the elementary waves considered are therefore the result of a double **spectral decomposition**, first over time and then over the two dimensions of this plane of invariance. We leave it to the reader to discover, in particular in the chapter on frequential wave packets, the consequences of this temporal Fourier transformation for describing interference phenomena in the harmonic regime, a regime where time is excluded and in which space dominates.

Now for the *How?*

This book is effectively divided into three parts: first there are five intro-

---

[2] CRC Press (2018).

ductory chapters dedicated to the sometimes quite detailed presentation of generic ideas related to **Maxwell's equations**. The use of both frequential and spatial wave packets, drawing on the formalism of distributions and their Fourier transforms, provides an anchor for a uniquely macroscopic description of all the classical concepts of electromagnetism (polarization of matter, inertia, dispersion and absorption, causality, spatial and temporal coherence linked with detection, resolution in time and space, polarization of light, etc.). These first five chapters are treated with the utmost care, since they provide the key to unambiguously tackling the detailed notions in the pivotal chapter that follows them, being entirely dedicated to a concise description of **optical interference filters**. Finally, we get to three chapters that are more thematic and that delve more deeply into the use of all the ideas tackled in the previous parts via three a priori unconnected questions (reading these chapters will demonstrate that these notions reveal a certain amount of overlap); these chapters deal first with the use in **total reflection** of multilayer stacks and the giant electromagnetic field enhancements that might result (Chapter 7), then with **light scattering** by stacks containing slightly rough interfaces (Chapter 8), and finally **planar multilayer microcavities** (Chapter 9).

Chapter 7 covers the often poorly understood notions of total reflection and evanescent waves and highlights the presence of absorption lines in the spectrum reflected by a planar multilayer component in this particular operational regime. The chapter also identifies the connection between these resonances in absorption and the guided modes of the multilayer structure considered in isolation. In this total reflection regime, in which admittances are purely imaginary, it is possible to make the substrate admittance zero by inserting a thin layer; for a long time, we have hesitated to call this a *magic* layer, though the effects of its presence have huge consequences for the local intensity of the electromagnetic field. Exploiting these field enhancements finds numerous applications in the domains of sensors and microsources.

The ideas set out in detail in Chapter 8 show how an extremely reliable prediction of the spatial and spectral distribution of the light scattered by a stack of thin layers may be obtained, however complex it may be. Experience shows that highly intense and very narrow angular or spectral lobes may appear in the scattering distributions of these complex filters (such as those used in Earth observation satellites), and it is therefore necessary to understand their precise characteristics to avoid the sometimes serious problems of ghosting between several detection spectral bands. This time in the monochromatic regime, however, these same notions can also help in assessing the restrictions imposed by the problems of scattered light interference on the ultimate performance of interferometers designed to detect gravitational waves.

Chapter 9 is dedicated to studying a configuration somewhat different from those in the other chapters, one in which the source of light, here an emissive surface, is located *within* the multilayer stack. The electromagnetic approach we develop shows that the emission into free space of the micro-cavity thus created is governed by satisfying two resonance conditions, one being related to the structure of the multilayer stack being considered, the other to the position of the emissive surface within the stack. Furthermore, this same approach allows the energy balance of this microcavity to be set out in detail, making it possible to make a comparative assessment of the amount of light emitted into free space and the light trapped in the component, either by total reflection within the substrate or by coupling in the guided modes associated with the multilayer structure.

Is the view that we propose in this book exhaustive? Clearly not, but that was never our intention even though, during the slow gestation of this book, we sometimes envisaged the addition of other chapters, especially in relation to metamaterials or the phase properties of stratified media. But in the next edition, who knows?

Finally, the reader may quite legitimately be surprised by the absence of bibliographic references within the body of this book, although it does end with a list of selected books for the reader to pursue his or her further interests. This was a deliberate choice on our part, as our aim was to emphasize the internal coherence of the argument at the expense of its context. Note that this rule applies equally to our own publications, particularly to those relating to the approach detailed in this book. Obviously, such an absence of references should not be seen as a lack of willingness on our part to include them.

In conclusion, we sincerely hope this book provides as much interest to you, the reader, as the satisfaction we derived from writing it.

# Acknowledgments

In particular, we would like to thank

- Chris Collister, for the great care he has taken in translating this entire book, originally written in French
- Thomas McGuinn from the translation agency Mondo Agit UK for the quality of our interaction and his swift response to our inquiries;
- Nicolas Bélivier from the company 6NK, for creating Figures 1.1, 1.3, 3.3, 5.1, 6.2, 6.3, 6.7, 6.8, and 6.10
- Catherine Grèzes-Besset for her meticulous proofreading of the final text

We would also like to thank the Institut Fresnel, Aix-Marseille University, the Centre National de la Recherche Scientifique, and the École Centrale Marseille for the quality of the day-to-day support they lent to our work.

# 1

# Fundamentals of Electromagnetic Optics

## 1.1 Introduction

Electromagnetic optics is essential when the dimensions of illuminated objects have the same order of magnitude as the wavelength of the illuminating source. In light of recent rapid developments in micro- and nanotechnologies, the burgeoning of modern photonics is not surprising, relying as it does on nanocomponents and microsystems. These great advances have allowed many concepts over the course of the past two decades to be revisited, including photonic crystals, microstructured fibers, the flat lens, and negative refractive indices; diffractive optics, luminescent microcavities, metamaterials, and invisibility cloaks; super-resolution and more generally imagery in diffusive media; and new microscopy techniques, plasmonics, nano-antennas, metasurfaces, slow light, etc.

All these concepts are based on the formalism (in the classical sense) of electromagnetic optics such as the well-known phenomena of reflection, transmission, absorption, diffraction, scattering, interference filtering, frequency generation, and birefringence. Furthermore, the marriage of new technologies and computing power has led to the synthesis of 1D, 2D, and 3D components capable of confining electromagnetic energy in volumes of the order of a fraction of a wavelength, or of drastically enhancing the optical field locally so as to augment the interaction between light and matter.

Given this myriad of concepts and applications, it seemed important that we should address the basic concepts of electromagnetic optics. Any physical theory is based on a familiarity with certain hypotheses, even before getting to grips with the mathematical formalism that is associated with or extends them. Indeed, this familiarity is very often at the heart of innovation or of questioning the validity of received wisdom. Finally, a good grasp of the fundamentals, making judicious use of any analogies, is a great help in understanding a discipline.

This first chapter puts the accent on the assumptions current within the macroscopic formalism of (classical) electromagnetic optics. As is often the case, experimentation generates hypotheses and validates approximations; this chapter does not claim to give an exact description of phenomena, but the authors have attempted to be as objective as possible.

## 1.2 Sources and Fields

### 1.2.1 Sources of Light

The existence of a light wave implicitly presupposes that, somewhere in space, there is a source of this light. This source may be *primary* (or *direct*), such as the sun, for example, or it may be *secondary* (or *indirect*), such as the scattering of this same light from the sun by the terrestrial atmosphere.

In all cases, for each of the atoms that constitute the primary source, emission of light is related to the radiative transition of an electron from a higher to a lower excitation level, such as the fundamental level, for example. Such an approach, invoking a description of phenomena at the atomic level, is said to be microscopic; by contrast, an approach in which only the ensemble behavior is of interest is said to be macroscopic. Electromagnetic optics here is concerned with this second approach.

Let us refer to the space triplet $(x, y, z)$ as $\vec{\rho}$, the coordinate doublet $(x, y)$ as $\vec{r}$, and time as $t$. Sources of light will be described by scalar charge densities $q(\vec{\rho}, t)$ and by vector current densities $\vec{J}(\vec{\rho}, t)$, both being space and time dependent.

### 1.2.2 Fields Produced by These Sources

If, within a region of space $\Omega$ and of boundary $S$, there is a multiplicity of charges and currents described by the spatiotemporal densities $q(\vec{\rho}, t)$ and $\vec{J}(\vec{\rho}, t)$, then this region $\Omega$ is the source of emission of light (more precisely, of electromagnetic waves), described by the four vector fields defined over $\mathbb{R}^3$, namely:

- An electric field $\vec{E}(\vec{\rho}, t)$
- A magnetic field $\vec{B}(\vec{\rho}, t)$
- An electric induction $\vec{D}(\vec{\rho}, t)$, also known as *electric displacement*
- A magnetic induction $\vec{H}(\vec{\rho}, t)$, sometimes referred to as *magnetic displacement* or *magnetic excitation*

These four vector fields therefore correspond to twelve unknowns and are mutually related through **Maxwell's equations**:

$$
\begin{cases}
\mathbf{curl}\,\vec{E} = -\dfrac{\partial \vec{B}}{\partial t} \\[2mm]
\mathbf{curl}\,\vec{H} = \vec{J} + \dfrac{\partial \vec{D}}{\partial t} \\[2mm]
\mathrm{div}\,\vec{D} = q \\[2mm]
\mathrm{div}\,\vec{B} = 0
\end{cases}
\tag{1.1}
$$

Maxwell's equations involve eight relationships (two vector and two scalar relations) in twelve unknowns. The system seems therefore underdetermined, but we shall see later how to handle the incomplete nature of this determination. We may remark at this point that current $\vec{J}$ and charge $q$ are not independent, since we can write:

$$
\mathrm{div}(\mathbf{curl}\,\vec{H}) = 0 = \mathrm{div}\left(\vec{J} + \frac{\partial \vec{D}}{\partial t}\right) \quad \Rightarrow \quad \mathrm{div}\vec{J} + \frac{\partial q}{\partial t} = 0.
\tag{1.2}
$$

This equation simply expresses conservation of charge. If we integrate this equation over a volume $\Omega$, we obtain

$$
\iiint_{\Omega}\left(\mathrm{div}\vec{J} + \frac{\partial q}{\partial t}\right) dv = 0,
\tag{1.3}
$$

or, equivalently, using the Green–Ostrogradsky theorem:

$$
\iint_{S} \vec{J}.\vec{n}\,dS + \frac{\partial Q}{\partial t} = 0,
\tag{1.4}
$$

where $S$ is the surface enclosing the volume $\Omega$ and $Q$ is the total value of the electric charge at some instant $t$. Hence the charge that escapes the region $\Omega$ is equal to the flux of the current vector crossing surface $S$.

## 1.3 Electrostatics in Vacuum

We propose in this section to show how the laws of electrostatics can be revisited using Maxwell's equations.

### 1.3.1 Notion of Potential

In electrostatics, charge and current are not dependent on time. The same is true of the vector fields, so Maxwell's equations take the following simplified form:

$$\begin{cases} \mathbf{curl}\ \vec{E} = \vec{0} \\ \mathbf{curl}\ \vec{H} = \vec{J}(\vec{\rho}) \\ \mathrm{div}\vec{D} = q(\vec{\rho}) \\ \mathrm{div}\vec{B} = 0 \end{cases} \qquad (1.5)$$

where conservation of charge means that $\mathrm{div}\vec{J} = 0$. We also note that the fields $\vec{E}$ and $\vec{H}$ are now decoupled, so the electrostatic and magnetostatic cases can be treated independently.

The curl of the electric field being identically 0 means that the field $\vec{E}$ derives from a scalar potential, such that we can always write

$$\mathbf{curl}\ \vec{E} = \vec{0} \quad \Rightarrow \quad \exists V,\ \text{such that}\ \vec{E} = -\mathbf{grad}\ V\,. \qquad (1.6)$$

The scalar variable $V$ is known as the *electrostatic potential*. Use of the $-$ sign is arbitrary and originates historically in the convention adopted for the direction of current flow with respect to the algebraic value of the potential difference. Note also that there is no uniqueness to this definition, since the value of the potential is determined to an additive constant.

Furthermore, in the case corresponding to in vacuum electrostatics, the electric induction $\vec{D}$ is proportional to the electric field $\vec{E}$, where the coefficient of proportionality, $\epsilon_v$, is known as the *permittivity of vacuum*:

$$\vec{D} = \epsilon_v \vec{E}\,. \qquad (1.7)$$

Hence, the third of Maxwell's equations can be written

$$\mathrm{div}\vec{D} = q = \epsilon_v\, \mathrm{div}\vec{E} = -\epsilon_v\, \mathrm{div}(\mathbf{grad}\ V) = -\epsilon_v\, \Delta V\,, \qquad (1.8)$$

leading to Poisson's equation, i.e.,

$$\Delta V + \frac{q}{\epsilon_v} = 0\,. \qquad (1.9)$$

Using classical methods, the general solution to this differential equation is somewhat lengthy and difficult. It is therefore desirable to make use of distributions and Green's function.

### 1.3.2 Solution by Green's Function

Consider the differential equation

$$\Delta V(\vec{\rho}) = -\frac{q(\vec{\rho})}{\epsilon_v}\,, \qquad (1.10)$$

and let $G(\vec{\rho})$ be the Green's function in the solution of the associated differential equation, where the second term has been replaced by a Dirac distribution:

$$\Delta G(\vec{\rho}) = \delta(\vec{\rho}).$$ (1.11)

If we are able to determine this Green's function $G(\vec{\rho})$, it will be straightforward to derive the general solution of the differential equation (1.10) by simply writing

$$V(\vec{\rho}) = -\left[G \star \frac{q}{\vec{\rho}\ \epsilon_v}\right](\vec{\rho}).$$ (1.12)

Indeed using the rules for differentiating a convolution, we get

$$\Delta V = -\Delta \left[G \star \frac{q}{\epsilon_v}\right] = -\Delta G \star \frac{q}{\epsilon_v} = -\delta \star \frac{q}{\epsilon_v} = -\frac{q}{\epsilon_v},$$ (1.13)

where we have implicitly used the fact that the Dirac distribution constitutes the neutral element of the convolution.

To solve the associated differential equation (1.11), we assume a sphere of radius $R$ centered at the origin, with $\Omega$ the volume defined by this sphere and $S$ its surface. Using the properties of the Dirac distribution we get

$$\iiint_\Omega \Delta G(\vec{\rho})\, dv = 1.$$ (1.14)

We now apply Green–Ostrogradsky theorem to this volume integral, not forgetting that the Laplacian of a scalar quantity is equal to the divergence of its gradient. Hence

$$\iiint_\Omega \Delta G(\vec{\rho})\, dv = \iiint_\Omega \text{div}[\textbf{grad } G(\vec{\rho})]\, dv = \iint_S \textbf{grad } G(\vec{\rho}).\vec{n}\, dS.$$ (1.15)

The problem corresponding to this Green's function has spherical symmetry, such that the scalar product of the gradient of the Green's function $G$ with the normal to the surface of the sphere is constant over the sphere and is given by

$$\textbf{grad } G(\vec{\rho}).\vec{n} = \left.\frac{\partial G}{\partial \rho}\right|_{\rho = R}.$$ (1.16)

Hence

$$\iint_S \textbf{grad } G(\vec{\rho}).\vec{n}\, dS = \left.\frac{\partial G}{\partial \rho}\right|_{\rho = R} \iint_S dS = 4\pi R^2 \left.\frac{\partial G}{\partial \rho}\right|_{\rho = R} = 1.$$ (1.17)

and thus

$$\frac{\partial G}{\partial \rho}\bigg|_{\rho=R} = \frac{1}{4\pi R^2} \; \forall R, \quad \text{i.e.,} \quad G(\vec{\rho}) = -\frac{1}{4\pi \rho}. \tag{1.18}$$

Finally, the analytical form of the potential $V(\vec{\rho})$ is obtained by using the equation (1.12), i.e.,

$$V(\vec{\rho}) = \left[\frac{1}{4\pi|\vec{\rho}|} \star \frac{q}{\epsilon_v}\right](\vec{\rho}) = \frac{1}{4\pi\epsilon_v} \int_{\vec{\rho}'} \frac{q(\vec{\rho}')}{|\vec{\rho} - \vec{\rho}'|} \, d\vec{\rho}', \tag{1.19}$$

where $d\vec{\rho}'$ denotes the volume element $dx'dy'dz'$, $\vec{\rho}'$ encompasses the region $\Omega$ containing the charges, and $S$ is the closed surface bounding the region, as shown in Figure 1.1.

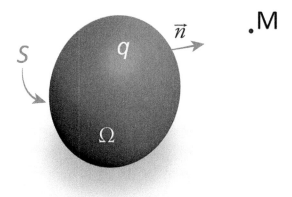

Figure 1.1 Electrostatics in vacuum.

Note that, by extension, the function $V(\vec{\rho})$ also allows an electrostatic potential to be defined within the charged region $\Omega$. Indeed, the volume integral in equation (1.19) is defined everywhere, even when the denominator goes to 0 (to convince yourself, it is sufficient to transform into spherical coordinates after having expanded the charge $q$ about $\vec{\rho}$).

We also show that on crossing the surface $S$ enclosing these charges, the variable $V$ remains continuous and differentiable with a continuous derivative. Consequently, the electric field $\vec{E}$ is also continuous across this surface $S$, such that we can write

$$\sigma[\vec{E}] = \vec{0}, \tag{1.20}$$

where $\sigma[\vec{E}]$ denotes the *jump*, i.e., the discontinuity in the field between the interior and the exterior, i.e.,

$$\sigma[\vec{E}] = \vec{E}_{\text{Exterior } \Omega} - \vec{E}_{\text{Interior } \Omega} . \tag{1.21}$$

However, it is important to bear in mind that this result is no longer true if the charges are surface or point charges (in contrast with volume charges), i.e., singular in the sense of distributions. We shall take these particular cases into account in the next section.

### 1.3.3 Singular Charges and Field Discontinuities

#### Surface Charges

Under certain particular situations (induced charges, fictitious charges, metallic materials, etc.) we will need to consider the case of *localized charges* on a surface $S$. This particular distribution of charges will be described by introducing the Dirac distribution $\delta_S(\vec{\rho})$, whose argument tends to 0 for all points $\vec{\rho}_S$ belonging to surface $S$. Hence we write

$$q_S(\vec{\rho}) = q_S \delta_S(\vec{\rho}) = q_S \delta(\vec{\rho} - \vec{\rho}_S) , \tag{1.22}$$

and the potential $V$ will again be calculated using equation (1.19), but taking account of the presence of these surface charges; this leads to

$$V(\vec{\rho}) = \frac{1}{4\pi\epsilon_v} \int_{\vec{\rho}'} \frac{q_S \delta(\vec{\rho}' - \vec{\rho}_S)}{|\vec{\rho} - \vec{\rho}'|} d\vec{\rho}' = \frac{1}{4\pi\epsilon_v} \int_S \frac{q_S}{R} dS , \tag{1.23}$$

where $R$ denotes the quantity $|\vec{\rho} - \vec{\rho}_S|$ in which $\vec{\rho}_S$ describes the surface $S$. It can be shown that, in this case,

$$\begin{cases} \vec{n} \wedge \sigma[\vec{E}] = \vec{0} \\ \vec{n}.\sigma[\vec{E}] = \dfrac{q_S}{\epsilon_v} \end{cases} , \tag{1.24}$$

which means that the tangential component of the electric field remains continuous across the surface $S$, but that its normal component has a discontinuity directly related to the value of the surface charge $q_S$.

#### Point Charge

The notion of a point charge localized at a point $\vec{\rho}_0$ in space is frequently encountered. This is represented by the distribution

$$q(\vec{\rho}) = q_0 \, \delta(\vec{\rho} - \vec{\rho}_0) , \tag{1.25}$$

and hence, using the properties of the Dirac distribution,

$$V(\vec{\rho}) = \frac{q_0}{4\pi\epsilon_v} \int_{\vec{\rho}'} \frac{\delta(\vec{\rho}' - \vec{\rho}_0)}{|\vec{\rho} - \vec{\rho}'|} d\vec{\rho}' = \frac{1}{4\pi\epsilon_v} \frac{q_0}{|\vec{\rho} - \vec{\rho}_0|} = \frac{1}{4\pi\epsilon_v} \frac{q_0}{R}, \qquad (1.26)$$

where $R$ denotes $|\vec{\rho} - \vec{\rho}_0|$. Using this formalism, the electrostatic potential is liable to diverge in the case of a point charge (though this was not true in the case of a volume charge).

### 1.3.4 *Potential Created by an Electrostatic Dipole*

We will also need to invoke the notion of a dipole to describe the effects of induced polarization in matter. To this end, consider two equal and opposite point charges $q$, separated by a distance $l$, as shown in Figure 1.2. This ensemble constitutes an electrostatic dipole characterized by a dipole moment $\vec{M} = q\vec{l}$, where $\vec{l}$ is a vector of modulus $l$ oriented in the direction of negative to positive charge.

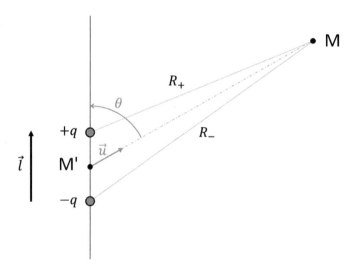

Figure 1.2 Electrostatic dipole.

The potential created at a point M results from the combined effect of these two charges and is expressed thus

$$V(\mathrm{M}) = \frac{+q}{4\pi\epsilon_v R_+} + \frac{-q}{4\pi\epsilon_v R_-} = \frac{q}{4\pi\epsilon_v} \left( \frac{1}{R_+} - \frac{1}{R_-} \right), \qquad (1.27)$$

where $R_+$ (or $R_-$ respectively) denotes the distance between point M and charge $+q$ (or $-q$ respectively). If the distance between this dipole and point

M is large compared with the distance $l$ separating the two charges, we can, to a first approximation, write

$$V(M) = \frac{q}{4\pi\epsilon_v}\left(\frac{R_- - R_+}{R_+ R_-}\right) \simeq \frac{q}{4\pi\epsilon_v}\cdot\frac{\Delta R}{R^2} = \frac{q}{4\pi\epsilon_v}\cdot\frac{l\cos\theta}{R^2}. \tag{1.28}$$

Let $\vec{u}$ be the unit vector along the half-line defined by the center M′ of the dipole and the point M. Under these conditions we can replace $ql\cos\theta$ by $\vec{\mathcal{M}}.\vec{u}$ in the expression for the potential, leading to

$$V(M) = \frac{1}{4\pi\epsilon_v}\cdot\frac{\vec{\mathcal{M}}.\vec{u}}{R^2}. \tag{1.29}$$

Distance $R$ can be expressed as $R = \sqrt{(x-x')^2 + (y-y')^2 + (z-z')^2}$, where $(x,y,z)$ denotes the coordinates of point M and $(x',y',z')$ denotes those of point M′.

Consider the quantity **grad** $R$. To calculate this correctly, we need to define over which triplet of coordinates the differential operator must act. We consider that this operator acts on the triplet $(x',y',z')$ and we shall initially consider differentiation with respect to $x'$:

$$\frac{\partial}{\partial x'}R = \frac{\partial}{\partial x'}\sqrt{(x-x')^2 + (y-y')^2 + (z-z')^2} = -\frac{x-x'}{R} = \frac{x'-x}{R}, \tag{1.30}$$

and hence

$$\mathbf{grad}_{M'}\,R = \frac{\overrightarrow{MM'}}{R} = -\vec{u}. \tag{1.31}$$

Consequently we can write

$$\mathbf{grad}_{M'}\left(\frac{1}{R}\right) = -\frac{1}{R^2}\mathbf{grad}_{M'}\,R = \frac{\vec{u}}{R^2}. \tag{1.32}$$

Using the result we have just established, we can put equation (1.29) in the following equivalent form:

$$V(M',M) = \frac{1}{4\pi\epsilon_v}\,\vec{\mathcal{M}}.\mathbf{grad}_{M'}\left(\frac{1}{R}\right), \tag{1.33}$$

where $V(M',M)$ is the potential at point M created by the dipole at M′.

## 1.4 Electrostatics in Matter

Having established an expression for the potential in vacuo, we now turn our attention to perturbation induced on this potential by the presence of matter.

Let us again consider a region $\Omega$ of space, bounded by a surface $S$ and occupied by a charge density $q(\vec{\rho}')$. At some point M in space, this ensemble of charges creates a potential $V_{\text{vacuo}}(\text{M}) = V_0(q, \text{M})$ that we can now calculate using the results obtained in Section 1.3.

### 1.4.1 Creation of Induced Dipoles

We now introduce into this space a volume $\mathcal{V}$ of dielectric material (electrically neutral) of boundary $\Sigma$, as shown in Figure 1.3.

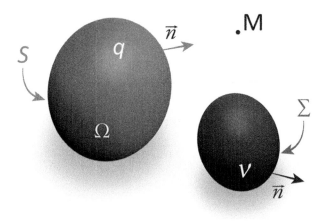

Figure 1.3 Electrostatics in matter.

This material is subject to an electric field $\vec{E} = -\mathbf{grad}\, V_0$ generated by the multiplicity of charges $q(\vec{\rho}')$ within the region $\Omega$. As a result the electric field induces a local polarization of the material in the volume $\mathcal{V}$, a polarization that can be thought of as a separation of the centroids of positive and negative charges under the action of the field (in the absence of a field, the dielectric material is electrically neutral and the centroids merge together).

### 1.4.2 Dipole Approximation

Thus, created within the volume $\mathcal{V}$ of material is an assembly of dipoles induced by the presence of the charged region $\Omega$, characterized using a dipole moment density $\vec{P}$, such that $d\vec{\mathcal{M}} = \vec{P}\, dv$.

Given this dipole approximation, we can therefore write the potential at any point M as the sum of the potentials created by the volumes $\Omega$ and $\mathcal{V}$, i.e.,

$$V_{\text{total}}(M) = V_{\text{vacuo}}(M) + V_{\text{matter}}(M) = V_0(q, M) + V_1(\vec{P}, M), \qquad (1.34)$$

with

$$V_0(\vec{\rho}) = \frac{1}{4\pi\epsilon_v} \int_\Omega \frac{q(\vec{\rho}')}{|\vec{\rho} - \vec{\rho}'|} \, d\vec{\rho}', \qquad (1.35)$$

and, in accordance with equation (1.33):

$$V_1(\vec{\rho}) = \frac{1}{4\pi\epsilon_v} \int_{\mathcal{V}} \vec{P}(\vec{\rho}').\mathbf{grad}_{\rho'} \left( \frac{1}{|\vec{\rho} - \vec{\rho}'|} \right) d\vec{\rho}'. \qquad (1.36)$$

### *1.4.3 Fictitious Charges*

There is one final stage of calculation before tackling the matter/vacuum analogy for the potential equations. The first step is to calculate the divergence of the ratio $\vec{P}/R$, where $R$ represents, as before, the quantity $|\vec{\rho} - \vec{\rho}'|$. We can write

$$\mathrm{div}\left(\frac{\vec{P}}{R}\right) = \frac{1}{R}\mathrm{div}\vec{P} + \vec{P}.\mathbf{grad}\left(\frac{1}{R}\right). \qquad (1.37)$$

Using this relationship, the potential $V_1(\vec{\rho})$ created by the electrical polarization $\vec{P}$ of the material can be put in the form

$$V_1(\vec{\rho}) = \frac{1}{4\pi\epsilon_v} \int_{\mathcal{V}} \left[ \mathrm{div}\left(\frac{\vec{P}}{R}\right) - \frac{1}{R}\mathrm{div}\vec{P} \right] d\vec{\rho}'. \qquad (1.38)$$

Let us consider the second term of the integral. We can write

$$V_{1,\mathcal{V}}(\vec{\rho}) = -\frac{1}{4\pi\epsilon_v} \int_{\mathcal{V}} \frac{\mathrm{div}\vec{P}}{|\vec{\rho} - \vec{\rho}'|} \, d\vec{\rho}'. \qquad (1.39)$$

By analogy with formula (1.35), it is as if a volume charge density $q_{\mathcal{V}}$ had been created within the volume $\mathcal{V}$ of matter, defined by

$$q_{\mathcal{V}} = -\mathrm{div}\vec{P}. \qquad (1.40)$$

The first term in equation (1.38) can also be transformed, this time by applying the Green–Ostrogradsky theorem, which says that the flux of a vector passing through a closed surface is equal to the integral of the divergence of the vector over the volume enclosed by that surface. Consequently,

$$\frac{1}{4\pi\epsilon_v}\int_{\mathcal{V}}\operatorname{div}\left(\frac{\vec{P}}{R}\right)d\rho' = \frac{1}{4\pi\epsilon_v}\int_{\Sigma}\frac{\vec{P}}{R}.\vec{n}\,d\Sigma, \tag{1.41}$$

where $\vec{n}$ is the normal to the surface $\Sigma$, directed toward the exterior of the region $\mathcal{V}$. By analogy with formula (1.23), it is as if the surface $\Sigma$ were carrying a surface charge density defined by

$$q_\Sigma = \vec{n}.\vec{P}. \tag{1.42}$$

Hence the introduction of matter into a potential causes two fictitious charge distributions to appear, volume and surface respectively. The existence of these additional charge densities in the presence of matter leads to a reformulation of the third of Maxwell's equations as follows:

$$\operatorname{div}(\epsilon_v\vec{E}) = q + q_\mathcal{V} = q - \operatorname{div}\vec{P}, \tag{1.43}$$

or, introducing the vector $\vec{D}$ defined by $\vec{D} = \epsilon_v\vec{E} + \vec{P}$,

$$\operatorname{div}\vec{D} = \operatorname{div}(\epsilon_v\vec{E} + \vec{P}) = q. \tag{1.44}$$

This relationship shows that, when the vector $\vec{D}$ is used, Maxwell's equations take the same form in both matter and vacuum, i.e., the divergence of the vector $\vec{D}$ is equal to the (noninduced) imposed charges.

### *1.4.4 Discontinuities*

We also need to verify this analogy at the boundary of region $\mathcal{V}$, i.e., over the surface $\Sigma$. In accordance with the two discontinuity relations introduced in Section 1.3.3 in the case of a surface charge distribution, the field discontinuity in the presence of matter must now be written to take account of the new induced charges:

$$\begin{cases} \vec{n} \wedge \sigma[\vec{E}] = \vec{0} \\ \vec{n}.\sigma[\vec{E}] = \dfrac{q_\Sigma}{\epsilon_v} \end{cases} \tag{1.45}$$

The first relationship, concerning the continuity of the tangential component of the electric field, remains unchanged. Conversely, for the normal component we need to take account of the presence of the surface charge $q_\Sigma$ induced in the material.

We may remark that, since no polarization is induced outside the volume $\mathcal{V}$, the jump in vector $\vec{P}$ may be written as

$$\sigma[\vec{P}] = \vec{0} - \vec{P} = -\vec{P}. \tag{1.46}$$

We immediately deduce that

$$\vec{n}.\sigma[\vec{D}] = \vec{n}.\sigma[\epsilon_v \vec{E} + \vec{P}] = q_\Sigma - \vec{n}.\vec{P} = 0. \tag{1.47}$$

Hence for the discontinuity in the normal component we find the same expression as that obtained in vacuo in the absence of any imposed surface charge, provided vector $\vec{D}$ is used in place of vector $\vec{E}$. In summary, we note that the respective tangential and normal components of fields $\vec{E}$ and $\vec{D}$ are continuous. We also keep in mind that the induced charges (in matter) no longer appear in Maxwell's equations, since they are taken into account by vector $\vec{D}$.

We have verified that Maxwell's equations are identical both in vacuum and in matter, for the field equations as well as the field discontinuities. We shall see in Chapter 2 how these properties (equations and discontinuities) can be condensed into a single expression using distributions.

For completeness, note that we have adopted a description of the polarization of matter that ignores the possibility of more complex charge distributions such as multipoles. It is for this reason that this approach is commonly referred to as the *dipole approximation*.

## 1.5 Linear Optics in an Isotropic Medium

We showed in the previous section that the introduction of matter leads to replacing the characteristic vector $\epsilon_v \vec{E}$ by the vector $\vec{D} = \epsilon_v \vec{E} + \vec{P}$, while continuing to use the same formulation for the Maxwell's equations.

At this stage, however, the number of unknowns is not reduced ($\vec{D}$ is a function of $\vec{P}$ and vice versa). We can go further, however, by expressing the polarization of matter in the form of an expansion of the applied field $\vec{E}$.

Taking some arbitrary component of this polarization vector we can write

$$P_i(\vec{E}) = P_i(\vec{0}) + \sum_{j=1}^{3} \frac{\partial P_i}{\partial E_j} E_j + \frac{1}{2} \sum_{j,k} \frac{\partial^2 P_i}{\partial E_j \partial E_k} E_j E_k + \cdots. \tag{1.48}$$

We first notice that the electric polarization in the absence of an exciting field (apart from exceptions such as ferroelectric materials, for example) is 0, i.e., $P_i(\vec{0}) = 0$.

Field component terms that have a quadratic dependence, as well as higher order terms, are responsible for **nonlinear** phenomena in optics such as frequency doubling, optical rectification, parametric amplification, or self-phase modulation. These will not be covered here and we shall remain within the domain of so-called of linear optics; in this domain processes are sometimes referred to as *elastic*, meaning that a material receiving a wave of arbitrary wavelength will not alter that wavelength despite such phenomena as reflection, transmission, absorption, diffraction, scattering, etc.; we shall study these later.

If we restrict ourselves to those terms in the expansion for which the components of the electric field have a linear dependence, then equation (1.48) can be cast in the following simplified form:

$$P_i(\vec{E}) \simeq \sum_j \frac{\partial P_i}{\partial E_j} E_j = \epsilon_v \sum_j \chi_{ij} E_j \quad \text{with} \quad \epsilon_v \chi_{ij} = \frac{\partial P_i}{\partial E_j}, \tag{1.49}$$

where $\chi_{ij}$ denotes the components of the **susceptibility** tensor.

In the general case, this leads to a matrix relation between induced polarization and field, namely

$$\vec{P} = \epsilon_v [\chi] \vec{E}, \tag{1.50}$$

and enables the properties of so-called anisotropic materials to be taken into account.

If the medium under consideration is isotropic, i.e., its microstructure does not depend on the direction of observation, then for reasons of symmetry no polarization can be generated that is not collinear with the excitation. As a result

$$\chi_{ij} = \chi \delta_{ij}, \tag{1.51}$$

where $\delta_{ij}$ is the Kronecker delta symbol ($\delta_{ij} = 1$, if $i = j$ and $\delta_{ij} = 0$, if $i \neq j$).

Under these conditions, the susceptibility matrix reduces to the identity matrix such that a single scalar $\chi_E$ links the vectors $\vec{P}$ and $\vec{E}$ thus:

$$\vec{P} = \epsilon_v \chi_E \vec{E}. \tag{1.52}$$

We then obtain for vector $\vec{D}$:

$$\vec{D} = \epsilon_v \vec{E} + \vec{P} = \epsilon_v \vec{E} + \epsilon_v \chi_E \vec{E} = \epsilon_v (1 + \chi_E) \vec{E} = \epsilon_v \epsilon_r \vec{E} = \epsilon \vec{E}, \tag{1.53}$$

where $\epsilon$ is the electric permittivity of the medium and $\epsilon_r$ its relative permittivity ($\epsilon = \epsilon_v \epsilon_r$).

Keep in mind that this expression has been established for linear isotropic media with no permanent polarization. Moreover, the number of unknowns has been reduced, since the three components of the vector $\vec{D}$ are given as a function of the field $\vec{E}$. Such reduction results from the introduction of the material physical parameter (the permittivity $\epsilon$).

## 1.6 Magnetostatics

Magnetostatics can be tackled by adopting a procedure similar to that just followed for the electrostatic case. We shall be content with a broad-brush description in this section.

### 1.6.1 Magnetostatics in Vacuum

We again start from the simplified form of Maxwell's equations when charge and current have no time dependence, i.e.,

$$\begin{cases} \mathbf{curl}\ \vec{H} = \vec{J}(\vec{\rho}) \\ \mathrm{div}\vec{B} = 0 \end{cases} \tag{1.54}$$

Since the divergence of vector $\vec{B}$ is 0, this means that the vector can be represented as the curl of a vector $\vec{A}$, i.e.,

$$\vec{B} = \mathbf{curl}\ \vec{A}. \tag{1.55}$$

Vector $\vec{A}$ is known as the *magnetostatic vector potential* and, as with the electrostatic scalar potential, its definition is not unique. Indeed we note that it is defined to an arbitrary gradient:

$$\vec{A}' = \vec{A} + \mathbf{grad}\ \Phi \quad \Rightarrow \quad \mathbf{curl}\ \vec{A}' = \mathbf{curl}\ \vec{A}, \tag{1.56}$$

where $\Phi$ is some arbitrary function.

Furthermore, in the case corresponding to the magnetostatics of a vacuum, the magnetic field $\vec{B}$ is proportional to the magnetic induction $\vec{H}$, where the coefficient of proportionality is known as the *magnetic permeability of vacuum* $\mu_v$, i.e.,

$$\vec{B} = \mu_v \vec{H}. \tag{1.57}$$

Consequently, Maxwell's second equation can be written

$$\mathbf{curl}\ \vec{H} = \frac{1}{\mu_v}\mathbf{curl}\ \vec{B} = \frac{1}{\mu_v}\mathbf{curl}\ (\mathbf{curl}\ \vec{A})$$

$$= \frac{1}{\mu_v}[\mathbf{grad}\ (\mathrm{div}\vec{A}) - \Delta\vec{A}] = \vec{J}(\vec{\rho}). \tag{1.58}$$

As previously mentioned, the choice of vector $\vec{A}$ is not unique, and it is always possible to consider one that satisfies the condition $\operatorname{div}\vec{A} = 0$, a condition usually referred to as the *Coulomb gauge*. If this particular gauge is adopted, we get

$$\Delta\vec{A} + \mu_v \vec{J} = 0 . \tag{1.59}$$

Apart from the fact that it is now a vector equation, this differential equation is similar to the Poisson equation (1.9) established in the electrostatic case. Hence its general solution can be written

$$\vec{A}(\vec{\rho}) = \left[\frac{1}{4\pi|\vec{\rho}|} \star \mu_v \vec{J}\right](\vec{\rho}) = \frac{\mu_v}{4\pi}\int_{\vec{\rho'}} \frac{\vec{J}(\vec{\rho'})}{|\vec{\rho} - \vec{\rho'}|}\, d\vec{\rho'} . \tag{1.60}$$

### 1.6.2 Magnetostatics in Matter

It can be seen that electrostatics and magnetostatics in vacuum are strongly analogous. This analogy can be extended in the presence of matter, the field generated by the imposed current volume density this time generating a *magnetization* $\vec{P}_H$ of the matter (sometimes referred to as magnetic polarization, by analogy with the electrostatic case). Hence for an isotropic linear medium and for materials with no permanent magnetization (such as paramagnetic or diamagnetic materials):

$$\vec{B} = \mu_v(\vec{H} + \vec{P}_H) \qquad \text{with} \qquad \vec{P}_H = \chi_H \vec{H} , \tag{1.61}$$

where $\chi_H$ is the magnetic susceptibility; then

$$\vec{B} = \mu_v(1 + \chi_H)\vec{H} = \mu_v \mu_r \vec{H} = \mu\vec{H} , \tag{1.62}$$

where $\mu$ is the magnetic permeability of the medium and $\mu_r$ is the relative permeability ($\mu = \mu_v \mu_r$).

Note that the materials commonly used in optics (such as dielectric materials) rarely demonstrate induced magnetization, in contrast with metals and alloys.

It is also important to point out at this stage that we have treated the electrostatic and magnetostatic cases separately, which led us to consider an electric polarization (or magnetic respectively) induced uniquely by an electric charge (or by an electric current respectively). We could nevertheless envisage some coupling between these induced *polarizations* in the case of a spatiotemporal regime (hence introducing chirality phenomena in optics).

Finally, before tackling electromagnetism in the next section, we note that the number of unknowns associated with Maxwell's equations in isotropic linear matter is just 6 (namely the components of the vectors $\vec{E}$ and $\vec{H}$), ultimately leaving 6 unknowns in 8 equations, which are clearly linked. The parameters $\epsilon$ and $\mu$, characteristic of the microstructure of matter, are not unknowns; they take account of the induced polarizations in matter. As a consequence, these induced charges and currents no longer appear in Maxwell's equations, providing the analogy between vacuum and matter formalism (keeping the imposed excitation sources).

Note that these are the assumptions of linearity and isotropy that allowed the problem to be simplified (we will not discuss the case of anisotropic or nonlinear materials). In conclusion, the fields $\vec{E}$ and $\vec{H}$ are obtained from scalar (electrostatics) and vector (magnetostatics) potentials, these being deduced from sources $(q, \vec{J})$, via equations (1.19) and (1.60).

## 1.7 Electromagnetism

Recall the equations obtained in the static regime:

$$\vec{D} = \epsilon_v \vec{E} + \vec{P}_E \qquad \vec{B} = \mu_v(\vec{H} + \vec{P}_H), \tag{1.63}$$

where electric polarization $\vec{P}_E$ and magnetic polarization $\vec{P}_H$ are expressed thus:

$$\vec{P}_E = \epsilon_v \chi_E \vec{E} \qquad \vec{P}_H = \chi_H \vec{H}. \tag{1.64}$$

### 1.7.1 Inertia of Matter

Let us now suppose that charge and current depend on time, something that differentiates electromagnetism from electrostatics and magnetostatics. We shall assume that the local properties previously defined remain true, except that we now need to take the temporal response (also referred to as *inertia*) of the medium into account, i.e., the time required by the material to *perceive* the temporal variations of the sources.

Put another way, the electric and magnetic polarizations induced in the material do not react **instantaneously** to the action of the exciting field emitted by the sources, such that it is **not possible** to write at each instant $t$:

$$\vec{P}_E(\vec{\rho}, t) = \epsilon_v \chi_E \vec{E}(\vec{\rho}, t) \qquad \vec{P}_H(\vec{\rho}, t) = \chi_H \vec{H}(\vec{\rho}, t). \tag{1.65}$$

This classical observation is true for all systems subject to some external action. In reality, the response of the system depends on the excitation throughout the duration it is applied, expressible through the following integral equations:

$$\vec{P}_E(\vec{\rho}, t) = \epsilon_v \int_\tau \chi_E(\tau) \vec{E}(\vec{\rho}, t - \tau) \, d\tau$$

$$\vec{P}_H(\vec{\rho}, t) = \int_\tau \chi_H(\tau) \vec{H}(\vec{\rho}, t - \tau) \, d\tau$$

(1.66)

where, in order to satisfy the principle of causality, the scalar functions $\chi_E(\tau)$ and $\chi_H(\tau)$ are 0 for negative values of time $\tau$.

With such relations, *polarization* appears as the result of a linear combination of the field at different instants $t - \tau$, weighted by the scalar function $\chi(\tau)$. Such behavior is classical and also said to be that of a linear filter. Note that the case of an *instantaneous* reaction (as in a vacuum or *perfect* material) corresponds to the case in which the susceptibility is written as a time-dependent Dirac distribution, i.e. for each polarization:

$$\chi_E(\tau) = \chi_0 \delta(\tau).$$

(1.67)

Retaining the linearity and isotropy assumptions made in Sections 1.5 and 1.6.2, we thus obtain for the spatiotemporal regime:

$$\vec{P}_E = \epsilon_v \left[ \chi_E \underset{t}{\star} \vec{E} \right] \quad \Rightarrow \quad \vec{D} = \epsilon_v \left[ (\delta + \chi_E) \underset{t}{\star} \vec{E} \right] = \left[ \epsilon \underset{t}{\star} \vec{E} \right]$$

(1.68)

$$\vec{P}_H = \left[ \chi_H \underset{t}{\star} \vec{H} \right] \quad \Rightarrow \quad \vec{B} = \mu_v \left[ (\delta + \chi_H) \underset{t}{\star} \vec{H} \right] = \left[ \mu \underset{t}{\star} \vec{H} \right],$$

(1.69)

where the suffix $t$ assigned to the convolutions shows that these apply only on the temporal variable (and not on spatial variables).

Equations (1.68) and (1.69) are the electromagnetic constitutive equations for the material. Note that the functions $\epsilon(t)$ and $\mu(t)$ appearing in these equations are not those of the static regime, these being connected with the integration of $\epsilon(t)$ and $\mu(t)$ over time.

Furthermore, fields $\vec{E}$ and $\vec{H}$ are now coupled in this temporal regime such that we could have considered an additional electric polarization induced by a magnetic field, and vice versa.

We highlight a final assumption underlying the developments in this chapter, in which we have always assumed that the media were **local**. This means that at each point M in space, the polarization depends only on the excitation at this **same point M**. In calculating the polarization while still within a macroscopic formalism, we could, however, consider the influence of excitation in the immediate neighborhood of this point M: we would therefore obtain a spatial convolution, analogous to the temporal convolution previously introduced. This approach, described under the generic term *nonlocal* *electromagnetism*, could be significant if we are interested in the development of nanomaterials or metamaterials.

### 1.7.2 Maxwell's Equations

Under these assumptions, Maxwell's equations, initially given in (1.1), may at last be written as follows:

$$
\begin{cases}
\mathbf{curl}\ \vec{E}(\vec{\rho}, t) = -\dfrac{\partial}{\partial t}[\mu \underset{t}{\star} \vec{H}](\vec{\rho}, t) \\[2mm]
\mathbf{curl}\ \vec{H}(\vec{\rho}, t) = \vec{J}(\vec{\rho}, t) + \dfrac{\partial}{\partial t}[\epsilon \underset{t}{\star} \vec{E}](\vec{\rho}, t) \\[2mm]
\mathrm{div}[\epsilon \underset{t}{\star} \vec{E}](\vec{\rho}, t) = q(\vec{\rho}, t) \\[2mm]
\mathrm{div}[\mu \underset{t}{\star} \vec{H}](\vec{\rho}, t) = 0
\end{cases}
\tag{1.70}
$$

Finally we note that, until now, we have always assumed dielectric materials, ignoring the case of metals in which additional electric currents are induced; this will be discussed in Chapter 2.

### 1.7.3 Introduction to Causality Relations

The notion of causality will allow us to close this chapter. On the basis of an apparently anodyne remark (*the response depends only on past excitation*), we assumed in the previous section that the support of the susceptibility functions was positive, i.e., they obeyed the condition

$$
\chi_E(t) = \chi_H(t) = 0 \quad \forall t < 0.
\tag{1.71}
$$

This can also be written

$$
\chi(t) = \frac{1}{2}[1 + \mathrm{Sgn}(t)]\,\chi(t),
\tag{1.72}
$$

where $\mathrm{Sgn}(t)$ is the function *sign of* $t$, defined by $\mathrm{Sgn}(t) = |t|/t$.

This property of **causality** is the basis of a well-known mathematical transformation, the so-called Kramers–Kronig relations, essential in understanding dispersion phenomena in electromagnetism. We shall return to this in Chapter 5.

# 2

# The Harmonic Regime

## 2.1 Introduction

Let us assume that the local medium is linear, isotropic, homogeneous, and not permanently polarized. The electromagnetic fields and current sources are assumed to be spatiotemporally dependent.

It is useful to recall the spatiotemporal Maxwell's equations that we established in the previous chapter:

$$
\begin{cases}
\mathbf{curl}\ \vec{E}(\vec{\rho}, t) = -\dfrac{\partial}{\partial t}[\mu \underset{t}{\star} \vec{H}](\vec{\rho}, t) \\[2mm]
\mathbf{curl}\ \vec{H}(\vec{\rho}, t) = \vec{J}(\vec{\rho}, t) + \dfrac{\partial}{\partial t}[\epsilon \underset{t}{\star} \vec{E}](\vec{\rho}, t) \\[2mm]
\mathrm{div}[\epsilon \underset{t}{\star} \vec{E}](\vec{\rho}, t) = q(\vec{\rho}, t) \\[2mm]
\mathrm{div}[\mu \underset{t}{\star} \vec{H}](\vec{\rho}, t) = 0
\end{cases}
\qquad (2.1)
$$

These provide us with a complete description of the electromagnetic field $[\vec{E}(\vec{\rho}, t), \vec{H}(\vec{\rho}, t)]$ within a material medium.

It is not hard to see that a direct solution of this system of equations is not straightforward, in particular because of the integral forms associated with the time-dependent convolutions. We shall show in the next few sections how we can get around this by using the Fourier transform.

## 2.2 The Particular Case of Perfect Matter

Nonetheless, it is worth noting that this system can be solved directly in the particular case of *perfect* matter, i.e., for a medium in which the response to an excitation is *instantaneous*. We saw in Section 1.7 that electric and magnetic susceptibilities can be described in terms of time-dependent Dirac distributions, and we recall that these constitute the neutral elements of the convolution. This gives us the means to tackle these convolutions, so that we can then apply the **curl** operator to the first equation:

$$\mathbf{curl}\left[\mathbf{curl}\ \vec{E}(\vec{\rho},t)\right] = -\mu\frac{\partial}{\partial t}\left[\mathbf{curl}\ \vec{H}(\vec{\rho},t)\right] \qquad (2.2)$$

and then use the second of Maxwell's equations to reformulate the second term:

$$\mathbf{curl}\left[\mathbf{curl}\ \vec{E}(\vec{\rho},t)\right] = -\mu\frac{\partial}{\partial t}\left[\vec{J}(\vec{\rho},t) + \epsilon\frac{\partial}{\partial t}\vec{E}(\vec{\rho},t)\right]. \qquad (2.3)$$

In a homogeneous medium the quantities $\epsilon$ and $\mu$ are independent of the space variable, and we end up with the wave equation:

$$\Delta\vec{E}(\vec{\rho},t) - \epsilon\mu\frac{\partial^2}{\partial t^2}\vec{E}(\vec{\rho},t) = \mu\frac{\partial}{\partial t}\vec{J}(\vec{\rho},t) + \frac{1}{\epsilon}\ \mathbf{grad}\ q(\vec{\rho},t). \qquad (2.4)$$

The solution $\vec{E}_0(\vec{\rho},t)$ to equation (2.4) without the second term (frequently called a *homogeneous* solution) is easy to find, given that any complex exponential of the form $\exp(\pm i\vec{\beta}.\vec{\rho} \pm i\omega t)$ satisfies the problem provided

$$\vec{\beta}.\vec{\beta} = k^2 = \omega^2\epsilon\mu, \qquad (2.5)$$

where $\vec{\beta}.\vec{\beta}$ is the scalar product in $\mathbb{R}^3$ and where $k$ is real in the case of our perfect medium.

As a result, the real field $\vec{E}_0(\vec{\rho},t)$ is proportional to any linear combination (integral or discrete) of elementary trigonometric functions of the form

$$\vec{E}_0 \cos(\pm\vec{\beta}(\omega).\vec{\rho} \pm \omega t), \qquad (2.6)$$

and we can therefore make use of the wave vector $\vec{\beta}$ and the pulsation $\omega$ to define a velocity, refractive index, wavelength, etc.

However, in these *perfect* media the dielectric and magnetic constants are real, so this highly specific case does not allow all the complex wave vectors in equation (2.5). Finally, under these same assumptions, the constants $\epsilon$ and $\mu$ are independent of frequency regardless of the spectral domain being considered; this absence of dispersion in the refractive index is unrealistic.

For completeness, we now turn to the particular solution $\vec{E}_1(\vec{\rho},t)$ of the wave equation (2.4). In a similar manner to the calculation developed in Section 1.3.2 for the electrostatic potential, the solution to this equation is given directly by the convolution of the Green's function $G(\vec{\rho},t)$ associated with the second member of that equation, i.e.,

$$\vec{E}_1(\vec{\rho},t) = [G \underset{\vec{\rho},t}{\star} \vec{S}](\vec{\rho},t) \quad \text{with} \quad \vec{S}(\vec{\rho},t) = \frac{1}{\epsilon}\ \mathbf{grad}\ q(\vec{\rho},t) + \mu\frac{\partial}{\partial t}\vec{J}(\vec{\rho},t). \qquad (2.7)$$

On this occasion, however, the convolution is defined jointly over space and time.

Furthermore, the Green's function is here specific to the spatiotemporal wave equation and must therefore satisfy

$$\Delta G(\vec{\rho}, t) - \epsilon\mu \frac{\partial^2}{\partial t^2} G(\vec{\rho}, t) = \delta(\vec{\rho})\delta(t) \,. \tag{2.8}$$

It can be shown that the solution to this equation takes the form

$$G(\vec{\rho}, t) = -\frac{1}{4\pi |\vec{\rho}|} \delta(t - \sqrt{\epsilon\mu} |\vec{\rho}|) \,, \tag{2.9}$$

which leads directly to an integral expression for the field $\vec{E}_1(\vec{\rho}, t)$:

$$\vec{E}_1(\vec{\rho}, t) = [G \underset{\vec{\rho}, t}{\star} \vec{S}](\vec{\rho}, t) = -\frac{1}{4\pi} \int_{\vec{\rho}'} \frac{\vec{S}(\vec{\rho}', t - \sqrt{\epsilon\mu} |\vec{\rho} - \vec{\rho}'|)}{|\vec{\rho} - \vec{\rho}'|} d\vec{\rho}' \,, \tag{2.10}$$

where it can be seen that information is transmitted with a delay related to the propagation speed $v$ of the radiated wave ($v = 1/\sqrt{\epsilon\mu}$).

## 2.3 Maxwell's Equations in the Harmonic Regime

### 2.3.1 Introduction to the Time-Dependent Fourier Transform

We now wish to define an exact general solution to the system of equations (2.1) in frequency space. To do so, we apply a time-domain Fourier transform to the electromagnetic fields $\vec{E}(\vec{\rho}, t)$ and $\vec{H}(\vec{\rho}, t)$, leading to the fields $\vec{\mathcal{E}}(\vec{\rho}, f)$ and $\vec{\mathcal{H}}(\vec{\rho}, f)$ in the frequency domain, defined respectively by

$$\vec{\mathcal{E}}(\vec{\rho}, f) = \int_t \vec{E}(\vec{\rho}, t) e^{2i\pi ft} \, dt \tag{2.11}$$

and

$$\vec{\mathcal{H}}(\vec{\rho}, f) = \int_t \vec{H}(\vec{\rho}, t) e^{2i\pi ft} \, dt \,, \tag{2.12}$$

where $f$ is the temporal frequency, the conjugate variable of time $t$ in the sense of the Fourier transform. Similarly, we introduce the quantities $\tilde{\epsilon}(f)$ and $\tilde{\mu}(f)$ defined by

$$\tilde{\epsilon}(f) = \int_t \epsilon(t) e^{2i\pi ft} \, dt \tag{2.13}$$

and

$$\tilde{\mu}(f) = \int_t \mu(t) e^{2i\pi ft} \, dt \,. \tag{2.14}$$

Note that we have here adopted the positive sign convention in expressing this temporal Fourier transform. We shall see that this choice will have consequences on the sign of the imaginary part of the refractive index for lossy media. Note that some texts define the temporal Fourier transform using the opposite sign convention.

### 2.3.2 Application to Maxwell's Equations

By way of example, we now apply a temporal Fourier transform to the first of Maxwell's equations. Though the properties of Fourier transform are well known, we will proceed step by step to obtain the harmonic formulation of Maxwell's equations. We then get

$$\int_t \mathbf{curl}\ \vec{E}(\vec{\rho},t)e^{2i\pi ft}\,dt = -\int_t \frac{\partial}{\partial t}[\mu \underset{t}{\star} \vec{H}](\vec{\rho},t)e^{2i\pi ft}\,dt\,. \qquad (2.15)$$

In calculating the first term in this equation, it is sufficient to note that the curl operator acts only on space variables, while the Fourier transform is an integral over time. This means that the time Fourier transformation of a curl is just the curl of this temporal Fourier transform, i.e.,

$$\int_t \mathbf{curl}\ \vec{E}(\vec{\rho},t)e^{2i\pi ft}\,dt = \mathbf{curl}\left[\int_t \vec{E}(\vec{\rho},t)e^{2i\pi ft}\,dt\right] = \mathbf{curl}\ \vec{\mathcal{E}}(\vec{\rho},f)\,. \qquad (2.16)$$

As far as the second term is concerned, the calculation has to be carried out in three steps. We first use the characteristic property of the derivative of a convolution, namely that the derivative operator can be applied to either one or the other of the two functions in the convolution, i.e.,

$$\frac{\partial}{\partial t}[\mu \underset{t}{\star} \vec{H}] = \mu \underset{t}{\star} \frac{\partial}{\partial t}\vec{H}\,. \qquad (2.17)$$

Second, we have to calculate the Fourier transform of the convolution of two functions; we know that this is equal to the simple product of the Fourier transforms of each of these two functions, i.e.,

$$\int_t [\mu \underset{t}{\star} \frac{\partial}{\partial t}\vec{H}](\vec{\rho},t)e^{2i\pi ft}\,dt = \tilde{\mu}(f)\int_t \frac{\partial}{\partial t}\vec{H}(\vec{\rho},t)e^{2i\pi ft}\,dt\,. \qquad (2.18)$$

The temporal Fourier transform of the temporal derivative of the field $\vec{H}(\vec{\rho},t)$ remains to be determined. Using the sign convention adopted in the definition of these Fourier transforms, we obtain

$$\int_t \frac{\partial}{\partial t}\vec{H}(\vec{\rho},t)e^{2i\pi ft}\,dt = -2i\pi f\vec{\mathcal{H}}(\vec{\rho},f)\,. \qquad (2.19)$$

In conclusion, applying a Fourier transform to the first of Maxwell's equations leads to the final relationship between temporal field Fourier transforms:

$$\mathbf{curl}\ \vec{\mathcal{E}}(\vec{\rho}, f) = 2i\pi f \tilde{\mu}(f) \vec{\mathcal{H}}(\vec{\rho}, f)\,. \tag{2.20}$$

Similarly, applying a Fourier transform to the second Maxwell's equation leads to the following relationship between frequency-domain fields:

$$\mathbf{curl}\ \vec{\mathcal{H}}(\vec{\rho}, f) = \vec{\mathcal{J}}(\vec{\rho}, f) - 2i\pi f \tilde{\epsilon}(f) \vec{\mathcal{E}}(\vec{\rho}, f)\,, \tag{2.21}$$

where $\vec{\mathcal{J}}(\vec{\rho}, f)$ is the temporal Fourier transform of the volume current density $\vec{J}(\vec{\rho}, t)$.

For the last two of Maxwell's equations, we similarly obtain, for a homogeneous medium:

$$\mathrm{div}[\tilde{\epsilon}(f)\vec{\mathcal{E}}(\vec{\rho}, f)] = \tilde{\epsilon}(f)\,\mathrm{div}\vec{\mathcal{E}}(\vec{\rho}, f) = \mathcal{Q}(\vec{\rho}, f) \tag{2.22}$$

$$\mathrm{div}[\tilde{\mu}(f)\vec{\mathcal{H}}(\vec{\rho}, f)] = \tilde{\mu}(f)\,\mathrm{div}\vec{\mathcal{H}}(\vec{\rho}, f) = 0\,, \tag{2.23}$$

where $\mathcal{Q}(\vec{\rho}, f)$ is the temporal Fourier transform of the volume charge distribution $q(\vec{\rho}, t)$.

### 2.3.3 Overview

These relationships between frequency variables resulting from application of a temporal Fourier transform to the time-dependent Maxwell's equations are customarily referred to as the time-harmonic Maxwell's equations. We shall see later that this harmonic regime corresponds to a single wavelength (and hence to a single color if this wavelength lies within the visible spectrum), and thus we also speak of the monochromatic regime.

For a homogeneous medium, these time-harmonic Maxwell's equations can finally be expressed as

$$\begin{cases} \mathbf{curl}\ \vec{\mathcal{E}}(\vec{\rho}, f) = 2i\pi f\,\tilde{\mu}(f)\vec{\mathcal{H}}(\vec{\rho}, f) \\ \mathbf{curl}\ \vec{\mathcal{H}}(\vec{\rho}, f) = \vec{\mathcal{J}}(\vec{\rho}, f) - 2i\pi f\,\tilde{\epsilon}(f)\vec{\mathcal{E}}(\vec{\rho}, f) \\ \tilde{\epsilon}(f)\,\mathrm{div}\vec{\mathcal{E}}(\vec{\rho}, f) = \mathcal{Q}(\vec{\rho}, f) \\ \mathrm{div}\vec{\mathcal{H}}(\vec{\rho}, f) = 0 \end{cases}\,, \tag{2.24}$$

or equivalently, invoking temporal pulsation $\omega$ in place of $2\pi f$:

$$\begin{cases} \mathbf{curl}\ \vec{\mathcal{E}}(\vec{\rho}, f) = i\omega\,\tilde{\mu}(f)\vec{\mathcal{H}}(\vec{\rho}, f) \\ \mathbf{curl}\ \vec{\mathcal{H}}(\vec{\rho}, f) = \vec{\mathcal{J}}(\vec{\rho}, f) - i\omega\,\tilde{\epsilon}(f)\vec{\mathcal{E}}(\vec{\rho}, f) \\ \tilde{\epsilon}(f)\,\mathrm{div}\vec{\mathcal{E}}(\vec{\rho}, f) = \mathcal{Q}(\vec{\rho}, f) \\ \mathrm{div}\vec{\mathcal{H}}(\vec{\rho}, f) = 0 \end{cases} \tag{2.25}$$

Note that in an **inhomogeneous medium**, the equations involving divergence must be kept in their initial form, i.e.,

$$div[\tilde{\epsilon}\vec{\mathcal{E}}] = \mathcal{Q} \quad \text{and} \quad div[\tilde{\mu}\vec{\mathcal{H}}] = 0, \tag{2.26}$$

since in this case $\tilde{\epsilon}$ and $\tilde{\mu}$ depend on space variables.

Furthermore, in the absence of sources, there is a certain redundancy in equations (2.25) since the last two can be deduced from the first two after applying the divergence operator:

$$\begin{array}{ll} \mathrm{div}[\mathbf{curl}\ \vec{\mathcal{E}}(\vec{\rho}, f)] = 0 & \Rightarrow \quad \mathrm{div}[\tilde{\mu}(f)\vec{\mathcal{H}}(\vec{\rho}, f)] = 0 \\ \mathrm{div}[\mathbf{curl}\ \vec{\mathcal{H}}(\vec{\rho}, f)] = 0 & \Rightarrow \quad \mathrm{div}[\tilde{\epsilon}(f)\vec{\mathcal{E}}(\vec{\rho}, f)] = 0 \end{array} \tag{2.27}$$

Otherwise the sources are connected as: $\mathrm{div}\vec{\mathcal{J}}(\vec{\rho}, f) = i\omega\mathcal{Q}(\vec{\rho}, f)$, which can be shown by application of the divergence operator to the second of Maxwell's equation, and taking into account the third.

We shall see that, in general, the first two equations (involving the curl operator) provide the tangential field components while the last two (using the divergence operator) provide the normal components of these same fields.

## 2.4 Reconstruction of the Temporal Field Dependence

The procedure outlined in Section 2.3 will enable us to calculate the values of fields $\vec{\mathcal{E}}(\vec{\rho}, f)$ and $\vec{\mathcal{H}}(\vec{\rho}, f)$ at any frequency $f$, provided we know the properties of the material medium defined by $\tilde{\epsilon}(f)$ and $\tilde{\mu}(f)$, as well as the nature of the sources described by $\vec{\mathcal{J}}(\vec{\rho}, f)$ and $\mathcal{Q}(\vec{\rho}, f)$.

Once these frequency field expressions have been calculated, we can deduce their corresponding temporal forms by applying an inverse Fourier transform to our result, namely

$$\vec{E}(\vec{\rho}, t) = \int_f \vec{\mathcal{E}}(\vec{\rho}, f)e^{-2i\pi ft}\, df \tag{2.28}$$

and

$$\vec{H}(\vec{\rho}, t) = \int_f \vec{\mathcal{H}}(\vec{\rho}, f)e^{-2i\pi ft}\, df. \tag{2.29}$$

This operation is known as **reconstruction** of the temporal field dependence. We can also take the complex exponential functions $\exp(-2i\pi ft)$ as a *continuous basis* over which the fields $\vec{E}(\vec{\rho}, t)$ or $\vec{H}(\vec{\rho}, t)$ can be decomposed, with the frequency fields $\vec{\mathcal{E}}(\vec{\rho}, f)$ and $\vec{\mathcal{H}}(\vec{\rho}, f)$ then appearing as *coefficients* of this spectral decomposition.

Note also that the field $\vec{E}(\vec{\rho}, t)$ is a real quantity defined in $\mathbb{R}^3$, since it is measurable. On the other hand, the associated frequency field $\vec{\mathcal{E}}(\vec{\rho}, f)$ is a complex quantity defined in $\mathbb{C}^3$. However, this field has Hermitian symmetry since it is the Fourier transform of a real field:

$$\vec{\mathcal{E}}^*(\vec{\rho}, f) = \int_t \vec{E}^*(\vec{\rho}, t) e^{-2i\pi ft} \, dt = \int_t \vec{E}(\vec{\rho}, t) e^{-2i\pi ft} \, dt = \vec{\mathcal{E}}(\vec{\rho}, -f), \quad (2.30)$$

where the symbol $*$ denotes the complex conjugate. If we use this property in (2.28) we get

$$\vec{E}(\vec{\rho}, t) = \int_{-\infty}^{+\infty} \vec{\mathcal{E}}(\vec{\rho}, f) e^{-2i\pi ft} \, df = \int_0^{+\infty} \left[ \vec{\mathcal{E}}(\vec{\rho}, f) e^{-2i\pi ft} + \vec{\mathcal{E}}^*(\vec{\rho}, f) e^{2i\pi ft} \right] df, \quad (2.31)$$

or equivalently

$$\vec{E}(\vec{\rho}, t) = 2 \int_0^{+\infty} \Re \left[ \vec{\mathcal{E}}(\vec{\rho}, f) e^{-2i\pi ft} \right] df. \quad (2.32)$$

If we arbitrarily take one of the components $j = x, y$ or $z$ of these fields, we can write

$$E_j(\vec{\rho}, t) = 2 \int_0^{+\infty} |\mathcal{E}_j(\vec{\rho}, f)| \cos[2\pi ft - \phi_j(f)] \, df, \quad (2.33)$$

where $|\mathcal{E}_j(\vec{\rho}, f)|$ is the modulus (in the complex number sense) of component $f$ of the frequency field and $\phi_j(f)$ is its argument.

This last relationship shows clearly that the reconstructed field is real and expressible as a linear combination (integral) of trigonometric functions, each function having a specific frequency and phase. This result is valid for any real signal that admits a Fourier transform. We shall now see how the moduli and phases of these variables depend on the spatial variable.

It is worth noting that (2.28) also enables a monochromatic wave to be described, provided the frequency spectrum is describable in terms of a Dirac distribution:

$$\vec{E}(\vec{\rho}, t) = \vec{E}_0(\vec{\rho}) \, \cos(2\pi f_0 t)$$

$$\Rightarrow \quad \vec{\mathcal{E}}(\vec{\rho}, f) = \frac{1}{2} \vec{E}_0(\vec{\rho}) \left[ \delta(f - f_0) + \delta(f + f_0) \right]. \tag{2.34}$$

## 2.5 Equation of Propagation in a Homogeneous Medium

Let us consider the special case in which charge and current are both 0 in a homogeneous medium. The time-harmonic Maxwell's equations then take the following simplified form:

$$\begin{cases} \mathbf{curl} \ \vec{\mathcal{E}}(\vec{\rho}, f) = 2i\pi f \, \tilde{\mu}(f) \vec{\mathcal{H}}(\vec{\rho}, f) = i\omega \, \tilde{\mu}(f) \vec{\mathcal{H}}(\vec{\rho}, f) \\ \mathbf{curl} \ \vec{\mathcal{H}}(\vec{\rho}, f) = -2i\pi f \, \tilde{\epsilon}(f) \vec{\mathcal{E}}(\vec{\rho}, f) = -i\omega \, \tilde{\epsilon}(f) \vec{\mathcal{E}}(\vec{\rho}, f) \\ \mathrm{div} \vec{\mathcal{E}}(\vec{\rho}, f) = 0 \\ \mathrm{div} \vec{\mathcal{H}}(\vec{\rho}, f) = 0 \end{cases} \tag{2.35}$$

Applying the curl operator to the first of Maxwell's equations, we get

$$\mathbf{curl}[\mathbf{curl} \ \vec{\mathcal{E}}(\vec{\rho}, f)] = \mathbf{grad}[\mathrm{div} \vec{\mathcal{E}}(\vec{\rho}, f)] - \Delta \vec{\mathcal{E}}(\vec{\rho}, f)$$

$$= i\omega \, \tilde{\mu}(f) \, \mathbf{curl} \ \vec{\mathcal{H}}(\vec{\rho}, f). \tag{2.36}$$

Now using the second and third equations, this expression becomes

$$- \Delta \vec{\mathcal{E}}(\vec{\rho}, f) = -i^2 \omega^2 \, \tilde{\epsilon}(f) \tilde{\mu}(f) \, \vec{\mathcal{E}}(\vec{\rho}, f) \tag{2.37}$$

and can be expressed in the classical form known as the Helmholtz equation, namely

$$\Delta \vec{\mathcal{E}}(\vec{\rho}, f) + k^2 \vec{\mathcal{E}}(\vec{\rho}, f) = 0 \qquad \text{with} \qquad k^2 = \omega^2 \tilde{\epsilon}(f) \tilde{\mu}(f), \tag{2.38}$$

where $k$ is a function of frequency $f$ (or angular frequency $\omega$) and assumes values in $\mathbb{C}$.

We can immediately check that any exponential of the form $\exp(i\vec{\beta}.\vec{\rho})$ is a solution of this equation, provided the vector $\vec{\beta}$ satisfies the equation

$$\vec{\beta}.\vec{\beta} = k^2 = \omega^2 \tilde{\epsilon}(f) \tilde{\mu}(f). \tag{2.39}$$

There exists an infinity of vectors $\vec{\beta}$ satisfying this condition at some given frequency $f$, most of them defining the directions of propagation of the frequency component for field $\vec{\mathcal{E}}(\vec{\rho}, f)$; the most general expression for this field will therefore have the form

$$\vec{\mathcal{E}}(\vec{\rho}, f) = \vec{\mathcal{A}}(f) \, e^{i\vec{\beta}.\vec{\rho}} \quad \text{with} \quad \vec{\beta}.\vec{\beta} = \omega^2 \tilde{\epsilon}(f) \tilde{\mu}(f). \tag{2.40}$$

This last expression corresponds to a monochromatic, unidirectional, and progressive or retrograde wave that may be propagative or evanescent under certain conditions to be described later.

## 2.6 Phase Velocity, Refractive Index, and Wavelength

We begin by assuming that the medium under consideration is transparent ($k$ real) over a certain frequency range, and we shall restrict ourselves here to a single direction of wave vector $\vec{\beta}(f)$, since we have yet to deal with an infinite range of frequencies. The case of several combined directions of wave vectors will be covered in Chapter 4. Though the medium is assumed transparent (real $k$), the wave vector can be complex, as for evanescent waves also discussed later in Chapter 4. In what follows, we shall consider only real values for this wave vector so that we can restrict our analysis to propagative waves only.

We can now proceed to reconstruction of the temporal dependence of the field by writing

$$\vec{E}(\vec{\rho}, t) = \int_f \vec{\mathcal{E}}(\vec{\rho}, f) e^{-2i\pi ft} df = \int_f \vec{A}(f) e^{i\vec{\beta}.\vec{\rho}} e^{-2i\pi ft} df . \qquad (2.41)$$

Making use of the Hermitian symmetry property introduced in Section 2.4, for each component $j$ of the vector field $\vec{E}(\vec{\rho}, t)$, we can write (with $j = x, y$ or $z$):

$$E_j(\vec{\rho}, t) = 2 \int_{f>0} |A_j(f)| \cos[\vec{\beta}.\vec{\rho} - 2\pi ft + \phi_j(f)] df . \qquad (2.42)$$

### 2.6.1 Phase Velocity

Consider the argument $\psi$ of the cosine function that appears in each elementary component of (2.42) and choose the direction defined by vector $\vec{\beta}$. This argument is then expressed as

$$\psi = |\vec{\beta}||\vec{\rho}| - 2\pi ft + \phi_j(f) = k\rho - \omega t + \phi_i(f) . \qquad (2.43)$$

The quantity being considered thus represents a plane wave propagating in the direction defined by vector $\vec{\beta}$ at a speed determined by the invariance condition of this phase $\psi$, i.e.,

$$d\psi = 0 = k\, d\rho - \omega\, dt \quad \Rightarrow \quad v = \frac{d\rho}{dt} = \frac{\omega}{k} . \qquad (2.44)$$

With this definition in mind, the quantity $v$ is known as the **phase velocity**. If we now insert into (2.44) the definition of the variable $k$ that appeared in equation (2.38), we get

$$v = \frac{\omega}{k} = \frac{1}{\sqrt{\tilde{\epsilon}(f)\tilde{\mu}(f)}}, \qquad (2.45)$$

which is often expressed in the condensed form $\tilde{\epsilon}(f)\tilde{\mu}(f)v^2 = 1$.

In the special case where the medium in which these waves are propagating is vacuum, we get $\epsilon_v\mu_v c^2 = 1$, where $c$ has the usual meaning of the speed of light in vacuo. We shall show that this phase velocity does not necessarily correspond to the speed at which energy is transported by the electromagnetic wave. This will be addressed in Chapter 3, where we shall introduce the idea of group velocity taking account of the frequency spectrum associated with the wave.

### 2.6.2 Refractive Index

We define the refractive index $n$ of the medium as the ratio of the speed of propagation of light in vacuum to that in the medium, i.e.,

$$n = \frac{c}{v} = \frac{\sqrt{\tilde{\epsilon}(f)\tilde{\mu}(f)}}{\sqrt{\epsilon_0\mu_0}} = \sqrt{\tilde{\epsilon}_r(f)\tilde{\mu}_r(f)}. \qquad (2.46)$$

Note also at this stage that this relationship does not necessarily show that the refractive index is greater than 1, or that the phase velocity is less than $c$. This point will be addressed when we introduce group velocity in Chapter 3.

### 2.6.3 Wavelength

If we now fix the time in (2.43), the spatial period $\lambda$ in the phase variations, known as the wavelength, is defined by

$$\lambda = \frac{2\pi}{k} \qquad (2.47)$$

$$\lambda = \frac{2\pi v}{\omega} = 2\pi\frac{c}{\omega}\cdot\frac{1}{n} = \frac{2\pi}{k_v}\cdot\frac{1}{n} = \frac{\lambda_v}{n}, \qquad (2.48)$$

where $\lambda_v$ and $k_v$ are the wavelength and modulus respectively of the wave vector in vacuum (at the same frequency $f$).

## 2.7 Absorbing Media

### 2.7.1 The General Case

The ideas introduced in the previous section were all defined assuming transparent media, i.e., those for which $\tilde{\epsilon}(f)$ and $\tilde{\mu}(f)$ are real. This assumption is clearly acceptable only over a certain range of frequencies, since the phenomena of dispersion and absorption are intrinsically linked, and only *perfect* matter (which reacts instantaneously to an excitation) can be nondispersive, at least in the absence of conduction currents in metals. This property results from the principle of causality and is directly illustrated through the Kramers–Kronig relations mentioned in the previous chapter, and to which we shall return later in this book (Chapter 5).

We now turn our attention to the case of absorbing media in a particular frequency domain. The variable $k$ is complex, but we can apply similar reasoning as before, noting that the complex exponential introduced in Section 2.5 can be written as

$$e^{i(\vec{\beta}.\vec{\rho}-\omega t)} = e^{-\vec{\beta}''.\vec{\rho}} e^{i(\vec{\beta}'.\vec{\rho}-\omega t)}, \tag{2.49}$$

where we have put $\vec{\beta} = \vec{\beta}' + i\vec{\beta}''$, with $\vec{\beta}.\vec{\beta} = k^2 \in \mathbb{C}$.

We see that the surfaces of equal phase and amplitude can be dissociated, since they are now determined by the real and imaginary parts of the wave vector. The amplitude of the wave decays as a result of the real exponent in $\exp(-\vec{\beta}''.\vec{\rho})$.

Furthermore, the equiphase surfaces can be used as before (i.e., in the case of transparent media) to arrive at the notion of phase velocity, this time given by

$$v = \frac{\omega}{|\vec{\beta}'|}. \tag{2.50}$$

We can elaborate on the expression of this variable. We start by defining a complex factor $n$ using the usual relation:

$$k = k_v n, \tag{2.51}$$

where $k_0 = \omega\sqrt{\epsilon_0\mu_0}$ is real.

If we assume that propagation (and attenuation) occurs parallel to the $z$-axis, we can write

$$\vec{\beta} = \beta\vec{z} \quad \Rightarrow \quad \vec{\beta}.\vec{\beta} = k^2 \quad \Rightarrow \quad \beta = \pm k, \tag{2.52}$$

and hence

$$|\vec{\beta}'| = |\beta'| = k' = k_v n' \quad \text{with} \quad n' = \Re[n] \quad \Rightarrow \quad v = \frac{\omega}{k_v n'} = \frac{c}{n'}. \tag{2.53}$$

The phase velocity therefore retains the same expression as in transparent media, provided we consider the real part of the complex refractive index defined by (2.51). Note that this complex factor can be written as

$$n = \frac{k}{k_v} = \frac{\omega\sqrt{\tilde{\epsilon}\tilde{\mu}}}{\omega\sqrt{\epsilon_v\mu_v}} \quad \Rightarrow \quad n = \sqrt{\tilde{\epsilon}_r\tilde{\mu}_r}\,. \tag{2.54}$$

So, introducing the electric and magnetic susceptibilities,

$$\tilde{\epsilon}_r = 1 + \tilde{\chi}_E \quad \text{and} \quad \tilde{\mu}_r = 1 + \tilde{\chi}_H \quad \Rightarrow \quad n = \sqrt{(1+\tilde{\chi}_E)(1+\tilde{\chi}_H)}\,. \tag{2.55}$$

In the most general situation, this relationship does not show that the real part of the complex refractive index is greater than 1 or that the phase velocity is bounded by $c$. The result will depend on the susceptibility parameters inherent in the microscopic structure of the medium. We shall see that there is no inherent paradox, as it is the group velocity (and not the phase velocity) that characterizes the transport of light energy.

### 2.7.2 The Special Case of Metals

Metals represent a particular case of absorbing media. We shall now see why the imaginary part of their refractive index is generally high and positive.

Up until now, we have considered photo-induced currents and charges resulting from the material (dielectric) polarization, and which are taken into account by the physical parameters (permittivity and permeability) of the material. For that reason these sources no longer appear in Maxwell's equations. We also notice that the electric charges resulted from a dissociation of the centroids of positive and negative charges under the action of the field. However, for specific materials, additional sources are created by the field excitation. The most common is that of metals, in which a conduction electric current must be introduced.

Here we shall relate this induced current to the excitation that it creates, and we shall assume that Ohm's law, written in the form

$$\vec{J} = \gamma \underset{t}{\star} \vec{E} \tag{2.56}$$

is satisfied and that the convolution is over time. Hence we obtain an expression similar to the one we obtained for the polarization of matter in Chapter 1. Note that the conductivity function $\gamma(t)$ is once again causal.

After applying a temporal Fourier transform, we get

$$\vec{\mathcal{J}}(\vec{\rho}, f) = \tilde{\gamma}(f)\vec{\mathcal{E}}(\vec{\rho}, f)\,, \tag{2.57}$$

so the second of Maxwell's equations assumes the following modified form:

$$\mathbf{curl}\ \vec{\mathcal{H}}(\vec{\rho}, f) = -i\omega\left[\tilde{\epsilon}(f) + i\frac{\tilde{\gamma}(f)}{\omega}\right]\vec{\mathcal{E}}(\vec{\rho}, f) = -i\omega\tilde{\epsilon}_{eq}(f)\vec{\mathcal{E}}(\vec{\rho}, f)\,. \quad (2.58)$$

Thus for metals Maxwell's equations remain unaltered provided they incorporate a complex permittivity $\tilde{\epsilon}_{eq}(f)$ defined by the equation

$$\tilde{\epsilon}_{eq}(f) = \tilde{\epsilon}(f) + i\frac{\tilde{\gamma}(f)}{\omega}\,. \quad (2.59)$$

This relationship explains why, by comparison with that of absorbing dielectrics, the imaginary permittivity of metals is high. We may also remark that for metals, and by contrast with what we have seen so far in dielectrics, the absence of inertia in the matter does not imply that there is neither dispersion nor loss. A regular or singular (time Dirac) conductivity always creates an imaginary part in the equivalent permittivity, and this imaginary part implies a dispersive behavior on account of the $1/\omega$ term.

To conclude this section, it should also be stressed that in arbitrary media, the complex index defined from (2.54) may reveal a real part with nonclassical values (negative of lower than unity). This a very specific case resulting from the definition of the square root in the complex plane.

## 2.8 Mutual Orthogonality

We wish to recall at this point the structure of the electromagnetic field and introduce the idea of mutual orthogonality. This will then enable us to define the state of polarization of light.

Consider a monochromatic plane wave propagating in free space with wave vector $\vec{\beta}$ within an homogeneous isotropic material. The equation of harmonic propagation showed that the electric field of this elementary component took the form

$$\vec{\mathcal{E}}(\vec{\rho}, f) = \vec{A}(f)e^{i\vec{\beta}\cdot\vec{\rho}}\,. \quad (2.60)$$

We now introduce this expression into the first of Maxwell's equations (harmonic form):

$$\mathbf{curl}\ \vec{\mathcal{E}}(\vec{\rho}, f) = i\omega\tilde{\mu}(f)\vec{\mathcal{H}}(\vec{\rho}, f)\,, \quad (2.61)$$

taking account of the particular form of the vector $\vec{\mathcal{E}}(\vec{\rho}, f)$, i.e.,

$$\mathbf{curl}\ \vec{\mathcal{E}}(\vec{\rho}, f) = e^{i\vec{\beta}\cdot\vec{\rho}}\,\mathbf{curl}\ \vec{A}(f) + \mathbf{grad}\ [e^{i\vec{\beta}\cdot\vec{\rho}}] \wedge \vec{A}(f)\,, \quad (2.62)$$

and hence

$$\mathbf{curl}\ \vec{\mathcal{E}}(\vec{\rho}, f) = \vec{0} + i\vec{\beta} \wedge \vec{A}(f)e^{i\vec{\beta}\cdot\vec{\rho}} = i\vec{\beta} \wedge \vec{\mathcal{E}}(\vec{\rho}, f)\,. \quad (2.63)$$

Consequently, we deduce that

$$\vec{\mathcal{H}}(\vec{\rho}, f) = \frac{1}{\omega \tilde{\mu}(f)} \vec{\beta} \wedge \vec{\mathcal{E}}(\vec{\rho}, f). \qquad (2.64)$$

Similarly the second of Maxwell's equations, in the absence of sources, gives

$$\mathbf{curl}\ \vec{\mathcal{H}}(\vec{\rho}, f) = -i\omega\tilde{\epsilon}(f)\vec{\mathcal{E}}(\vec{\rho}, f) \ \Rightarrow\ \vec{\mathcal{E}}(\vec{\rho}, f) = -\frac{1}{\omega\tilde{\epsilon}(f)} \vec{\beta} \wedge \vec{\mathcal{H}}(\vec{\rho}, f). \quad (2.65)$$

In the case of a transparent medium ($\vec{\beta} \in \mathbb{R}^3$), these expressions show that the vectors $\vec{\beta}$, $\vec{\mathcal{E}}$, and $\vec{\mathcal{H}}$ are mutually orthogonal.

In conclusion, for a plane wave propagating within a transparent isotropic medium, the electric and magnetic fields vibrate in a plane perpendicular to the wave vector and remain mutually orthogonal during these vibrations.

## 2.9 The Polarization State of Light

Consider the simple case in which the frequency component of the field reduces to a plane wave of frequency $f$ propagating in the direction defined by the vector $\vec{\beta}$ ($\vec{\beta} \in \mathbb{R}^3$ ), i.e.,

$$\vec{\mathcal{E}}(\vec{\rho}, f) = \vec{A}(f)e^{i\vec{\beta}.\vec{\rho}}. \qquad (2.66)$$

We have shown that, in a transparent isotropic medium, this field $\vec{\mathcal{E}}$ is perpendicular to the vector $\vec{\beta}$ and may therefore be decomposed in a coordinate system $Ouvw$ associated with the plane perpendicular to this vector, where $Ow$ is parallel to $\vec{\beta}$. In this reference frame we can write

$$\begin{cases} \mathcal{E}_u(\vec{\rho}, f) = A_u(f)e^{i\vec{\beta}.\vec{\rho}} \\ \mathcal{E}_v(\vec{\rho}, f) = A_v(f)e^{i\vec{\beta}.\vec{\rho}} \end{cases}, \qquad (2.67)$$

where these two fields are often called the **polarization modes**; we will see further (Chapter 6) how to define the S and P polarization modes in regard to the geometry of optical components.

To get back to a temporal dependence, it is sufficient to use (2.32) taking account of the monochromatic nature of the wave being considered:

$$\begin{cases} E_u(\vec{\rho}, t) = 2\Re\left\{A_u(f)e^{i(\vec{\beta}.\vec{\rho}-\omega t)}\right\} = 2\,|A_u|\cos(\vec{\beta}.\vec{\rho} - \omega t + \phi_u) \\ E_v(\vec{\rho}, t) = 2\Re\left\{A_v(f)e^{i(\vec{\beta}.\vec{\rho}-\omega t)}\right\} = 2\,|A_v|\cos(\vec{\beta}.\vec{\rho} - \omega t + \phi_v) \end{cases}, \qquad (2.68)$$

where the right-hand terms have deliberately not indicated any frequency dependence of the quantities $\beta$, $|A_u|$, $|A_v|$, $\phi_u$, and $\phi_v$.

We now observe the angle $\eta$ that this field $\vec{E}$ forms with the axis $Ou$ in the plane $Ouv$. We get

$$\tan \eta = \frac{E_v}{E_u} = \frac{|\mathcal{A}_u| \cos(\vec{\beta}.\vec{\rho} - \omega t + \phi_u)}{|\mathcal{A}_v| \cos(\vec{\beta}.\vec{\rho} - \omega t + \phi_v)}. \qquad (2.69)$$

This expression shows that, in general, the angle $\eta$ is not constant and hence that the direction of the field varies with time. The speed of rotation can be expressed as

$$\frac{d\eta}{dt} = -\omega \frac{|\mathcal{A}_u|.|\mathcal{A}_v|}{|\mathcal{A}_u|^2 \cos^2(\omega t - \phi_u) + |\mathcal{A}_v|^2 \cos^2(\omega t - \phi_v)} \sin(\phi_v - \phi_u). \qquad (2.70)$$

We first observe that the sign of this speed is not modified versus time, which means that the ellipse is browsed in a unique direction. Furthermore, this speed is 0 when the phases are identical on the two axes ($\phi_u = \phi_v$), the condition for which the ratio $E_v/E_u$ is effectively constant. In this particular case, we say that the light is **linearly polarized** or that it exhibits **rectilinear polarization**.

In general, polarization will be not linear, but elliptical. This can easily be proved with the parametrized curve $u(t) = E_u(t)$ and $v(t) = E_v(t)$. Then since every ellipse is characterized with three parameters, it is common to use spherical coordinates to assign each ellipse a point at the surface of a sphere of radius 1 (the Poincaré sphere). Bear in mind here that, to emphasize the complexity of the vectorial nature of light (whose components remain mutually orthogonal over time), the speed of rotation is in the order of $10^{14}$ rads/s.

To conclude this short section, note also that light polarization (not to be confused with material polarization) has been introduced for a plane wave, that is, a wave with a unique frequency and a unique wave vector. The general elliptical temporal behavior of the field locus is characteristic of fully polarized light. We will see further (Chapter 5) how to define partial or total polarization of light, a property that can be associated only with a frequency packet, as opposed to a monochromatic wave.

## 2.10 Absorption, Flux, and Energy

We are interested in the energy balance in the harmonic regime. The case of the general regime (spatiotemporal) will be tackled in Chapter 4. Since optical power is provided by sources, we make the assumption that vector current densities $\vec{\mathcal{J}}(\vec{\rho}, f)$ and scalar charges $\mathcal{Q}(\vec{\rho}, f)$ are present in Maxwell's equations. Let $\Omega$ be the volume containing these sources, $\Sigma$ be the closed surface that delimits this volume, and $\vec{n}$ be the outward directed local normal to this surface $\Sigma$.

In order to simplify the expressions in what follows, we shall not be constantly recalling the parameters on which each of the variables considered depend, regardless of whether they are scalar or vector.

The first two Maxwell's equations are therefore written (in the harmonic regime)

$$\begin{cases} \mathbf{curl}\ \vec{\mathcal{E}} = iw\ \tilde{\mu}\vec{\mathcal{H}} \\ \mathbf{curl}\ \vec{\mathcal{H}} = -iw\ \tilde{\epsilon}\vec{\mathcal{E}} + \vec{\mathcal{J}} \end{cases} \tag{2.71}$$

Consider the vector $\vec{\Pi}$, which is known as the **Poynting vector** and defined by

$$\vec{\Pi} = \frac{1}{2}(\vec{\mathcal{E}}^* \wedge \vec{\mathcal{H}})\,, \tag{2.72}$$

where, as before, the asterisk notation implies the complex conjugate. We now calculate the divergence of the Poynting vector:

$$\operatorname{div}\vec{\Pi} = \frac{1}{2}\operatorname{div}(\vec{\mathcal{E}}^* \wedge \vec{\mathcal{H}}) = \frac{1}{2}\vec{\mathcal{H}}.\mathbf{curl}\ \vec{\mathcal{E}}^* - \frac{1}{2}\vec{\mathcal{E}}^*.\mathbf{curl}\ \vec{\mathcal{H}}$$
$$= -\frac{i}{2}w\tilde{\mu}^*|\vec{\mathcal{H}}|^2 + \frac{i}{2}w\tilde{\epsilon}|\vec{\mathcal{E}}|^2 - \frac{1}{2}\vec{\mathcal{J}}.\vec{\mathcal{E}}^*\,. \tag{2.73}$$

We now consider the general case of an absorbing medium and put

$$\tilde{\epsilon} = \tilde{\epsilon}' + i\tilde{\epsilon}'' \qquad \tilde{\mu} = \tilde{\mu}' + i\tilde{\mu}''\,. \tag{2.74}$$

Hence the real part of the divergence of the Poynting vector can be expressed in the following form

$$\Re[\operatorname{div}\vec{\Pi}] = -\frac{1}{2}w\tilde{\epsilon}''|\vec{\mathcal{E}}|^2 - \frac{1}{2}w\tilde{\mu}''|\vec{\mathcal{H}}|^2 - \frac{1}{2}\Re[\vec{\mathcal{J}}.\vec{\mathcal{E}}^*]\,. \tag{2.75}$$

If we integrate this expression over the volume $\Omega$ enclosing the sources, we get

$$\int_\Omega \Re[\operatorname{div}\vec{\Pi}]\, dV = -\int_\Omega \frac{1}{2}w\tilde{\epsilon}''|\vec{\mathcal{E}}|^2\, dV$$
$$-\int_\Omega \frac{1}{2}w\tilde{\mu}''|\vec{\mathcal{H}}|^2\, dV - \int_\Omega \frac{1}{2}\Re[\vec{\mathcal{J}}.\vec{\mathcal{E}}^*]dV. \tag{2.76}$$

Applying the Green–Ostrogradsky theorem to the first term in this equation and slightly rearranging its structure we finally obtain

$$-\int_\Omega \frac{1}{2}\Re[\vec{\mathcal{J}}.\vec{\mathcal{E}}^*]dV = \int_\Sigma \Re[\vec{\Pi}.\vec{n}]\, dS + \int_\Omega \left[\frac{1}{2}w\tilde{\epsilon}''|\vec{\mathcal{E}}|^2 + \frac{1}{2}w\tilde{\mu}''|\vec{\mathcal{H}}|^2\right] dV\,. \tag{2.77}$$

This equation can be written in the following compact form:

$$F = \Phi + A, \tag{2.78}$$

where

$$F = - \int_\Omega \frac{1}{2} \Re[\vec{\mathcal{J}}.\vec{\mathcal{E}}^*] dV \tag{2.79}$$

is the optical power provided by the sources contained within the region $\Omega$,

$$\Phi = \int_\Sigma \Re[\vec{\Pi}.\vec{n}] \, dS \tag{2.80}$$

is the power that escapes by radiation (*flux*) through the closed surface, and

$$A = \int_\Omega \left[ \frac{1}{2} \omega \tilde{\epsilon}'' |\vec{\mathcal{E}}|^2 + \frac{1}{2} \omega \tilde{\mu}'' |\vec{\mathcal{H}}|^2 \right] dV \tag{2.81}$$

is the amount of light absorbed in volume $\Omega$ in electric and magnetic form.

Equations (2.77) and (2.78) simply express the very general principle of conservation of energy in the harmonic regime.

It is also useful at this point to introduce the densities defined by

$$\frac{dF}{dV} = -\frac{1}{2} \Re[\vec{\mathcal{J}}.\vec{\mathcal{E}}^*], \text{ power volume density}$$

$$\frac{d\Phi}{dS} = \Re[\vec{\Pi}.\vec{n}], \text{ flux surface density} \tag{2.82}$$

$$\frac{dA}{dV} = \frac{1}{2} \omega \left[ \tilde{\epsilon}'' |\vec{\mathcal{E}}|^2 + \tilde{\mu}'' |\vec{\mathcal{H}}|^2 \right], \text{ absorption volume density},$$

noting that these variables are **local** functions, in contrast with the variables $F$, $\Phi$ and $A$, which express an energy balance in some particular volume $\Omega$.

Bear in mind also that this energy balance has been established for a monochromatic wave without reference to its spatial distribution. In the case of a progressive unidirectional wave in nonabsorbing free space ($\vec{\beta}$ in $\mathbb{R}^3$), the Poynting vector can be further developed using the orthogonality relations, so

$$\vec{\Pi} = \frac{1}{2} \left\{ \vec{\mathcal{E}}^* \wedge [\frac{1}{\omega \tilde{\mu}} (\vec{\beta} \wedge \vec{\mathcal{E}})] \right\} = \frac{1}{2 \omega \tilde{\mu}} |\vec{\mathcal{E}}|^2 \vec{\beta}. \tag{2.83}$$

As a result, the surface flux density through a surface perpendicular to the direction of propagation can be written

$$\frac{d\Phi}{dS} = \frac{\beta}{2 \omega \tilde{\mu}} |\vec{\mathcal{E}}|^2 = \frac{n}{2 \eta_v \mu_r} |\vec{\mathcal{E}}|^2, \tag{2.84}$$

where $\eta_v = \sqrt{\frac{\mu_v}{\epsilon_v}}$ is the impedance of vacuum. We shall see in Chapter 4 how these ideas can be linked to those of irradiance, intensity, or radiance.

## 2.11  The Equation of Harmonic Propagation with Source Term

In contrast with the procedure followed in Chapter 1 in which we considered the particular solution of Poisson's equation given by the associated Green's function, in the harmonic regime we are for the moment restricted to examining the homogeneous solution of the Helmholtz equation, since the sources were all at an infinite distance from a transparent medium. This procedure revealed the analytical form of the fields propagating in free space far from any sources.

However, in particular when studying near-field phenomena, it is useful to know the particular solution of the harmonic wave equation in the presence of a source term $\vec{\mathcal{S}}(\vec{\rho}, f)$, i.e.,

$$\Delta\vec{\mathcal{E}}(\vec{\rho}, f) + k^2\vec{\mathcal{E}}(\vec{\rho}, f) = \vec{\mathcal{S}}(\vec{\rho}, f) = -i\omega\tilde{\mu}\vec{\mathcal{J}}(\vec{\rho}, f) + \mathbf{grad}\left[\frac{1}{\tilde{\epsilon}(f)}\mathcal{Q}(\vec{\rho}, f)\right],$$
(2.85)

whose general solution is given by

$$\vec{\mathcal{E}}(\vec{\rho}, f) = [G \star \vec{\mathcal{S}}](\vec{\rho}, f),$$
(2.86)

where the convolution is over the space variable $\vec{\rho}$ and the associated Green's function must satisfy

$$\Delta G(\vec{\rho}) + k^2 G(\vec{\rho}) = \delta(\vec{\rho}).$$
(2.87)

It can be shown that this Green's function can be expressed as

$$G(\vec{\rho}) = \pm\frac{1}{4\pi}\frac{e^{i\vec{\beta}.\vec{\rho}}}{|\rho|} \quad \text{with} \quad \vec{\beta}.\vec{\beta} = k^2.$$
(2.88)

This expression is useful when dealing with volume sources in the harmonic regime or when tackling the notion of a point source or spherical wave. Notice that the homogeneous solution must be added to this particular solution so as to satisfy the boundary conditions and arrive at the correct field.

## 2.12  Maxwell's Equations in the Sense of Distributions

It is important to conclude this chapter with a reformulation of Maxwell's equations in terms of distributions (generalized functions).

In any electromagnetic optics problem, the analytical form of the field is expressible as a linear combination of *homogeneous* solutions (i.e., without a source term), and *particular* solutions (with source terms) of the propagation equation. Since the particular solutions are specifically defined (often as the source convolved with the associated Green's function), these are therefore the homogeneous solutions that are the (unknown) constants to be determined. This determination is made via the continuity equations (boundary conditions), and it is for this reason that discontinuities in the electromagnetic field are often described as sources of this field.

A useful technique for remembering these continuity relations is to consider that Maxwell's equations, until now written in the sense of functions, remain true in the sense of distributions. In the harmonic regime and in the presence of sources, these distribution-type equations can be written in the usual form

$$
\begin{cases}
\mathbf{curl}\ \vec{\mathcal{E}}(\vec{\rho}, f) = i\omega\tilde{\mu}(f)\vec{\mathcal{H}}(\vec{\rho}, f) + \vec{\mathcal{M}}(\vec{\rho}, f) \\
\mathbf{curl}\ \vec{\mathcal{H}}(\vec{\rho}, f) = -i\omega\tilde{\epsilon}(f)\vec{\mathcal{E}}(\vec{\rho}, f) + \vec{\mathcal{J}}(\vec{\rho}, f) \\
\mathrm{div}[\tilde{\epsilon}(f)\vec{\mathcal{E}}(\vec{\rho}, f)] = \mathcal{Q}(\vec{\rho}, f) \\
\mathrm{div}[\tilde{\mu}(f)\vec{\mathcal{H}}(\vec{\rho}, f)] = \mathcal{P}(\vec{\rho}, f)
\end{cases}, \tag{2.89}
$$

but, in order to make the problem as general as possible, electric and magnetic currents $\vec{\mathcal{J}}$ and $\vec{\mathcal{M}}$ respectively are included, as well as electric and magnetic charges $\mathcal{Q}$ and $\mathcal{P}$. These current and charge distributions can be real or fictitious (i.e., resulting from a mathematical operation).

To preserve the greatest possible generality for this problem, we have to assume that these variables possess parts that are *regular* (*bijective* with functions) but also singular (Dirac distributions or limits of functions), depending on whether or not they can be associated with common functions. Hence, where the discontinuities occur on a surface S, we can write

$$
\begin{cases}
\vec{\mathcal{M}} = \vec{\mathcal{M}}_R + \vec{\mathcal{M}}_S\delta_S \\
\vec{\mathcal{J}} = \vec{\mathcal{J}}_R + \vec{\mathcal{J}}_S\delta_S \\
\mathcal{Q} = \mathcal{Q}_R + \mathcal{Q}_S\delta_S \\
\mathcal{P} = \mathcal{P}_R + \mathcal{P}_S\delta_S
\end{cases}. \tag{2.90}
$$

We now analyze what happens to the continuity relations when these equations are expanded in the sense of distributions. For the first equation we get

$$
\mathbf{curl}\ \vec{\mathcal{E}} = (\mathbf{curl}\ \vec{\mathcal{E}}) + \vec{n} \wedge \sigma[\vec{\mathcal{E}}].\,\delta_S = i\omega\tilde{\mu}\vec{\mathcal{H}} + \vec{\mathcal{M}}_R + \vec{\mathcal{M}}_S\delta_S, \tag{2.91}
$$

where $\sigma[\vec{\mathcal{E}}]$ denotes the discontinuity in the vector $\vec{\mathcal{E}}$ on crossing surface $S$, $\vec{n}$ is the normal to the surface $S$; the term in brackets $(\mathbf{curl}\ \vec{\mathcal{E}})$ is known as the *derivative without precaution* for this field and corresponds to a derivative in the sense of functions, while the term that contains the discontinuity (the jump) is a Dirac surface distribution. The support of this distribution is the surface S of the field discontinuity.

Identifying the regular parts of the distributions, this immediately gives us an equation identical to that used until now in the sense of functions:

$$(\mathbf{curl}\ \vec{\mathcal{E}}) = i\omega\tilde{\mu}\vec{\mathcal{H}} + \vec{\mathcal{M}}_R. \tag{2.92}$$

Furthermore, identifying the singular parts,

$$\vec{n} \wedge \sigma[\vec{\mathcal{E}}].\delta_S = \vec{\mathcal{M}}_S\delta_S \quad \Rightarrow \quad \vec{n} \wedge \sigma[\vec{\mathcal{E}}] = \vec{\mathcal{M}}_S, \tag{2.93}$$

which shows that any discontinuity in the tangential component of the electric field implies the presence of a magnetic surface current.

The same procedure can obviously be applied to the other three Maxwell's equations. In all cases, the regular parts lead to the same equations as those already established in the sense of functions. As for the singular parts, these enable the discontinuities in the vectors $\vec{\mathcal{E}}$ and $\vec{\mathcal{H}}$ to be expressed thus:

• For the tangential component of the magnetic field:

$$\mathbf{curl}\ \vec{\mathcal{H}} = (\mathbf{curl}\ \vec{\mathcal{H}}) + \vec{n} \wedge \sigma[\vec{\mathcal{H}}].\delta_S = -i\omega\tilde{\epsilon}\vec{\mathcal{E}} + \vec{\mathcal{J}}_R + \vec{\mathcal{J}}_S\delta_S$$
$$\Rightarrow \quad \vec{n} \wedge \sigma[\vec{\mathcal{H}}] = \vec{\mathcal{J}}_S. \tag{2.94}$$

• For the normal component of the electric field:

$$\mathrm{div}[\tilde{\epsilon}\vec{\mathcal{E}}] = (\mathrm{div}[\tilde{\epsilon}\vec{\mathcal{E}}]) + \vec{n}.\sigma[\tilde{\epsilon}\vec{\mathcal{E}}].\delta_S = \mathcal{Q}_R + \mathcal{Q}_S\delta_S$$
$$\Rightarrow \quad \vec{n}.\sigma[\tilde{\epsilon}\vec{\mathcal{E}}] = \mathcal{Q}_S. \tag{2.95}$$

• For the normal component of the magnetic field:

$$\mathrm{div}[\tilde{\mu}\vec{\mathcal{H}}] = (\mathrm{div}[\tilde{\mu}\vec{\mathcal{H}}]) + \vec{n}.\sigma[\tilde{\mu}\vec{\mathcal{H}}].\delta_S = \mathcal{P}_R + \mathcal{P}_S\delta_S$$
$$\Rightarrow \quad \vec{n}.\sigma[\tilde{\mu}\vec{\mathcal{H}}] = \mathcal{P}_S. \tag{2.96}$$

Remember that in the absence of singular sources (carried by a surface distribution), the fields $\vec{\mathcal{E}}$ and $\vec{\mathcal{H}}$ have continuous tangential components, likewise for the normal components of $\tilde{\epsilon}\vec{\mathcal{E}}$ and $\tilde{\mu}\vec{\mathcal{H}}$. Note that the singular (surface) currents are tangential to the discontinuity surface S. In another way, note also that the electric or magnetic volume currents create no discontinuities in the tangential field components, and that the electric and magnetic volume charges create no discontinuities in the normal components of $\tilde{\epsilon}\vec{\mathcal{E}}$ and $\tilde{\mu}\vec{\mathcal{H}}$.

Finally, in the case in which there are no currents or charges, the tangential components of the electric and magnetic fields are continuous.

To conclude, writing Maxwell's equations in the sense of distributions immediately gives access to the discontinuity relations. We also remark that the equations governing the $\vec{\mathcal{E}}$ and $\vec{\mathcal{H}}$ fields are identical from a mathematical standpoint, provided permittivity and permeability behave appropriately ($\tilde{\epsilon} = -\tilde{\mu}$), and that electric and magnetic sources are identical (give or take a sign for the charge).

# 3
# Frequency Wave Packet

## 3.1 Introduction

Since its invention in 1960, the **Laser** (*Light Amplification by Stimulated Emission of Radiation*) has seen such extraordinary development that it is now part of everyday life in such varied objects as barcode readers; printers; high-speed telecommunications systems; CD, DVD, and Blu-Ray players; optical levels; devices for measuring speed and distance; and others.

In the collective unconscious the laser is a narrow directional monochromatic beam: this is by no means untrue, but it does need to be substantiated rather more scientifically. Recall for the moment that, when the output power is continuous, i.e., constant over time, most of these sources can be taken as a monochromatic wave of the type we studied previously in the harmonic regime. By way of example, today's He–Ne lasers deliver a power of a few milliwatts in the visible spectrum over a few bands in the order of a picometer wide.

However, there are also many pulsed laser sources whose mode of operation means that they deliver a light energy $\Delta W$ over some characteristic time $\Delta t$. An order of magnitude for the peak power of the emitted pulse of light can be defined by

$$P \simeq \frac{\Delta W}{\Delta t}. \tag{3.1}$$

Various operational modes can be envisaged for these pulsed lasers, giving extremely varied pulse durations:

1. First, *relaxed* operation, in which the amplifier medium is continuously pumped; the emitted pulses are associated with relaxation oscillations of this amplifying medium. Pulse widths are very variable and can be anything from a few microseconds to around a hundred milliseconds.

2. *Triggered* (or Q-switched) operation, in which the energy stored in the optical cavity is suddenly converted into the form of light, the duration of the pulses thus obtained being from a few nanoseconds to a few hundreds of nanoseconds.

3. Finally *mode-locking*, where the different oscillatory modes in the amplifying cavity are phase-locked; this method leads to the shortest pulse lengths, between ten femtoseconds and roughly one hundred picoseconds.[1]

To consolidate our ideas, let us assume that the energy $\Delta W$ emitted per pulse is on the order of one millijoule; this means that the peak power in a ten femtosecond pulse is close to a hundred gigawatts!

Furthermore, it is also possible to focus this beam using an optical system with the appropriate properties onto transverse dimensions of the order of a wavelength and hence to generate a luminous spot of extreme brightness. The greatest possible surface power density (or irradiance) will be in the order of $P/\lambda^2$, which gives for our example nearly $15 \times 10^{18}$ W per square centimeter!

These very considerable energy densities can be put to good use in drilling, cutting or welding, just by using light alone. They have also led to major scientific projects such as

1. **Thermonuclear fusion by inertial confinement** (*National Ignition Facility* [NIF] in the USA, *Laser MegaJoule* [LMJ] in France), in which 240 laser beams of wavelength 351 nm are isotropically focused onto a 2 mm diameter target containing a mix of deuterium and tritium. When the target absorbs the 1.8 megajoules delivered by the lasers in 20 nanoseconds, this causes the target's internal temperature to rise to 200 million Kelvin at a pressure of several terabar.

2. The **creation of elementary particles** simply by focusing a very high-power laser beam in a vacuum, which can be seen as one of the consequences of the mass–energy equivalence discovered by Einstein. This is the ultimate purpose of the European *Extreme Light Infrastructure* (ELI) project, whose aim is also to produce ultrashort pulses of high-energy particles such as photons, electrons, protons, neutrons, or neutrinos, or to develop new processes for radiotherapy or medical imaging.

Within such a context several questions arise: can a light wave characterized by such short pulse durations (e.g., 50 fs) still be considered to be

---

[1] 1 nanosecond = $10^{-9}$ second; 1 picosecond = $10^{-12}$ second; 1 femtosecond = $10^{-15}$ s.

monochromatic? When passing through a material, are the temporal properties of these ultrashort pulses affected? And how do we model the response of a two-beam interferometer when the light source is this particular type? We shall now attempt to answer these different questions.

## 3.2 Line Width and Pulse Duration

### 3.2.1 The General Case

At some point $\vec{\rho}_0$ in space arbitrarily chosen as the origin, we consider a pulse of light of *duration* $\Delta t$ and of *mean* angular frequency $\omega_0$. To keep things simple, we shall focus on just one algebraic component of the associated field projected onto one of the polarization axes. This component can then be put into the following form:

$$E(\vec{\rho}_0, t) = A_0(t) \cos \omega_0 t\,, \tag{3.2}$$

where we shall assume that $A_0(t)$ is a function that varies quite slowly over a characteristic length of time $t_c$ corresponding to the period of oscillation of the carrier ($t_c = 2\pi/\omega_0$). In other words, $A_0(t)$ corresponds to the envelope of the harmonic signal $\cos \omega_0 t$, with the signal oscillating at angular frequency $\omega_0$ within it.

The frequency spectrum of this pulse of light can be determined by calculating the Fourier transform of its temporal profile, i.e.,

$$\mathcal{E}(\vec{\rho}_0, f) = \int_{-\infty}^{+\infty} E(\vec{\rho}_0, t) e^{2i\pi ft} dt = \int_{-\infty}^{+\infty} A_0(t) \cos \omega_0 t\, e^{2i\pi ft} dt\,. \tag{3.3}$$

We replace the temporal pulsation $\omega_0$ by $2\pi f_0$, and then express the cosine as a half-sum of conjugate imaginary complex exponentials, i.e.,

$$\mathcal{E}(\vec{\rho}_0, f) = \frac{1}{2} \left[ \int_{-\infty}^{+\infty} A_0(t)\, e^{2i\pi(f - f_0)t} dt + \int_{-\infty}^{+\infty} A_0(t)\, e^{2i\pi(f + f_0)t} dt \right] \tag{3.4}$$

$$= \frac{1}{2} [\tilde{A}_0(f - f_0) + \tilde{A}_0(f + f_0)]\,,$$

where $\tilde{A}_0(f)$ is the Fourier transform of the envelope function $A_0(t)$.

In the equation describing this mechanism [equation (2.32)], integration can be carried out only over positive values of frequency:

$$\vec{E}(\vec{\rho}_0, t) = 2 \int_0^{+\infty} \Re \left[ \vec{\mathcal{E}}(\vec{\rho}_0, f) e^{-2i\pi ft} \right] df\,.$$

We are interested in the behavior of the function $\mathcal{E}(\vec{\rho}_0, f)$ in the frequency domain and, especially, in that of the support of the function $\tilde{A}_0(f)$ around the mean frequency $f_0$, which we shall now refer to as the *line width*.

Before going any further, we need a more rigorous definition of the two ideas we just introduced qualitatively, namely those of pulse duration $\Delta t$ and line width $\Delta f$. Let us start with the time and frequency widths defined as

$$(\Delta t)^2 = \frac{\displaystyle\int_{-\infty}^{+\infty} t^2 |A_0(t)|^2\, dt}{\displaystyle\int_{-\infty}^{+\infty} |A_0(t)|^2\, dt} \quad \text{and} \quad (\Delta f)^2 = \frac{\displaystyle\int_{-\infty}^{+\infty} f^2 |\tilde{A}_0(f)|^2\, df}{\displaystyle\int_{-\infty}^{+\infty} |\tilde{A}_0(f)|^2\, df} . \tag{3.5}$$

Consider the integral in the numerator of the expression that defines the line width $\Delta f$. This can be written as

$$\int_{-\infty}^{+\infty} f^2 |\tilde{A}_0(f)|^2\, df = \frac{1}{4\pi^2} \int_{-\infty}^{+\infty} |2i\pi f \tilde{A}_0(f)|^2\, df . \tag{3.6}$$

Then, using both the definition of the Fourier transform of the derivative of a function and Parseval's theorem:

$$\int_{-\infty}^{+\infty} |2i\pi f \tilde{A}_0(f)|^2\, df = \int_{-\infty}^{+\infty} \left| \frac{dA_0}{dt} \right|^2 dt . \tag{3.7}$$

Applying the same Parseval's theorem to the denominator of the expression that defines the line width, we immediately deduce that

$$(\Delta f)^2 = \frac{1}{4\pi^2} \frac{\displaystyle\int_{-\infty}^{+\infty} \left| \frac{dA_0}{dt} \right|^2 dt}{\displaystyle\int_{-\infty}^{+\infty} |A_0(t)|^2\, dt} . \tag{3.8}$$

We now form the product $(\Delta t)^2 (\Delta f)^2$, often known as the *time–bandwidth product*:

$$(\Delta t)^2 (\Delta f)^2 = \frac{1}{4\pi^2} \frac{\displaystyle\int_{-\infty}^{+\infty} t^2 |A_0(t)|^2\, dt}{\displaystyle\int_{-\infty}^{+\infty} |A_0(t)|^2\, dt} \frac{\displaystyle\int_{-\infty}^{+\infty} \left| \frac{dA_0}{dt} \right|^2 dt}{\displaystyle\int_{-\infty}^{+\infty} |A_0(t)|^2\, dt} . \tag{3.9}$$

Using the **Cauchy–Schwarz inequality**, we can write

$$\left| \int_{-\infty}^{+\infty} t A_0(t) \frac{dA_0^*}{dt}\, dt \right|^2 \leq \left( \int_{-\infty}^{+\infty} t^2 |A_0(t)|^2\, dt \right) \left( \int_{-\infty}^{+\infty} \left| \frac{dA_0}{dt} \right|^2 dt \right), \tag{3.10}$$

which allows us to define a lower bound of the numerator $N$ in equation (3.9). Making use of the fact that the envelope function $A_0(t)$ is real, we can write

$$N \geq \left[ \int_{-\infty}^{+\infty} t A_0(t) \frac{dA_0^*}{dt} dt \right]^2 = \left[ \frac{1}{2} \int_{-\infty}^{+\infty} t \left\{ \frac{d}{dt} A_0^2(t) \right\} dt \right]^2 . \qquad (3.11)$$

The last integral in (3.11) can be calculated by integration by parts, i.e.,

$$\frac{1}{2} \int_{-\infty}^{+\infty} t \left\{ \frac{d}{dt} A_0^2(t) \right\} dt = \frac{1}{2} \left[ t A_0^2(t) \right]_{-\infty}^{+\infty} - \frac{1}{2} \int_{-\infty}^{+\infty} A_0^2(t) \, dt = -\frac{1}{2} \int_{-\infty}^{+\infty} A_0^2(t) \, dt . \quad (3.12)$$

The zero value of the first term in the integration by parts is directly related to the finite nature of the length of the pulse in question, and hence its strong decay at infinity. Putting this result into (3.9) and (3.10), we finally get

$$(\Delta t)^2 (\Delta f)^2 \geq \frac{1}{16\pi^2} , \qquad (3.13)$$

This can also be written as

$$\Delta t . \Delta f \geq \frac{1}{4\pi} \quad \text{or} \quad \Delta t . \Delta \omega \geq \frac{1}{2} \quad \text{or} \quad \frac{\Delta \lambda}{\lambda^2} \geq \frac{1}{4\pi c \, \Delta t} . \qquad (3.14)$$

This shows that, the shorter the duration of a pulse of light, the greater its line width. To illustrate this point, a 40-fs pulse centered around a central frequency $f_0$ of 375 THz (i.e. $\lambda_0 = 800$ nm) will have a line width $\Delta f$ of at least 20 THz (i.e., $\Delta \lambda > 40$ nm).

To conclude this short section, we can stress that these inequalities result directly from the properties of the Fourier transform and from the definition that we used for the time and frequency widths.

### 3.2.2 Light Pulse with a Gaussian Profile

We now consider the special case for which the temporal envelope of the pulse is represented by a Gaussian function, namely

$$A_0(t) = A_0 e^{-\left(\frac{t}{\tau}\right)^2} , \qquad (3.15)$$

where $\tau$ is the temporal half-width of the pulse at $1/e$.

*Calculating the Pulse Duration*

Using the formalism introduced in Section 3.2.1, we can calculate the duration $\Delta t$ of this pulse:

$$(\Delta t)^2 = \frac{\displaystyle\int_{-\infty}^{+\infty} t^2 A_0^2(t)\, dt}{\displaystyle\int_{-\infty}^{+\infty} A_0^2(t)\, dt} = \frac{\displaystyle\int_{-\infty}^{+\infty} t^2 e^{-2\left(\frac{t}{\tau}\right)^2}\, dt}{\displaystyle\int_{-\infty}^{+\infty} e^{-2\left(\frac{t}{\tau}\right)^2}\, dt}. \tag{3.16}$$

We start by calculating the derivative of the function $F(t) = \exp[-2(t/\tau)^2]$, i.e.,

$$\frac{d}{dt} F(t) = -\frac{4}{\tau^2} t\, F(t). \tag{3.17}$$

Consequently, the numerator in equation (3.16) can be put in the form

$$\int_{-\infty}^{+\infty} t^2 e^{-2\left(\frac{t}{\tau}\right)^2}\, dt = -\frac{\tau^2}{4} \int_{-\infty}^{+\infty} t \frac{d}{dt}[F(t)]\, dt$$

$$= -\frac{\tau^2}{4} \left[ t e^{-2\left(\frac{t}{\tau}\right)^2} \right]_{-\infty}^{+\infty} + \frac{\tau^2}{4} \int_{-\infty}^{+\infty} e^{-2\left(\frac{t}{\tau}\right)^2}\, dt \tag{3.18}$$

$$= \frac{\tau^2}{4} \int_{-\infty}^{+\infty} e^{-2\left(\frac{t}{\tau}\right)^2}\, dt.$$

Substituting this result in (3.16), we immediately get

$$\Delta t = \frac{\tau}{2}. \tag{3.19}$$

*Calculating the Spectral Width*

To determine the line width associated with such a pulse we first need to calculate the Fourier transform $\tilde{A}_0(f)$ of the envelope function $A_0(t)$. To this end, we shall use the following memory aid, that allows the Fourier transform of a Gaussian function to be calculated in the general case:

$$e^{-\pi a t^2} \xrightarrow{\text{FT}} \frac{1}{\sqrt{a}}\, e^{-\frac{\pi}{a} f^2}.$$

A simple identification to (3.15) leads to $\frac{1}{a} = \pi \tau^2$, and hence to the expression we seek:

$$\tilde{A}_0(f) = A_0 \int_{-\infty}^{+\infty} e^{-\left(\frac{t}{\tau}\right)^2} e^{2i\pi f t}\, dt = A_0 \tau \sqrt{\pi}\, e^{-\pi^2 \tau^2 f^2}. \tag{3.20}$$

It remains to calculate the line width $\Delta f$ by applying the general formula given in (3.5); hence

$$(\Delta f)^2 = \frac{\displaystyle\int_{-\infty}^{+\infty} f^2 |\tilde{A}_0(f)|^2 \, df}{\displaystyle\int_{-\infty}^{+\infty} |\tilde{A}_0(f)|^2 \, df} = \frac{\displaystyle\int_{-\infty}^{+\infty} f^2 e^{-2\pi^2\tau^2 f^2} \, df}{\displaystyle\int_{-\infty}^{+\infty} e^{-2\pi^2\tau^2 f^2} \, df} \, , \qquad (3.21)$$

and then adopting a procedure that is entirely similar to that used in calculating the pulse duration $\Delta t$. It can easily be shown that

$$\Delta f = \frac{1}{2\pi\tau} \, . \qquad (3.22)$$

### Calculating the Time-Bandwidth Product

The time–bandwidth product has the value

$$\Delta t . \Delta f = \frac{\tau}{2}\frac{1}{2\pi\tau} = \frac{1}{4\pi} \qquad (3.23)$$

and corresponds to the minimum value imposed by the existence of a Fourier transform type of relationship between these two quantities. This property is an important feature of pulses of Gaussian profile, in accordance with the lower bound defined by the Cauchy–Schwarz inequality.

Figure 3.1 shows the temporal profile of such a pulse of light for which the parameter $\tau$ has been taken to be 40 fs and whose wavelength is assumed centered on 800 nm (upper graph); its frequency spectrum (lower graph) clearly shows the Hermitian symmetry whose generic character we have already mentioned.

It is often convenient to consider an alternative definition for the time–bandwidth product in which each of the two parameters is taken to be equal to the total width at the mid-height of the corresponding profile. This total mid-height width is commonly referred to by the acronym FWHM, or Full-Width at Half-Maximum.

In the special case of a pulse with a Gaussian temporal profile, it is easy to calculate the value taken by these two FWHM widths:

$$(\Delta t)_{\text{FWHM}} = 2\tau\sqrt{\ln 2} \quad \text{and} \quad (\Delta f)_{\text{FWHM}} = \frac{2\sqrt{\ln 2}}{\pi\tau} \qquad (3.24)$$

and then form the associated time–bandwidth product, i.e.,

$$(\Delta t)_{\text{FWHM}} . (\Delta f)_{\text{FWHM}} = \frac{4}{\pi}\ln 2 \simeq 0,88 \, . \qquad (3.25)$$

For the time–bandwidth product in the case of an arbitrary temporal profile

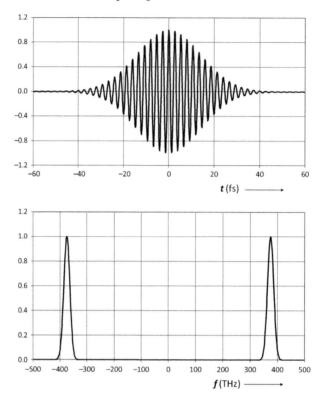

Figure 3.1 Temporal profile and frequency spectrum of a Gaussian pulse (central wavelength 800 nm; pulse duration 40 fs).

we shall frequently use an alternative formulation to (3.14), namely

$$(\Delta t)_{\text{FWHM}} \cdot (\Delta f)_{\text{FWHM}} \geq 1 . \tag{3.26}$$

### 3.2.3 Overview

To conclude this section, we emphasize the fact that the previously demonstrated results are intrinsically related to the properties of the Fourier transform: the product of the supports of a function and its Fourier transform has a lower bounded defined by a constant whose order of magnitude is 1. This down-factoring that we used here for time resolution in optics also underlies the limit of optical resolution in space; we shall cover this in Chapter 4, which is dedicated to the spatial wave packet.

We understand that this time–bandwidth product satisfies an inequality that we simply have to accept. The significance of this is that, although a short-duration pulse must necessarily imply an extended spectrum, a broad-spectrum source is not necessarily impulsive because in this second case, the inequality leads simply to $\Delta t > 0$. To find the impulsive character of the wave, we must have complementary phase spectral matching.

It is also interesting to note that the time–frequency equivalence is useful in the design of many experiments. For example, the first step in the *Chirped Pulse Amplification* (CPA) method shapes the pulse frequency spectrum, broadening it temporally to reduce the peak transported power $(P \simeq \frac{\Delta W}{\Delta t})$ before recompressing it in the second step to recover the initial temporal form. To a great extent this procedure allows problems of damage to optical elements subject to an intense luminous flux to be avoided.

In a more general sense, we often get the same information from an experiment by measuring a system's frequency or temporal response, provided we subsequently adjust the corresponding spectral or temporal windows.

## 3.3 Group Velocity

### 3.3.1 Centroid of a Pulse of Arbitrary Temporal Profile

Given that the *time–bandwidth product* has been identified and understood, we shall now study how the energy associated with the light wave propagates. First we shall return to the case of an **arbitrary temporal profile**, defining the idea of group velocity in the general case. We assume that the wave travels along the $Oz$ axis in a transparent homogeneous medium with wave vector $\vec{\beta}(f) = \beta(f)\,\vec{z}$, $\beta(f) \in \mathbb{R}$. Recall that this means we are dealing with a unidirectional wave with a range of frequencies around a central frequency.

We know that the power of a light pulse varies with the square of the electric field $E(z,t)$. To identify the position in $z$ of the energy centroid of this pulse, we form the product of this square $E^2(z,t)$ with a $z$-variable function having a symmetrical profile and a minimum at $z = z_0$, then we integrate the result over the space variable $z$. The result obtained represents the overlap between the two quantities whose product was calculated, and will obviously be a function of $z_0$. The particular $z_0$ corresponding to the minimum of this product $P(z_0)$ enables the position of the centroid of the pulse in question to be determined.

For example, if we choose a quadratic form of type $(z - z_0)^2$ as our symmetrical function, then the position $z_0$ of this centroid will be defined by

$$\frac{\partial}{\partial z_0} P(z_0) = \frac{\partial}{\partial z_0}\left[ \int_{-\infty}^{+\infty} (z - z_0)^2 E^2(z,t)\, dz \right] = 0\,; \tag{3.27}$$

so

$$- 2 \int_{-\infty}^{+\infty} (z - z_0) E^2(z, t) \, dz = 0 \,, \tag{3.28}$$

and hence

$$z_0(t) = \frac{\int_{-\infty}^{+\infty} z E^2(z, t) \, dz}{\int_{-\infty}^{+\infty} E^2(z, t) \, dz} \,. \tag{3.29}$$

### 3.3.2 Propagation Speed of the Centroid

Calculation of the propagation speed of the centroid requires the quantity $\frac{\partial z_0}{\partial t}$ to be evaluated; in the absence of more precise information about the field $E(z, t)$, this does not seem very simple.

As is often the case, a good solution here involves passing into Fourier space where the calculation of derivatives is much simpler; this also involves the use of Parseval's theorem.

We write

$$\frac{\int_{-\infty}^{+\infty} z E(z, t) . E^*(z, t) \, dz}{\int_{-\infty}^{+\infty} E(z, t) . E^*(z, t) \, dz} = \frac{\int_{-\infty}^{+\infty} z \widetilde{E}(\beta, t) . \widetilde{E}^*(\beta, t) \, d\beta}{\int_{-\infty}^{+\infty} \widetilde{E}(\beta, t) . \widetilde{E}^*(\beta, t) \, d\beta} \,, \tag{3.30}$$

where $\beta$ is the conjugate $z$ variable in the Fourier sense.

Note here that equation (3.30) involves a Fourier transformation with respect to the space variable $z$, a transformation that we have not yet defined. In writing this Fourier space transformation we shall adopt the negative sign convention (this will be justified when addressing the spatial packet in Chapter 4), in contrast to the temporal Fourier transforms that we have defined using the positive sign convention. Hence

$$\widetilde{E}(\beta, t) = \int_z E(z, t) e^{-2i\pi\beta z} \, dz \,. \tag{3.31}$$

### *Finding an Expression for the FT Along the Axis of Propagation*

We introduce the expression for the frequency wave packet that describes the propagation of the progressive field $E(z, t)$, i.e.,

$$E(z, t) = \int_f A(f) e^{ik(f)z} e^{-2i\pi f t} \, df \,. \tag{3.32}$$

Substituting this expression into (3.31) and changing the order of the integrals, we get

$$\tilde{E}(\beta,t) = \int_z \int_f A(f)e^{ik(f)z}e^{-2i\pi ft}e^{-2i\pi\beta z}\, df\, dz$$

$$= \int_f A(f)e^{-2i\pi ft}\left\{\int_z e^{ik(f)z}e^{-2i\pi\beta z}\, dz\right\}df\,, \qquad (3.33)$$

or, introducing the Dirac distribution $\delta$,

$$\tilde{E}(\beta,t) = 2\pi \int_f A(f)e^{-2i\pi ft}\delta[k(f) - 2\pi\beta]\, df\,. \qquad (3.34)$$

We need to introduce a change of variable at this stage; the Dirac distribution would allow the integral to be tackled provided the new variable is $k$ and not $f$. However, we shall see in Chapter 5 that the Kramers–Kronig relations predict that, in a transparent zone (as per our assumption), the real refractive index $n(\omega)$ is an increasing function of frequency. As a result, the function $k(\omega)$ is monotonic, since

$$\frac{\partial}{\partial\omega}k(\omega) = \frac{\partial}{\partial\omega}\left[\frac{n(\omega)\omega}{c}\right] = \frac{1}{c}\left\{n + \omega\frac{\partial n}{\partial\omega}\right\} > 0 \quad \text{if} \quad \frac{\partial n}{\partial\omega} > 0\,, \qquad (3.35)$$

and we are justified in considering the inverse function $f = s(k)$ giving frequency $f$ as a function of $k$. Under these conditions, equation (3.34) can be put in the form

$$\tilde{E}(\beta,t) = 2\pi \int_k A[s(k)]e^{-2i\pi s(k)t}\delta(k - 2\pi\beta)\left[\frac{\partial s}{\partial k}(k)\right] dk\,. \qquad (3.36)$$

Using the properties of the Dirac distribution, we get

$$\tilde{E}(\beta,t) = 2\pi A[s(2\pi\beta)]e^{-2i\pi s(2\pi\beta)t}\left[\frac{\partial s}{\partial k}(2\pi\beta)\right]. \qquad (3.37)$$

The following notation:

$$C(\beta) = 2\pi A[s(2\pi\beta)]\left[\frac{\partial s}{\partial k}(2\pi\beta)\right] \quad \text{and} \quad \omega(\beta) = 2\pi s(2\pi\beta) \qquad (3.38)$$

allows the Fourier transform of the field to be written in the form

$$\tilde{E}(\beta,t) = C(\beta)e^{-i\omega(\beta)t}\,. \qquad (3.39)$$

## Calculating the Centroid

We can invoke the various properties of the Fourier transform in the last expression (3.39) to write

$$\widetilde{zE}(\beta) = \frac{i}{2\pi} \frac{\partial}{\partial \beta} \left[ \tilde{E}(\beta) \right] = \frac{i}{2\pi} e^{-i\omega(\beta)t} \left\{ \frac{\partial C}{\partial \beta} - it \frac{\partial \omega}{\partial \beta} C(\beta) \right\}, \qquad (3.40)$$

and finally

$$z_0(t) = \frac{i}{2\pi} \frac{\int_\beta \frac{\partial C}{\partial \beta} C^* \, d\beta}{\int_\beta |C(\beta)|^2 \, d\beta} + \frac{1}{2\pi} \frac{\int_\beta \frac{\partial \omega}{\partial \beta} |C(\beta)|^2 \, d\beta}{\int_\beta |C(\beta)|^2 \, d\beta} \ t = a + v_g \, t. \qquad (3.41)$$

Hence we see that in a transparent homogeneous medium the centroid of the pulse moves at a constant speed $v_g$ known as the **group velocity**, defined by

$$v_g = \frac{1}{2\pi} \frac{\int_\beta \frac{\partial \omega}{\partial \beta} |C(\beta)|^2 \, d\beta}{\int_\beta |C(\beta)|^2 \, d\beta} = \frac{\int_k \frac{\partial \omega}{\partial k} |\tilde{C}(k)|^2 \, dk}{\int_k |\tilde{C}(k)|^2 \, dk}. \qquad (3.42)$$

For a large number of so-called *quasi-monochromatic* beams, the line width $\Delta\omega$ is small compared with the central frequency $\omega_0$. If we assume that, over this spectral range, variations in the derivative $\frac{\partial \omega}{\partial k}$ can be neglected, then the expression for the group velocity becomes

$$v_g = \left. \frac{\partial \omega}{\partial k} \right|_{\omega_0}. \qquad (3.43)$$

## Consequence

It is of interest to compare this group velocity with the phase velocity defined in the harmonic regime, where the angular frequency is assumed to be $\omega_0$. Let us expand on the expression for $v_g$:

$$\frac{1}{v_g} = \frac{\partial k}{\partial \omega} = \frac{\partial}{\partial \omega} \left[ \frac{n\omega}{c} \right] = \frac{n}{c} + \frac{\omega}{c} \frac{\partial n}{\partial \omega} \quad \Rightarrow \quad v_g = \frac{c}{n + \omega \frac{\partial n}{\partial \omega}}. \qquad (3.44)$$

We see that, in the absence of dispersion, the phase and group velocities are identical:

$$\frac{\partial n}{\partial \omega} = 0 \quad \Rightarrow \quad v_g = \frac{c}{n} = v. \qquad (3.45)$$

On the other hand, this is no longer true for dispersive media, a property used to slow light down in cases in which the dispersion is very high (this is the concept of *slow light*). Bearing in mind the property of $n(\omega)$ already mentioned, namely that the derivative $\frac{\partial n}{\partial \omega}$ is positive in a transparent window, note that the group velocity is always less than the phase velocity:

$$\frac{\partial n}{\partial \omega} > 0 \quad \Rightarrow \quad v_g < \frac{c}{n} = v \, . \tag{3.46}$$

We conclude by recalling that the group velocity is associated with the transmission of information (detecting the passage of a pulse): this is not the case for phase velocity.

## 3.4 Propagation of a Gaussian Pulse

### 3.4.1 Introduction

In many experiments or applications such as (in particular) long distance communication using optical fiber (*long-haul fiber optic network*), it is important that the pulses of light should propagate with the least amount of distortion. Hence, after having established the notion of group velocity as applied to the centroid of the pulse, we now turn our attention to the temporal shape of the beam that accompanies this centroid as it propagates.

Consider the special case of a pulse whose temporal envelope is a Gaussian profile, i.e.,

$$A_0(t) = A_0 \, e^{-\left(\frac{t}{\tau}\right)^2} \, . \tag{3.47}$$

We saw in Section 3.2.2 that the behavior of this particular case constitutes a limiting case in terms of the *time–bandwidth product*. This will enable us to obtain, with relative ease, analytical results whose scope is in fact quite generic.

As before, the beam is assumed unidirectional, propagating along the $Oz$ axis. Recall that we saw that its frequency spectrum on some arbitrary abscissa $z$ was deduced from that on the abscissa $z_0 = 0$ by

$$\mathcal{E}(z, f) = \mathcal{E}(0, f) e^{ik(f)z} \, . \tag{3.48}$$

We have only to use the temporal reconstruction theorem to deduce the spatiotemporal dependence of the field with which it is associated, i.e.,

$$E(z, t) = \int_{-\infty}^{+\infty} \mathcal{E}(z, f) e^{-2i\pi ft} \, df = 2\Re \left[ \int_{0}^{+\infty} \mathcal{E}(0, f) \, e^{i[k(f)z - 2\pi ft]} \, df \right] . \tag{3.49}$$

### *3.4.2 Nondispersive Medium*

If the material medium in which the pulse is propagating is **nondispersive**, i.e., if the refractive index $n$ does not depend on the frequency $f$, the calculation is very simple and can continue to follow a generic procedure, i.e.,

$$k(\omega) = n\frac{\omega}{c} = \frac{2\pi n f}{c}$$

$$\Rightarrow \ E(z,t) = \int_{-\infty}^{+\infty} \mathcal{E}(0,f)\, e^{-2i\pi f\left(t - \frac{nz}{c}\right)}\, df = E\left(0, t - \frac{nz}{c}\right). \qquad (3.50)$$

We see that the shape of the pulse has not been altered, since we have just a simple translation related to the speed of propagation of light in this medium (recall that the phase and group velocities are identical in nondispersive media).

For a pulse with a Gaussian temporal profile, we therefore have

$$E(z,t) = A_0\, e^{-\left(\frac{nz/c - t}{\tau}\right)^2} \cos[k(\omega_0)z - \omega_0 t]. \qquad (3.51)$$

Note here that in absence of dispersion, the group velocity and the phase velocity are identical and are given by $v = v_g = \omega_0/k(\omega_0) = c/n$.

### *3.4.3 Linear Dispersion Medium*

We now assume that the material medium in which this pulse of light is propagating is **dispersive** and that the spectral dependence of the modulus of the wave vector $k(\omega)$ around the central frequency $\omega_0$ is **linear** in nature, i.e., that

$$k(\omega) = k(\omega_0) + (\omega - \omega_0)\left.\frac{dk}{d\omega}\right|_{\omega_0} = k_0 + \alpha_0(\omega - \omega_0). \qquad (3.52)$$

In the expression in (3.49), let $\mathcal{J}$ be the integral in square brackets. This integral is

$$\mathcal{J} = \int_{0}^{+\infty} \mathcal{E}(0,f)\, e^{i[k_0 z + \alpha_0(\omega - \omega_0)z - 2\pi f t]}\, df$$

$$= \frac{A_0 \tau \sqrt{\pi}}{2} e^{i(k_0 z - 2\pi f_0 t)} \int_{0}^{+\infty} e^{-\pi^2 \tau^2 (f - f_0)^2}\, e^{2i\pi[\alpha_0 z - t](f - f_0)}\, df. \qquad (3.53)$$

After a change of variable $u = f - f_0$, we get

$$\mathcal{J} = \frac{A_0 \tau \sqrt{\pi}}{2} e^{i(k_0 z - 2\pi f_0 t)} \int\limits_{-f_0}^{+\infty} e^{-\pi^2 \tau^2 u^2} e^{2i\pi u[\alpha_0 z - t]} \, du. \tag{3.54}$$

Exploiting the rapid decay of the exponential spectral profile, we can replace the lower integration limit $-f_0$ by $-\infty$ and get the inverse Fourier transform of the spectral profile $\mathcal{E}(0, f)$, taken at the point $\alpha_0 z - t$, i.e.,

$$\mathcal{J} = \frac{A_0}{2} e^{i(k_0 z - 2\pi f_0 t)} \, e^{-\left(\frac{\alpha_0 z - t}{\tau}\right)^2}. \tag{3.55}$$

It just remains to take the real part of this quantity to end up with an expression for the spatiotemporal dependence of the electric field $E$ on abscissa $z$ and at time $t$:

$$E(z, t) = E_0 \, e^{-\left(\frac{\alpha_0 z - t}{\tau}\right)^2} \cos[k_0 z - \omega_0 t]. \tag{3.56}$$

We conclude that in a medium characterized by **linear spectral dispersion**, a light pulse of duration $\tau$ again propagates **without distortion**, but at a speed this time defined by

$$\alpha_0 dz - dt = 0 \quad \Rightarrow \quad \frac{dz}{dt} = \frac{1}{\alpha_0} = \left.\frac{d\omega}{dk}\right|_{\omega_0}, \tag{3.57}$$

which equates with the group velocity $v_g$ introduced in Section 3.3.2.

### *3.4.4 Quadratic Dispersion Medium*

Still with a dispersive medium, we now assume that an accurate description of its spectral dependence requires a second-order term to be included in the truncated Taylor expansion of $k(\omega)$ about the central frequency $\omega_0$, i.e.,

$$k(\omega) = k(\omega_0) + (\omega - \omega_0) \left.\frac{dk}{d\omega}\right|_{\omega_0} + \frac{1}{2}(\omega - \omega_0)^2 \left.\frac{d^2 k}{d\omega^2}\right|_{\omega_0}$$

$$= k_0 + \alpha_0(\omega - \omega_0) + \frac{\gamma_0}{2}(\omega - \omega_0)^2. \tag{3.58}$$

We say that this time the medium in question has a quadratic spectral dispersion. The integral $\mathcal{J}$ involved in the reconstruction process is expressed here as

$$J = \int\limits_{0}^{+\infty} \mathcal{E}(0, f) \, e^{i[k(\omega)z - 2\pi ft]} \, df$$

$$= \frac{E_0 \tau \sqrt{\pi}}{2} e^{i(k_0 z - 2\pi f_0 t)} \int\limits_{0}^{+\infty} e^{-\pi^2 \tau^2 \left[1 - \frac{2i\gamma_0 z}{\tau^2}\right](f - f_0)^2} \, e^{2i\pi[\alpha_0 z - t](f - f_0)} \, df \, . \quad (3.59)$$

After the usual change of variable $(u = f - f_0)$, followed by replacement of the lower limit of integration $-f_0$ by $-\infty$, it remains for us to calculate the integral:

$$\mathcal{G} = \int\limits_{-\infty}^{+\infty} e^{-\pi^2 \tau^2 \left[1 - \frac{2i\gamma_0 z}{\tau^2}\right] u^2} \, e^{2i\pi[\alpha_0 z - t] u} \, du \, .$$

To do this we shall use the reciprocal version of the *aide memoire* we used in Section 3.2.2, namely

$$e^{-\frac{\pi}{a} f^2} \xrightarrow{\text{FT}^{-1}} \sqrt{a} \, e^{-\pi a t^2} \, . \quad (3.60)$$

In our case, $a$ is a complex:

$$a = \frac{1}{\pi \tau^2 \left[1 - \frac{2i\gamma_0 z}{\tau^2}\right]} = \frac{1 + \frac{2i\gamma_0 z}{\tau^2}}{\pi \tau^2 \left[1 + \left(\frac{2\gamma_0 z}{\tau^2}\right)^2\right]} \, , \quad (3.61)$$

but since its real part is positive, use of such a relationship is justified, and enables us to write

$$\mathcal{G} = \frac{1}{\sqrt{\pi} \tau \sqrt{1 - \frac{2i\gamma_0 z}{\tau^2}}} e^{-\frac{(\alpha_0 z - t)^2}{\tau^2 \left[1 - \frac{2i\gamma_0 z}{\tau^2}\right]}} \, . \quad (3.62)$$

By simplifying and altering the representation of the complex numbers in (3.62),

$$\mathcal{G} = \frac{1}{\sqrt{\pi} \tau(z)} e^{i\zeta(z)} e^{-\frac{(\alpha_0 z - t)^2}{\tau^2(z)}} e^{-\frac{2i\gamma_0 z(\alpha_0 z - t)^2}{\tau^2 \tau^2(z)}} \, , \quad (3.63)$$

where

$$\tau(z) = \tau \sqrt{1 + \left(\frac{2\gamma_0 z}{\tau^2}\right)^2} \qquad \zeta(z) = \frac{1}{2} \arctan\left(\frac{2\gamma_0 z}{\tau^2}\right) \, . \quad (3.64)$$

Hence the spatiotemporal dependence of the field $E$ can finally be expressed as

$$E(z,t) = E_0 \frac{\tau}{\tau(z)} e^{-\frac{(\alpha_0 z - t)^2}{\tau^2(z)}} \cos\left[k_0 z - \omega_0 t + \phi(z,t)\right], \qquad (3.65)$$

where $\phi(z,t)$ is the quantity

$$\phi(z,t) = \zeta(z) - \frac{2\gamma_0 z}{\tau^2} \frac{(\alpha_0 z - t)^2}{\tau^2(z)}. \qquad (3.66)$$

We conclude that the temporal profile of a pulse of light after propagation over a distance $z$ in a quadratic dispersive medium always has a Gaussian shape: this propagates at a speed equal to the group velocity $v_g$, but its duration $\Delta t = \tau(z)/2$ is now an increasing function of the propagation distance $z$. This **temporal broadening** is associated with a reduction in the pulses peak amplitude, in accordance with the principle of conservation of energy.

These two effects (temporal broadening and reduction in maximum amplitude) are clearly demonstrated in Figure 3.2, where we have shown the temporal profile of the initial pulse for two different values of propagation distance, namely $z_1 = 0$ (pulse duration $2\tau$ is 40 fs) and $z_2 = \tau^2\sqrt{3}/2\gamma_0$ (pulse duration $2\tau$ is 80 fs).

### *3.4.5 Transmission of Information*

To a first approximation, the transmission of digital information over a single-mode optical fiber can be taken to be a succession of *pulses* and *absence of pulses* of duration $\tau$, for which the arrangement in time reproduces the sequence of bits 1 and 0 in the binary message; the time shift between two consecutive pulses is the inverse of the data rate $D_b$ (expressed in bits/s). A proper configuration of the message on transmission naturally requires that

$$\tau < \frac{1}{D_b}. \qquad (3.67)$$

Broadening of the pulse due to the dispersion of the material in the fiber core introduces an intrinsic limit to this data rate, since detection of the message after propagating over a distance $Z$ requires that, as before, we have

$$\tau(z) < \frac{1}{D_b} \quad \text{and hence} \quad D_b < \frac{1}{\tau\sqrt{1 + \left(\frac{2\gamma_0 z}{\tau^2}\right)^2}}. \qquad (3.68)$$

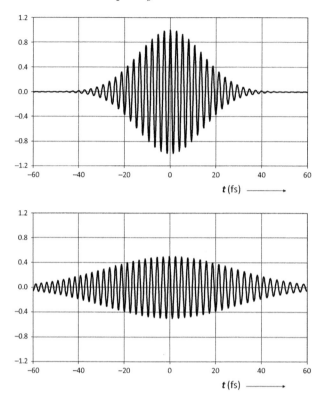

Figure 3.2 Evolution of the temporal profile of a Gaussian pulse propagating in a medium of quadratic spectral dispersion [$z_1 = 0$ (*upper graph*); $z_2 = \tau^2 \sqrt{3}/2\gamma_0$ (*lower graph*)].

Actually the situation is a bit more complicated than the approach developed here might suggest, since in this overview we must also take account of the spectral dispersion of the propagation constant in the guided mode within the fiber; such dispersion can be modified by a suitable choice of profile for the index used.

## 3.5 Modeling the Response of a 2-Beam Interferometer

In this section we propose to model the response of a 2-beam interferometer when illuminated by a very short duration pulse, and then cast some light on the effect of a spectral dispersion in the path difference on the response of such an interferometer.

The instrument we shall consider here is the Mach–Zehnder interferometer, and has the ability to sweep through optical path differences (OPD) around the zero-OPD, as seen in Figure 3.3.

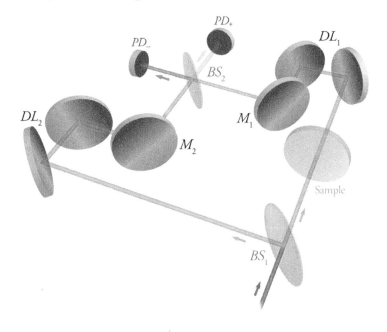

Figure 3.3 Mach–Zehnder interferometer.

The incident beam of light is separated into two sub-beams of the same power using a balanced lossless beam splitter $BS_1$ ($R_1 = T_1 = 50\%$):

- The transmitted beam, here used as the measuring arm, passes through the sample to be characterized and then, using a perfect plane mirror (reflection coefficient 100% and no phase shift on reflection), through a delay line $DL_1$ before being sent to a second beam splitter ($BS_2$) with the same properties as the first.
- The reflected beam, here used as the reference arm, follows a similar path (delay line, mirror, beam splitter) but without passing through the sample.

The interference patterns resulting from the superposition of these two beams are detected using two photodiodes $PD_+$ and $PD_-$, the + sign being associated with the interferometer's symmetric port (each beam undergoes a reflection and a transmission at the beam splitters), while the − sign is associated with the antisymmetric port of the same interferometer (two reflections for the reference arm to be compared with two transmissions for the measuring arm). Unless otherwise stated, we shall henceforth assume the photodiode $PD_+$ to be the nominal detector.

The pulse of light sent into this interferometer will be assumed identical to that described in Section 3.2.2 (central wavelength 800 nm, pulse length 40 fs).

To model the response of this interferometer in the presence of a sample consisting of a glass slab with plane parallel faces, we have to adopt the following procedure:

1. First we carry out a **spectral decomposition** of the incident pulse by applying a Fourier transform to its temporal profile, i.e.,

$$\mathcal{E}_0(f) = \int_{-\infty}^{+\infty} E_0(t)\, e^{2i\pi f t}\, dt = \int_{-\infty}^{+\infty} A_0(t) \cos 2\pi f_0 t\, e^{2i\pi f t}\, dt. \qquad (3.69)$$

2. Then we describe the propagation in each of the interferometer's arms by calculating the **phase shift accumulated** by each of the spectral components of the pulse, i.e.,

$$\mathcal{E}_1(f) = t_1 \mathcal{E}_0(f)\, e^{ik(f)\{L_1 + [n(f)-1]e\}} r_2$$
$$\mathcal{E}_2(f) = r_1 \mathcal{E}_0(f)\, e^{ik(f)L_2} t_2, \qquad (3.70)$$

   where $e$ is the mechanical thickness of the sample and $n$ its refractive index.

3. Before proceeding to the **temporal reconstruction** of the electric field resulting from their superposition,

$$E(t) = 2\Re \left[ \int_0^{+\infty} \{\mathcal{E}_1(f) + \mathcal{E}_2(f)\} e^{-2i\pi f t}\, df \right].$$

4. And finally to the **detection** of this resultant field by the photodiode located in the symmetric port of the interferometer; the electric current $I$ output by this detector is proportional to the temporal mean of the square of the electric field (we shall return to this particular point in greater detail in Chapter 5):

$$I = K \langle E^2(t) \rangle_T. \qquad (3.71)$$

The characteristic time $T$ over which the mean is taken is large compared with the duration $\tau$ of the incident pulse (and hence compared with the oscillations of the electric field).

Note that the phase differences between the two beams are here taken into account in frequency space, on elementary components that are perfectly monochromatic. In other words, this means that in this working space, **the**

**two sub-beams are always interfering**, regardless of their own temporal characteristics and regardless of the optical path difference to which they are subject.

We shall illustrate this mechanism using numerical integration to calculate the interferometer's response when it is empty (i.e., with no sample) and when it includes a slab of silica 10 mm thick. Before doing so, we introduce the variable $\Delta$ to quantify the difference in length between the two arms of the interferometer, i.e.,

$$\Delta = L_2 - L_1 . \tag{3.72}$$

**Case 1 Response of the interferometer when empty** Figure 3.4 shows various ways in which this response can be displayed. The graph on the left corresponds to the variation in electric current output by the detector photodiode as a function of the path difference $\Delta$.

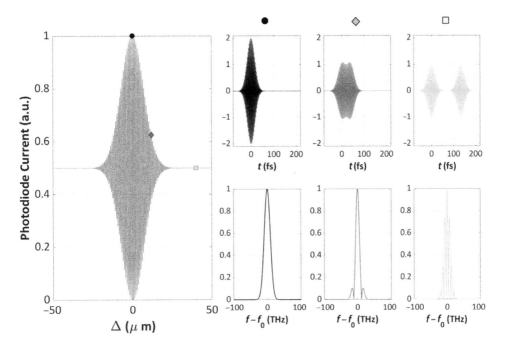

Figure 3.4 Response of the empty interferometer.

There are three various markers on this interferogram:

- The first, a black dot, corresponds to a zero OPD.
- The second, a gray diamond, corresponds to an OPD of 12 μm.
- Finally the third, a light gray square, corresponds to an OPD of 40 μm.

At a zero-OPD (black) the two signals are rigorously identical and lie perfectly on top of each other: the frequency spectrum and temporal profile therefore correspond to that of a single pulse, except that the amplitude is multiplied by a factor of 2.

If the OPD increases, the two pulses separate in time, and the greater the OPD, the greater this separation. If this gets to 40 microns, the current from the photodetector becomes almost constant, and the two pulses have no temporal overlap; it is as if the spectral components had not interfered, as pointed out earlier. In fact, however, this interference does not affect the amplitude of the field, but its **spectrum**, as the lower graphs in Figure 3.4 demonstrate.

**Case 2   Response of the interferometer in the presence of a 10 mm thick slab of silica** Figure 3.5 shows the same information as that used in Figure 3.4. We note a slight broadening of the interferogram, due to the spectral dispersion of the silica slab's refractive index.

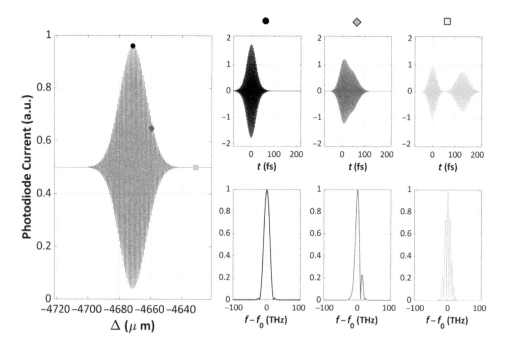

Figure 3.5 Response of the interferometer in the presence of a 10 mm thick slab of silica.

This is confirmed by examining the temporal profiles, in particular that corresponding to the greatest optical path difference (light gray): the pulse centered on $t = 0$ corresponds to the one passing down the reference arm, while the one centered on $t = \Delta/c = 130$ fs passes through the silica slab and, as a result, shows a temporal broadening comparable to that analyzed in Section 3.4.4.

The second important piece of information concerns the value of the optical path difference for which the signal output by the detector is a maximum: this maximum occurs when the path difference is $-4671.6$ μm (see left graph of Figure 3.5). The refractive index of silica at 800 nm is 1.4533. The path difference introduced by the silica slab is therefore $-e(n - 1) = -4533$ μm, which is different from the value observed. In fact, here we have to consider not the refractive index at the central wavelength of the pulse, but rather the group index at this same wavelength, i.e.,

$$n_g(f) = n(f_0) + f_0 \left.\frac{\partial n}{\partial f}\right|_{f_0} , \qquad (3.73)$$

which is 1.467192 at this same wavelength of 800 nm (375 THz). So, if we replace $n$ by $n_g$ in the computation of the path difference introduced by the silica slab, we obtain $\Delta = -e(n_g - 1) = -4671.9$ μm, in very good agreement with the expected value ($-4671.6$ μm).

# 4

# Spatial Wave Packet

## 4.1 Introduction

In Chapter 3 we considered monochromatic unidirectional waves and the superposition of such waves at different frequencies: these we gave the general description of frequency wave packets.

However, we were able to observe that all these monochromatic waves had parallel wave vectors, meaning that we considered a unidirectional beam of white light (broad spectrum) propagating along the $Oz$ axis.

This approach is clearly not complete, since it only allows representation of fields that fill the whole of space, and whose modulus is spatially constant. This is somewhat limiting; indeed, we observe that, if the wave vector of a unidirectional monochromatic component $(\vec{A}\,e^{i\vec{\beta}\cdot\vec{\rho}})$ is real, then its modulus is constant over all space.

This is why we now consider purely monochromatic fields corresponding to more general spatial distributions, such as those that describe, for example, divergent or convergent focused beams. Hence, at some given angular frequency $\omega$, we have to take account of the coexistence of several real wave vectors $\vec{\beta}(\omega)$ having the same modulus but different directions, this time resulting in a spatial wave packet. Furthermore, a new Fourier transform with respect to a space variable will be required for these developments.

The fields that we shall consider from this point can be described by restricted dependencies of the form $\vec{\mathcal{E}}(\vec{\rho}, f) = \vec{\mathcal{E}}(\vec{\rho}) = \vec{\mathcal{E}}(\vec{r}, z)$, since the frequency $f$ is fixed. Furthermore, we shall consider only the case of homogeneous, isotropic, linear media containing neither charge nor current.

Finally, we shall restrict ourselves initially to analyzing just the 2D $(x, z)$ problem, since this will allow all the ideas needed later on to be introduced. Hence this situation can be described by a simplified form of the Helmholtz equation:

$$\Delta\vec{\mathcal{E}} + k^2\vec{\mathcal{E}} = \frac{\partial^2\vec{\mathcal{E}}}{\partial x^2} + \frac{\partial^2\vec{\mathcal{E}}}{\partial z^2} + k^2\vec{\mathcal{E}} = \vec{0} \quad \text{with} \quad k^2 = \omega^2\tilde{\epsilon}(f)\tilde{\mu}(f), \ k \in \mathbb{C}. \ (4.1)$$

## 4.2 Spatial Fourier Transform

The one-dimensional spatial Fourier transform $\vec{\mathbb{E}}(\nu, z)$ is defined using the equation

$$\vec{\mathbb{E}}(\nu, z) = \int_{-\infty}^{+\infty} \vec{\mathcal{E}}(x, z)e^{-2i\pi\nu x}\, dx, \qquad (4.2)$$

where we have adopted the negative sign convention (this is opposite to the choice we made in the temporal case).

In equation (4.2), it is usual to denote the spatial frequency $\nu$ as the conjugate Fourier variable for the space variable $x$. For convenience, we also often use the notion of spatial pulsation given by $\sigma = 2\pi\nu$. By comparison with the previous chapter where the temporal time/frequency pair $(t, \omega = 2\pi f)$ was involved and in which $\omega$ was the temporal pulsation, we are now concerned with the spatial space/frequency pair $(x, \sigma = 2\pi\nu)$.

We now have to find the equation of propagation that satisfies the spatial Fourier transform of the field; this is obtained by rewriting the Helmholtz equation in the spatial frequency plane. This calculation is generally made easier if the mathematical tools related to the Fourier transformation have been mastered (convolution, differentiation, etc.). However, we prefer here to describe the various stages. Hence, applying the one-dimensional Fourier transform to the reduced Helmholtz equation (4.1), we get

$$\int_{-\infty}^{+\infty} \frac{\partial^2\vec{\mathcal{E}}}{\partial x^2}e^{-2i\pi\nu x}\, dx + \frac{\partial^2}{\partial z^2}\int_{-\infty}^{+\infty}\vec{\mathcal{E}}(x, z)e^{-2i\pi\nu x}\, dx$$

$$+ k^2\int_{-\infty}^{+\infty}\vec{\mathcal{E}}(x, z)e^{-2i\pi\nu x}\, dx = \vec{0} \qquad (4.3)$$

or

$$\int_{-\infty}^{+\infty} \frac{\partial^2\vec{\mathcal{E}}}{\partial x^2}e^{-2i\pi\nu x}\, dx + \frac{\partial^2\vec{\mathbb{E}}(\nu, z)}{\partial z^2} + k^2\vec{\mathbb{E}}(\nu, z) = \vec{0}. \qquad (4.4)$$

To calculate the first term of this equation we shall use the rule for the Fourier transform of the derivative of a function. We get

$$\int\limits_{-\infty}^{+\infty} \frac{\partial^2 \vec{\mathcal{E}}}{\partial x^2} e^{-2i\pi\nu x} \, dx = 2i\pi\nu \int\limits_{-\infty}^{+\infty} \frac{\partial \vec{\mathcal{E}}}{\partial x} e^{-2i\pi\nu x} \, dx = (2i\pi\nu)^2 \int\limits_{-\infty}^{+\infty} \vec{\mathcal{E}}(x, z) e^{-2i\pi\nu x} \, dx$$

$$= -4\pi^2 \nu^2 \vec{\mathbb{E}}(\nu, z) = -\sigma^2 \vec{\mathbb{E}}(\nu, z) \,. \tag{4.5}$$

The reduced Helmholtz equation then takes the following *frequential* form

$$- \sigma^2 \vec{\mathbb{E}}(\nu, z) + \frac{\partial^2 \vec{\mathbb{E}}(\nu, z)}{\partial z^2} + k^2 \vec{\mathbb{E}}(\nu, z) = \vec{0} \tag{4.6}$$

or

$$\frac{\partial^2 \vec{\mathbb{E}}(\nu, z)}{\partial z^2} + \alpha^2 \vec{\mathbb{E}}(\nu, z) = \vec{0} \qquad \text{with} \qquad \alpha^2 = k^2 - \sigma^2 \,. \tag{4.7}$$

The general solution of this differential equation is well known, and is given by

$$\vec{\mathbb{E}}(\nu, z) = \vec{\mathbb{A}}^+(\nu) e^{i\alpha(\nu)z} + \vec{\mathbb{A}}^-(\nu) e^{-i\alpha(\nu)z} \,. \tag{4.8}$$

We then proceed to reconstruction of the spatial dependence of the field using the inverse Fourier transform, written here as

$$\vec{\mathcal{E}}(x, z) = \int\limits_{-\infty}^{+\infty} \vec{\mathbb{E}}(\nu, z) e^{2i\pi\nu x} \, d\nu \,, \tag{4.9}$$

and finally,

$$\vec{\mathcal{E}}(x, z) = \int\limits_{-\infty}^{+\infty} \vec{\mathbb{A}}^+(\nu) e^{i[\sigma x + \alpha(\nu)z]} \, d\nu + \int\limits_{-\infty}^{+\infty} \vec{\mathbb{A}}^-(\nu) e^{i[\sigma x - \alpha(\nu)z]} \, d\nu \,. \tag{4.10}$$

This general solution corresponds to a sum of two spatial wave packets, progressive and retrograde respectively, with amplitudes $\vec{\mathbb{A}}^+(\nu)$ and $\vec{\mathbb{A}}^-(\nu)$ with wave vectors expressed as

$$\vec{\beta}^+ = \begin{bmatrix} \sigma \\ 0 \\ \alpha(\sigma) \end{bmatrix} \qquad \text{and} \qquad \vec{\beta}^- = \begin{bmatrix} \sigma \\ 0 \\ -\alpha(\sigma) \end{bmatrix} \,. \tag{4.11}$$

Note that the spatial pulsation is simply the tangential component of the progressive and retrograde wave vectors. Moreover, and as mentioned in the introduction, it can be verified at this stage that the wave packet is of limited spatial support. Indeed, equation (4.10) is used in the text that follows to calculate the field on a plane whose abscissa is $z = 0$. We get

$$\vec{\mathcal{E}}(x,0) = \int\limits_{-\infty}^{+\infty} \vec{\mathbb{A}}^+(\nu) e^{2i\pi\nu x}\, d\nu + \int\limits_{-\infty}^{+\infty} \vec{\mathbb{A}}^-(\nu) e^{2i\pi\nu x}\, d\nu$$

$$= \mathrm{TF}^{-1}\left[\vec{\mathbb{A}}^+(\nu) + \vec{\mathbb{A}}^-(\nu)\right]_x . \tag{4.12}$$

The spatial distribution of the field in the $z = 0$ plane is thus described by inverse Fourier transforms of amplitudes $\vec{\mathbb{A}}^+(\nu)$ and $\vec{\mathbb{A}}^-(\nu)$; these are of bounded support because they correspond to an actual (experimental) illumination.

## 4.3 Plane Waves and Evanescent Waves

Spatial frequency is a variable that varies along the real axis. We shall now identify a cutoff frequency on this axis, leading to the concepts of plane wave and evanescent wave. First we shall assume that the medium is transparent ($k$ real).

### 4.3.1 Low Spatial Frequencies and Propagative Waves

This first regime corresponds to the condition $|\sigma| \leqslant k$. Hence, for each frequency, we can define a *normal* angle $\theta$ using the equation

$$\exists\, \theta, \quad \sigma = k \sin\theta , \tag{4.13}$$

where $\theta$ takes values between $-\pi/2$ and $+\pi/2$, corresponding to $-k \leqslant \sigma \leqslant k$.

Hence the normal component of the wave vector can be deduced:

$$\alpha = \sqrt{k^2 - \sigma^2} = k\cos\theta . \tag{4.14}$$

Consequently, the wave vectors of the progressive ($\vec{\beta}^+$) and retrograde ($\vec{\beta}^-$) components respectively are real, and are written as (see Figure 4.1):

$$\vec{\beta}^\pm = \begin{bmatrix} k\sin\theta \\ 0 \\ \pm k\cos\theta \end{bmatrix} = k\begin{bmatrix} \sin\theta \\ 0 \\ \pm\cos\theta \end{bmatrix}. \tag{4.15}$$

We shall define this set of solutions generically as *propagative* components of the field, and we shall see that they transport energy in the far field.

We must also bear in mind at this stage that an angle has been associated with each frequency, meaning that the spatial wave packet can be viewed as

*Spatial Wave Packet*

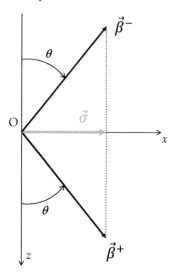

Figure 4.1 Schematic representation of the wave vectors and spatial pulsation.

a superposition of plane waves of different directions. We recall that though each plane wave shows an infinite support (their modulus is constant), their superposition has a finite support, following (4.12). The same integral expression (4.12) can also be used to describe the particular case of a single plane wave; for that it is sufficient to assume that the amplitude distributions $\mathbb{A}^{\pm}(\nu)$ are Dirac distributions $\delta(\nu - \nu_0)$, centered on the required frequency $\nu_0$.

The reader can verify that, for the propagative field, the integral then reduces to

$$\vec{\mathcal{E}}(x, z) = \vec{\mathbb{A}}^{+}(\nu_0) \, e^{i\vec{\beta}(\nu_0) \cdot \vec{\rho}}. \tag{4.16}$$

Nevertheless, it is important to stress here that the use of a singular distribution to describe the amplitude of the field changes the dimension of the physical quantities. To be convinced of this, the reader might attempt a field reconstruction as described in Section 2.4 and apply a dimensional analysis.

### 4.3.2 High Spatial Frequencies and Evanescent Waves

This second regime corresponds to the condition $|\sigma| > k$.

In this case we are not looking to introduce an angle, but we note that the normal component of the wave vector is pure imaginary:

$$a(\sigma) = \sqrt{k^2 - \sigma^2} = i\sqrt{\sigma^2 - k^2} = \pm i\alpha'' , \tag{4.17}$$

where the sign of the root must be chosen to prevent the amplitude of the field from diverging. If we are interested in the modulus of the progressive component of these plane waves, we can write for $\alpha = i\alpha''$:

$$|\vec{\mathcal{E}}(x, z)| = |\vec{\mathbb{A}}^+(\sigma)e^{i[\sigma x + i\alpha'' z]}| = |\vec{\mathbb{A}}^+(\sigma)|\, e^{-\alpha'' z} . \tag{4.18}$$

This shows that the amplitude of the field decreases exponentially in the $z$ direction, and that associated with this mechanism is a characteristic distance $z_c$ for which this amplitude is multiplied by $1/e$, i.e.,

$$z_c = \frac{1}{\alpha''} = \frac{1}{\sqrt{\sigma^2 - k^2}} . \tag{4.19}$$

Consider, for example, the particular case where $\sigma = k\sqrt{2}$; the associated characteristic distance is then of the order of the wavelength:

$$z_c = \frac{1}{\alpha''} = \frac{1}{k} = \frac{\lambda_v}{2\pi n} . \tag{4.20}$$

However, it must also be noted that $z_c$ takes values between $+\infty$ and $0$ when $\sigma$ varies from $k$ to $+\infty$.

   This set of solutions is usually given the generic description of *evanescent* components of the field, and the associated mechanism for the rapid fall in amplitude means that these components are generally observable only in the immediate neighborhood of interfaces or sources: hence we shall also speak of the *near field*, in contrast with the *far field* associated with the propagative component of the field and corresponding to low spatial frequencies.

   We shall also see that these evanescent waves transport no energy, independently of the arbitrarily defined characteristic distance $z_c$ associated with them. In other words, it is not because the field is nonzero in a region of space (and hence with a local electromagnetic volume density) that energy is transported into it. On the other hand, an object introduced into the region excited by the evanescent wave will cause a decoupling of energy resulting from this change in geometry.

   Note also that, had we sought to introduce the notions of refractive index, wavelength and phase velocity for an evanescent wave, the results would have been different. We would find a phase velocity $v = \frac{\omega}{\sigma}$ that depends on frequency and which is bounded by the phase velocity of plane waves: $0 < v = \frac{\omega}{\sigma} < \frac{\omega}{k}$. Similarly for the spatial period, which would be written as $\lambda = \frac{2\pi}{\sigma} \ldots$ Hence these quantities would not be characteristic of the medium.

In conclusion, it can be stated that all the calculations carried out here remain true in a complex (absorbing) medium, though with a less clear dissociation between plane and evanescent waves.

## 4.4 Flux Transported by a Packet of Plane Waves

We now turn our attention to the optical power transported by a wave packet that is then intercepted by a detector whose collecting area is assumed large compared with the spatial spread of the beam. To do this, we shall calculate the flux of the Poynting vector through a plane of arbitrary $z$ abscissa.

Assume a wave packet of progressive plane waves (in the $z > 0$ direction; see Figure 4.2), described by the following integral representation:

$$\vec{\mathcal{E}}(x, z) = \int_{-\infty}^{+\infty} \vec{\mathbb{A}}(\nu)\, e^{i(2\pi\nu x + \alpha z)} d\nu \,. \tag{4.21}$$

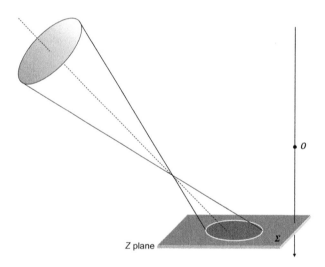

Figure 4.2 Schematic representation of a packet of plane waves propagating in the $Z > 0$ direction.

We saw in Chapter 2 that the flux transported by this packet of plane waves at a plane of abscissa $z$ is equal to the real part of the quantity $\Phi$ defined by

$$\Phi(z) = \int_{\Sigma} \vec{\Pi}(z).\vec{n}\, dS = \int_{x} \vec{\Pi}(z).\vec{z}\, dx \,. \tag{4.22}$$

Note, however, that the variable $\Phi$ denotes a line flux (per unit length $y$), insofar as we are here dealing with a 2D problem. In equation (4.22), $\vec{\Pi}$ is the Poynting vector associated with the wave packet; recall that, in the harmonic regime, this is given as

$$\vec{\Pi} = \frac{1}{2}(\vec{\mathcal{E}}^* \wedge \vec{\mathcal{H}}). \tag{4.23}$$

The fields $\vec{\mathcal{E}}$ and $\vec{\mathcal{H}}$ in this expression are harmonic, but no assumptions are made as to their spatial distribution. Hence they must again be decomposed into the form of spatial packets.

If for the vector $\vec{\mathcal{H}}$ we use a decomposition into a packet of plane waves similar to that used for the vector $\vec{\mathcal{E}}$, namely

$$\vec{\mathcal{H}}(x, z) = \int_{-\infty}^{+\infty} \vec{\mathbb{B}}(\nu) \, e^{i(2\pi\nu x + \alpha z)} d\nu, \tag{4.24}$$

then the expression for the flux $\Phi$ becomes

$$\Phi = \frac{1}{2} \int_x \vec{z}. \left\{ \int_\nu \vec{\mathbb{A}}^*(\nu) \, e^{-i[2\pi\nu x + \alpha^*(\nu)z]} d\nu \right\} \wedge \left\{ \int_{\nu'} \vec{\mathbb{B}}(\nu') \, e^{i[2\pi\nu' x + \alpha(\nu')z]} d\nu' \right\} dx.$$

Changing the order of integration, we get

$$\Phi = \frac{1}{2} \int_\nu \int_{\nu'} \vec{z}. \left[ \vec{\mathbb{A}}^*(\nu) \wedge \vec{\mathbb{B}}(\nu') \right] e^{-i\alpha^*(\nu)z} e^{i\alpha(\nu')z} \, d\nu \, d\nu' \int_x e^{-2i\pi(\nu - \nu')x} dx$$

or

$$\Phi = \frac{1}{2} \int_\nu \int_{\nu'} \vec{z}. \left[ \vec{\mathbb{A}}^*(\nu) \wedge \vec{\mathbb{B}}(\nu') \right] e^{-i\alpha^*(\nu)z} e^{i\alpha(\nu')z} \delta(\nu - \nu') \, d\nu \, d\nu'. \tag{4.25}$$

Then, using the properties of the Dirac distribution:

$$\Phi = \frac{1}{2} \int_\nu \vec{z}. \left[ \vec{\mathbb{A}}^*(\nu) \wedge \vec{\mathbb{B}}(\nu) \right] e^{i[\alpha(\nu) - \alpha^*(\nu)]z} \, d\nu. \tag{4.26}$$

Furthermore, we showed in Chapter 2 that in free space the vectors $\vec{\mathcal{E}}$ et $\vec{\mathcal{H}}$ associated with a monochromatic plane wave were linked by the orthogonality equation (2.64), which leads here to

$$\vec{\mathbb{B}}(\nu) = \frac{1}{\omega\tilde{\mu}} \vec{\beta}^+ \wedge \vec{\mathbb{A}}(\nu). \tag{4.27}$$

Consequently,

$$\vec{\mathbb{A}}^*(\nu) \wedge \vec{\mathbb{B}}(\nu) = \frac{1}{\omega\tilde{\mu}} \left\{ \vec{\mathbb{A}}^*(\nu) \wedge [\vec{\beta}^+ \wedge \vec{\mathbb{A}}(\nu)] \right\}. \tag{4.28}$$

Using the vector cross product formula, we eventually get

$$\vec{\mathbb{A}}^*(\nu) \wedge \vec{\mathbb{B}}(\nu) = \frac{1}{\omega\tilde{\mu}} \left\{ |\vec{\mathbb{A}}(\nu)|^2 \vec{\beta}^+ - [\vec{\mathbb{A}}^*(\nu).\vec{\beta}^+]\vec{\mathbb{A}}(\nu) \right\}, \qquad (4.29)$$

and hence

$$\vec{z}.\left[\vec{\mathbb{A}}^*(\nu) \wedge \vec{\mathbb{B}}(\nu)\right] = \frac{1}{\omega\tilde{\mu}} \left\{ |\vec{\mathbb{A}}(\nu)|^2[\vec{z}.\vec{\beta}^+] - [\vec{\mathbb{A}}^*(\nu).\vec{\beta}^+][\vec{z}.\vec{\mathbb{A}}(\nu)] \right\}. \qquad (4.30)$$

Using the third of Maxwell's equations, we can write

$$\mathrm{div}[\vec{\mathbb{A}}(v)e^{i\vec{\beta}^+ \cdot \vec{\rho}}] = 0 \;\Rightarrow\; \vec{\beta}^+.\vec{\mathbb{A}}(\nu) = 0$$

$$\Rightarrow\; \sigma[\vec{x}.\vec{\mathbb{A}}(\nu)] + \alpha[\vec{z}.\vec{\mathbb{A}}(\nu)] = 0, \qquad (4.31)$$

and hence

$$\sigma[\vec{x}.\vec{\mathbb{A}}^*(\nu)] = -\alpha^*[\vec{z}.\vec{\mathbb{A}}^*(\nu)] \qquad (4.32)$$

because $\sigma$ is real. As a consequence

$$[\vec{\mathbb{A}}^*(\nu).\vec{\beta}^+][\vec{z}.\vec{\mathbb{A}}(\nu)] = \{\sigma[\vec{x}.\vec{\mathbb{A}}^*(\nu)] + \alpha[\vec{z}.\vec{\mathbb{A}}^*(\nu)]\}[\vec{z}.\vec{\mathbb{A}}(\nu)]$$
$$= (\alpha - \alpha^*)|\vec{z}.\vec{\mathbb{A}}(\nu)|^2. \qquad (4.33)$$

Finally, noting that $\alpha - \alpha^* = 2i\alpha''$, and substituting into (4.26) the partial results corresponding to (4.30) and (4.33), we obtain

$$\Phi(z) = \frac{1}{2\omega\tilde{\mu}} \int_\nu [\alpha(\nu)|\vec{\mathbb{A}}(\nu)|^2 + 2i\alpha''|\vec{z}.\vec{\mathbb{A}}(\nu)|^2] e^{-2\alpha''(\nu)z}\, d\nu, \qquad (4.34)$$

where the real part is the transported flux, defined by

$$\Re[\Phi(z)] = \frac{1}{2\omega\tilde{\mu}} \int_\nu \Re[\alpha(\nu)]\,|\vec{\mathbb{A}}(\nu)|^2\, e^{-2\alpha''(\nu)z}\, d\nu. \qquad (4.35)$$

This equation first shows that the flux carried by a spatial wave packet is equal to the sum of the elementary fluxes carried by each spatial frequency; in other words, we do not have to consider beats between waves in different directions, and this results from the integration over an infinite receiver (with respect to the size beam at the receiver plane).

At this stage we observe that the flux is attenuated naturally in an absorbing medium with abscissa $z$. On the other hand, in a transparent medium ($\alpha'' = 0$), it no longer depends on this abscissa, corresponding to the fact that the receiver plane is assumed infinite (all the transported optical power is intercepted, regardless of whether the beam is divergent).

It is useful to introduce the **flux spectral density** transported per unit of spatial frequency $\nu$, written

$$\frac{d}{d\nu}\Re[\Phi(z)] = \frac{\Re[\alpha(\nu)]}{2\omega\tilde{\mu}}|\vec{\mathbb{A}}(\nu)|^2\,e^{-2\alpha''(\nu)z}\,, \qquad (4.36)$$

and which, notably, is seen to be driven by the value taken by the real part of parameter $\alpha$. In the case of an evanescent wave in particular, this last quantity is 0 in a transparent medium, showing that evanescent waves transport no energy (in contrast with propagative waves).

Equation (4.36) recalls the previously introduced cutoff frequency; for a transparent medium we can again state this as

- $|\sigma| < k \Rightarrow \Re[\alpha] = \alpha \Rightarrow \Phi \neq 0$ (plane waves, or low frequencies)
- $|\sigma| > k \Rightarrow \Re[\alpha] = 0 \Rightarrow \Phi = 0$ (evanescent waves, or high frequencies).

Thus, evanescent waves do not transport energy in a transparent medium. However, equation (4.36) shows that in an absorbing medium ($k$ complex), since $\Re[\alpha]$ is nonzero at any frequency, evanescent waves do contribute to the energy balance (this occurs mainly in the near vicinity of boundaries).

Note that the cut-off frequency $k$ depends on the refractive index and its dispersion, so that any modification of medium or wavelength may transform an evanescent wave into a plane wave (this is often called a *decoupling effect*), and conversely. Such an effect can be observed when a probe approaches the near field of a device, or when the surrounding medium (air, vacuum) is modified by the introduction of gas or other contaminants.

From a general point of view, evanescent waves can be addressed in a more concrete way if we consider total reflection occurring in devices; this will be covered in Chapter 7.

## 4.5 Optical Resolution

The notion of the spatial wave packet leads immediately to the limit of resolution, more commonly known as the diffraction limit.

Consider the spatial wave packet associated with the component propagating in the positive $z$ direction. According to equation (4.10), we can write

$$\vec{\mathcal{E}}(x,z) = \int_{\nu} \vec{\mathbb{A}}^+(\nu)e^{i[\sigma x + \alpha(\nu)z]}\,d\nu\,. \qquad (4.37)$$

As we have already seen, in the $z = 0$ plane and for some arbitrary component of this field:

$$\mathcal{E}_j(x,0) = \int_{\nu} \mathbb{A}_j^+(\nu)e^{2i\pi\nu x}\,d\nu\,. \qquad (4.38)$$

This shows that the two quantities $\mathcal{E}_j(x,0)$ and $\mathbb{A}_j^+(\nu)$ are related by a Fourier transformation. As a result, just as was shown for the time–bandwidth product, their mid-height widths $\Delta x$ and $\Delta\nu$ are not independent, since they are related by the inequality $\Delta x.\Delta\nu \geqslant 1$.

In the case of an experiment that involves illuminating a sample (e.g., under a microscope), the light comes from the far field (with respect to the wavelength of the radiation) such that only incident plane waves or low frequencies take part in the illumination. Under these conditions, each spatial frequency of the packet corresponds to an angle $\theta$ defined by $\sigma = k\sin\theta$. As a result, the frequency is necessarily contained within the interval $[-k, +k]$, the extremal bounds being obtained for a beam that is the most open possible, i.e., for $\theta = 90$ degree. We deduce that the support $\Delta\sigma$ is limited to $2k$; this imposes a minimal spatial extension on the wave packet being considered, defined by

$$\Delta x \geqslant \frac{2\pi}{2k} = \frac{\lambda_0}{2n}. \tag{4.39}$$

It is therefore impossible to obtain an optical beam whose lateral dimensions in air ($n = 1$) are less than half a wavelength. This sets the limit of resolution for propagative optics.

Recall that this diffraction limit was established for far-field illumination in the absence of evanescent waves. This does not, however, preclude attempts to increase the resolution by imagining a near-field illumination that might take advantage of the contribution of evanescent waves, making use of the proximity of nano-objects designed for this purpose. We then speak of super-resolution.

In conclusion, we again emphasize the lower bound in the products of the supports for the field and its Fourier transform. If the illumination is a plane wave, its spatial Fourier amplitude is a Dirac distribution, and hence its spatial support is infinite; conversely, if the light is highly focused its frequential support is large and the spatial extension is therefore reduced.

Eventually and generally speaking, since evanescent waves do not carry energy, far field optical measurements will most often deliver information related only to low spatial frequencies ($\sigma < k$). This is why it is currently said that light cannot see spatial periods less than the wavelength (or pulsations greater than the inverse $k$ of the medium). However, this does not mean that far field data do not allow to detect sub-wavelength particles or defects; actually only the low spatial frequencies of these defects will be recovered with far field data.

## 4.6 Generalization to the 3D Case

We now generalize the previous results to the 3D case, i.e., for a field

$$\vec{\mathcal{E}}(x, y, z) = \vec{\mathcal{E}}(\vec{r}, z)$$

dependent on the three space variables; this is the most widespread geometry and covers most of the experimental configurations encountered. Furthermore, this generalization is essential when introducing such commonly used notions as radiance, intensity, or irradiance. We shall therefore proceed as directly as possible by analogy with the 2D case.

### 4.6.1 The Helmholtz Equation, Spatial FT, and Reconstruction

The Helmholz equation is written in 3D as

$$\Delta \vec{\mathcal{E}} + k^2 \vec{\mathcal{E}} = \frac{\partial^2 \vec{\mathcal{E}}}{\partial x^2} + \frac{\partial^2 \vec{\mathcal{E}}}{\partial y^2} + \frac{\partial^2 \vec{\mathcal{E}}}{\partial z^2} + k^2 \vec{\mathcal{E}} = \vec{0}. \tag{4.40}$$

The spatial Fourier Transform (FT) is again defined in terms of the transverse coordinate $\vec{r} = (x, y)$, but which this time is a vector in $\mathbb{R}^2$. This choice comes from the fact that in the geometries we shall consider (sources at $z = -\infty$), most physical variables are *tempered*, i.e., of limited support with respect to the transverse variable, thus guaranteeing the existence of a Fourier transform. Note that the conjugate variable $\vec{\nu}$ (in the Fourier sense) is also a vector in $\mathbb{R}^2$; we can write this spatial frequency $\vec{\nu} = (\nu_x, \nu_y)$, though we can also use the spatial pulsation defined by $\vec{\sigma} = 2\pi\vec{\nu}$.

After applying the Fourier transform, (4.40) becomes

$$(2i\pi\nu_x)^2 \vec{\mathbb{E}}(\vec{\nu}, z) + (2i\pi\nu_y)^2 \vec{\mathbb{E}}(\vec{\nu}, z) + \frac{\partial^2}{\partial z^2} \vec{\mathbb{E}}(\vec{\nu}, z) + k^2 \vec{\mathbb{E}}(\vec{\nu}, z) = \vec{0} \tag{4.41}$$

or

$$\frac{\partial^2}{\partial z^2} \vec{\mathbb{E}}(\vec{\nu}, z) + \alpha^2 \vec{\mathbb{E}}(\vec{\nu}, z) = \vec{0} \quad \text{with} \quad \alpha^2(\nu) = k^2 - \sigma^2 = k^2 - (\sigma_x^2 + \sigma_y^2), \tag{4.42}$$

where $\sigma$ is the modulus of the spatial pulsation.

These equations are therefore identical to those obtained in the 2D case, provided the algebraic frequency $\sigma$ is replaced by its modulus $|\sigma|$. As before, the solution to equation (4.42) is immediate:

$$\vec{\mathbb{E}}(\vec{\nu}, z) = \vec{\mathbb{A}}^+(\vec{\nu}) e^{i\alpha(\nu)z} + \vec{\mathbb{A}}^-(\vec{\nu}) e^{-i\alpha(\nu)z}, \tag{4.43}$$

and reconstruction of the field eventually leads to

$$\vec{\mathcal{E}}(\vec{r}, z) = \int_{\vec{\nu}} \vec{\mathbb{A}}^+(\vec{\nu}) e^{i\alpha(\nu)z} e^{2i\pi\vec{\nu}.\vec{r}} \, d^2\vec{\nu} + \int_{\vec{\nu}} \vec{\mathbb{A}}^-(\vec{\nu}) e^{-i\alpha(\nu)z} e^{2i\pi\vec{\nu}.\vec{r}} \, d^2\vec{\nu}. \tag{4.44}$$

Note that these integrals are carried out over the variable $\vec{\nu}$ in $\mathbb{R}^2$, where $d^2\vec{\nu} = d\nu_x \, d\nu_y$. We again see the sum of two progressive and retrograde wave packets, since we can write

$$\vec{\mathcal{E}}(\vec{r}, z) = \int_{\vec{\nu}} \vec{\mathbb{A}}^+(\vec{\nu}) e^{i\vec{\beta}^+(\nu) \cdot \vec{\rho}} \, d^2\vec{\nu} + \int_{\vec{\nu}} \vec{\mathbb{A}}^-(\vec{\nu}) e^{i\vec{\beta}^-(\nu) \cdot \vec{\rho}} \, d^2\vec{\nu}, \tag{4.45}$$

where the wave vectors $\vec{\beta}^+(\vec{\nu})$ and $\vec{\beta}^-(\vec{\nu})$ are defined by

$$\vec{\beta}^+ = \begin{bmatrix} \sigma_x \\ \sigma_y \\ \alpha(\nu) \end{bmatrix} \quad \text{and} \quad \vec{\beta}^- = \begin{bmatrix} \sigma_x \\ \sigma_y \\ -\alpha(\nu) \end{bmatrix}. \tag{4.46}$$

We again find that the spatial pulsation is the tangential component of the wave vector.

### 4.6.2 Normal Angle, Polar Angle

*Plane Waves in a Transparent Medium*

We first consider the low frequency case ($\sigma < k$). This time each spatial frequency gives rise to the definition of two angles; the *normal* angle is defined as before in 2D:

$$0 \leqslant \sigma \leqslant k \quad \Rightarrow \quad \exists \theta \,, \sigma = k \sin\theta \quad \Rightarrow \quad \alpha = k \cos\theta. \tag{4.47}$$

Note, however, that since the modulus of $\sigma$ is positive, we have $0 \leqslant \theta \leqslant \frac{\pi}{2}$, which is different from the 2D geometry $(-\frac{\pi}{2} \leqslant \theta \leqslant \frac{\pi}{2})$.

We then introduce a second angle $\phi$, the *polar or azimuthal angle*, defined as follows:

$$\vec{\sigma} \in \mathbb{R}^2 \quad \exists \phi \,, \vec{\sigma} = \sigma \, (\cos\phi, \sin\phi) \quad \text{with} \quad 0 \leqslant \phi \leqslant 2\pi. \tag{4.48}$$

Finally, making use of spherical coordinates (as illustrated in Figure 4.3), the wave vectors are written as

$$\vec{\beta}^+ = \begin{bmatrix} k \sin\theta \cos\phi \\ k \sin\theta \sin\phi \\ k \cos\theta \end{bmatrix} \quad \text{and} \quad \vec{\beta}^- = \begin{bmatrix} k \sin\theta \cos\phi \\ k \sin\theta \sin\phi \\ -k \cos\theta \end{bmatrix}. \tag{4.49}$$

Keep in mind here that the physical variable which appears naturally is the spatial frequency, and that the angles are then defined from this frequency. This remark lies at the origin of the Snell–Descartes relations which we shall encounter in Chapter 6 dedicated to planar components.

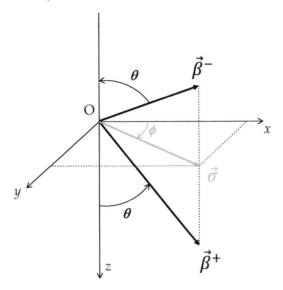

Figure 4.3 Schematic representation of the two angles (*normal* and *polar*) defining wave vectors and spatial pulsation.

### Evanescent Waves

The evanescent wave case is barely altered because the exponential in $z$ is still expressed in the same way ($e^{i\alpha z}$, with $\alpha = \sqrt{k^2 - \sigma^2}$). Consequently, it is now the modulus $\sigma$ of the spatial pulsation that is driving the cutoff on the real axis. Beyond this frequency ($\sigma > k$), the waves are evanescent and decay exponentially with no transport of energy.

### 4.6.3 Photometric Variables

#### Spectral Flux Density

Calculating the flux of the Poynting vector through an infinite plane of abscissa $z$ for a 3D wave packet is identical to that for the 2D case. The reader can quickly verify that we still have

$$\Phi(z) = \frac{1}{2\omega\tilde{\mu}} \int_{\vec{\nu}} \Re[\alpha(\nu)] \, |\vec{A}(\vec{\nu})|^2 \, e^{-2\alpha"(\nu)z} \, d^2\vec{\nu}. \tag{4.50}$$

To calculate the spectral flux density, we need to make the following change of variables:

$$(\nu_x, \nu_y) \to (\nu, \phi) \quad \text{with} \quad \nu = |\vec{\nu}|.$$

The Jacobian $J$ of this transformation is simple to calculate $(J = \nu \, d\nu \, d\phi)$, and we obtain

$$\Phi(z) = \frac{1}{2\omega\tilde{\mu}} \int_{\nu,\phi} \Re[\alpha(\nu)] \, |\vec{\mathbb{A}}(\vec{\nu})|^2 \, e^{-2\alpha''(\nu)z} \, \nu \, d\nu \, d\phi \,. \qquad (4.51)$$

Moreover, the **spectral flux density per unit polar angle** in a transparent medium $(\alpha'' = 0)$ is defined by

$$\frac{d^2\Phi}{d\nu \, d\phi} = \frac{1}{2\omega\tilde{\mu}} \Re[\alpha(\nu)] \, \nu \, |\vec{\mathbb{A}}(\vec{\nu})|^2 \,. \qquad (4.52)$$

Recall that this variable is 0 for evanescent waves in a transparent medium:

$$\sigma = 2\pi\nu > k \quad \Rightarrow \quad \frac{d^2\Phi}{d\nu \, d\phi} = 0 \,.$$

Note also that, though this function is 0 at 0 frequency (in contrast to the 2D case), its integral around this particular value of frequency is not 0; this reflects the case of an experimental measurement.

Eventually we stress on the fact that (4.52) is a definition, and not the consequence of (4.51). Further details about the connection between these two relations will be given in Chapter 7, Section 7.10.

### Intensity and Radiance

It is also usual to introduce the concept of solid angle. We restrict ourselves to plane waves in order to make a change of variables as follows:

$$\nu = \frac{k}{2\pi} \sin\theta \quad \Rightarrow \quad d\nu = \frac{k}{2\pi} \cos\theta \, d\theta \,,$$

leading to

$$d^2\Phi = \frac{1}{2\omega\tilde{\mu}} k \cos\theta \left( \frac{k \sin\theta}{2\pi} \right) |\vec{\mathbb{A}}(\vec{\nu})|^2 \frac{k}{2\pi} \cos\theta \, d\theta \, d\phi$$

$$= \frac{1}{2\omega\tilde{\mu}} \left( \frac{k^3}{4\pi^2} \right) |\vec{\mathbb{A}}(\vec{\nu})|^2 \cos^2\theta \, \sin\theta \, d\theta \, d\phi \,. \qquad (4.53)$$

Observing that the differential expression for the solid angle $(d^2\Omega = \sin\theta \, d\theta \, d\phi)$ appears in this last term, we obtain an expression for the **flux density per unit solid angle** in a transparent medium:

$$\frac{d^2\Phi}{d^2\Omega} = \frac{1}{2\omega\tilde{\mu}} \left( \frac{k^3}{4\pi^2} \right) |\vec{\mathbb{A}}(\vec{\nu})|^2 \cos^2\theta \,. \qquad (4.54)$$

This variable is called **intensity** in photometry and is written as $I$. We also use **radiance** $L$, defined by

$$\frac{d^2\Phi}{d^2\Omega} = L\cos\theta \,, \tag{4.55}$$

which leads to

$$L = \frac{1}{2\omega\tilde{\mu}} \left(\frac{k^3}{4\pi^2}\right) |\vec{\mathbb{A}}(\vec{\nu})|^2 \cos\theta \,. \tag{4.56}$$

*Irradiance*

We conclude this chapter with the concept of **irradiance**. Coming back to the expression for flux in a transparent medium, and assuming that the normal component $\alpha$ of the wave vector is more or less constant over the support of the amplitude distribution (as for a weakly diverging beam), the flux expression given in (4.50) becomes

$$\Phi = \frac{1}{2\omega\tilde{\mu}} \Re[\alpha(\nu_0)] \int_{\vec{\nu}} |\vec{\mathbb{A}}(\vec{\nu})|^2 \, d^2\vec{\nu} \,. \tag{4.57}$$

Using Parseval's theorem, we can also write

$$\Phi = \frac{1}{2\omega\tilde{\mu}} \Re[\alpha(\nu_0)] \int_{\vec{r}} |\vec{\mathcal{E}}(\vec{r},0)|^2 \, d^2\vec{r} \,. \tag{4.58}$$

From this we deduce an expression to define flux density in the plane $z = 0$, commonly referred to as irradiance:

$$\frac{d^2\Phi}{d^2S} = \frac{1}{2\omega\tilde{\mu}} \Re[\alpha(\nu_0)]|\vec{\mathcal{E}}(\vec{r},0)|^2 \,, \tag{4.59}$$

where $d^2S$ is the element of area $dx\,dy$.

This tells us that, for a weakly divergent beam, the irradiance is proportional to the modulus of the square of the field.

# 5

# Energy, Causality, and Coherence

In this chapter we have chosen to tackle three particularly important concepts:

1. The definition of an energy balance associated with an electromagnetic field in the spatiotemporal regime and its connection with that of the harmonic regime
2. The consequences of the principle of causality on the dispersion behavior (frequency dependence) of electric or magnetic susceptibility of a material medium
3. The definition of coherence of electromagnetic radiation and the study of those variables in particular that govern its ability to interfere with itself

## 5.1 Energy Balance in the Spatiotemporal Regime

### 5.1.1 Introduction

In Chapter 2 we saw how to establish an energy balance in the harmonic regime and then refined this approach in Chapter 4 in the case of a spatial wave packet. However, until now the ideas introduced have not concerned the spatiotemporal regime. We therefore propose to express the energy associated with broad-spectrum light as a function of the energy transported by each of its monochromatic components. We can immediately state that the intrinsic quadratic nature of these energy variables means that we cannot appeal to the principle of superposition, a principle we have used frequently where only the field amplitudes mattered. Furthermore, we will not be revisiting the fundamental concepts of energy and work, but will work by simple analogy to end up directly with a balance that can be used in our particular case, namely that of electromagnetic optics. Here we are concerned with the general case of the spatiotemporal regime, in which Maxwell's equations can be written as

$$
\begin{cases}
\mathbf{curl}\ \vec{E}(\vec{\rho},t) = -\dfrac{\partial}{\partial t}\vec{B}(\vec{\rho},t) + \vec{M}(\vec{\rho},t) \\[2mm]
\mathbf{curl}\ \vec{H}(\vec{\rho},t) = \vec{J}(\vec{\rho},t) + \dfrac{\partial}{\partial t}\vec{D}(\vec{\rho},t)
\end{cases}
, \qquad (5.1)
$$

with the various introduced fields satisfying the following constitutive equations:

$$
\begin{cases}
\vec{B}(\vec{\rho},t) = [\mu \star \vec{H}](\vec{\rho},t) \\[2mm]
\vec{D}(\vec{\rho},t) = [\epsilon \star \vec{E}](\vec{\rho},t)
\end{cases}
, \qquad (5.2)
$$

where the convolution is over time.

The source vectors [electric current $\vec{J}(\vec{\rho},t)$ and magnetic current $\vec{M}(\vec{\rho},t)$] are introduced to describe the most general case corresponding to real or fictitious sources. Figure 5.1 shows that the support for these sources is the domain $\Omega$ and that these sources are the origin of the fields.

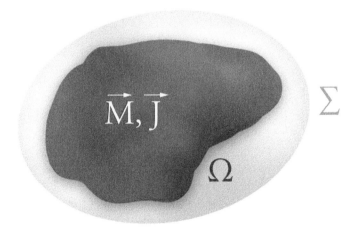

Figure 5.1 Currents $\vec{J}$ and $\vec{M}$ in a domain $\Omega$ surrounded by a fictitious closed surface $\Sigma$.

### 5.1.2 Definitions

To identify a relation that might describe this energy balance, we shall be using mathematical formulations involving products of sources and fields that reveal the power provided by these sources. Intuitively, we might want to show that the power provided by these sources in the domain $\Omega$ gives rise to a radiative flux in the far field through any closed surface $\Sigma$ enclosing this domain, in addition to a loss term within the domain $\Omega$.

We therefore see that the desired formula must involve surfaces and volumes together, leading us to consider a divergence term for which the integral over a volume can be expressed as a surface integral. Hence, using a real Poynting vector defined by $\vec{\mathbb{P}} = \vec{E} \wedge \vec{H}$, we obtain

$$\mathrm{div}\vec{\mathbb{P}} = \mathrm{div}[\vec{E} \wedge \vec{H}] = \vec{H}.\mathbf{curl}\ \vec{E} - \vec{E}.\mathbf{curl}\ \vec{H}\,, \tag{5.3}$$

or, using Maxwell's equations,

$$\mathrm{div}\vec{\mathbb{P}} = (\vec{M}.\vec{H} - \vec{J}.\vec{E}) - \left(\vec{H}.\frac{\partial}{\partial t}\vec{B} + \vec{E}.\frac{\partial}{\partial t}\vec{D}\right)\,. \tag{5.4}$$

After integration over the volume $\Omega$, we get

$$\iint_\Sigma \vec{\mathbb{P}}.\vec{n}\, dS = \iiint_\Omega (\vec{M}.\vec{H} - \vec{J}.\vec{E})\, dV - \iiint_\Omega \left(\vec{H}.\frac{\partial}{\partial t}\vec{B} + \vec{E}.\frac{\partial}{\partial t}\vec{D}\right) dV\,, \tag{5.5}$$

where $\vec{n}$ is the local normal to the surface $\Sigma$ directed outward from the domain $\Omega$, $dS$ is a surface element, and $dV$ is a volume element.

Hence, we have an instantaneous relationship that, by analogy, allows us to introduce

- $\vec{\mathbb{P}}.\vec{n} = \frac{d\Phi}{dS}$, the *surface density of flux* $\Phi$ radiated by sources through the closed surface $\Sigma$ surrounding the domain $\Omega$,
- $\vec{M}.\vec{H} - \vec{J}.\vec{E} = \frac{dF}{dV}$, the *volume power density* $F$ provided by sources within the domain $\Omega$,
- $\vec{H}.\frac{\partial}{\partial t}\vec{B} + \vec{E}.\frac{\partial}{\partial t}\vec{D} = \frac{dA}{dV}$, a *volume density of "losses"* $A$ in the domain $\Omega$.

Note that, taking account of the spatiotemporal regime of interest to us and in which all variables are real, we have defined a real Poynting vector. We shall see later how this vector is related to the vector introduced in the case of the harmonic regime, which is complex and preceded by a factor of $1/2$. For the moment, note that equation (5.5) can be rewritten as

$$\iint_\Sigma d\Phi = \iiint_\Omega dF - \iiint_\Omega dA \quad \Rightarrow \quad F(\Omega, t) = \Phi(\Sigma, t) + A(\Omega, t)\,. \tag{5.6}$$

Thus, at each instant the power provided in the domain $\Omega$ is equal to the outgoing flux through the limiting surface $\Sigma$, in addition to a loss term generated within this same volume $\Omega$. Note that this relation (5.6) is instantaneous and nonlocal. It is problematic in the general case to elaborate further on notions of energy without increasing the complexity of the formalism. However, in the absence of dispersion, i.e., where the medium reacts instantaneously, we saw that the temporal convolutions disappeared in the constitutive relations, so that the loss term reduces to

$$\frac{dA}{dV} = \mu\vec{H}.\frac{\partial}{\partial t}\vec{H} + \epsilon\vec{E}.\frac{\partial}{\partial t}\vec{E} = \frac{1}{2}\frac{\partial}{\partial t}(\vec{H}^2 + \vec{E}^2). \tag{5.7}$$

Introducing the variable $w$, defined by

$$w = \frac{1}{2}(\vec{H}^2 + \vec{E}^2), \tag{5.8}$$

(5.7) becomes

$$\frac{dA}{dV} = \frac{\partial w}{\partial t}, \tag{5.9}$$

and we finally obtain, at each instant $t$:

$$F(\Omega, t) = \Phi(\Sigma, t) + \frac{\partial W}{\partial t}(\Omega) \quad \text{with} \quad W(\Omega) = \iiint_{\Omega} w\, dV. \tag{5.10}$$

After temporal integration over some interval $\tau$, we get for the domain $\Omega$:

$$\int_0^\tau F(\Omega, t)\, dt = \int_0^\tau \Phi(\Sigma, t)\, dt + [W(\tau) - W(0)]. \tag{5.11}$$

Hence during this time interval $\tau$, the difference between the power provided and the outgoing flux is equal to the variation $\Delta W$ of electromagnetic energy in the volume $\Omega$. This then defines $w = \frac{dW}{dv}$ as a volume density of electromagnetic energy whose instantaneous variation is equal to the volume density of losses ($\frac{dA}{dV} = \frac{\partial w}{\partial t}$).

However, note that if dispersion is present, the integration we used in (5.7) is no longer immediate since we would have

$$\frac{dA}{dV} = \frac{\partial w}{\partial t} = \vec{H}.\left[\mu \star \frac{\partial}{\partial t}\vec{H}\right] + \vec{E}.\left[\epsilon \star \frac{\partial}{\partial t}\vec{E}\right]. \tag{5.12}$$

### 5.1.3 Harmonic Components

We shall now introduce into the foregoing relations a general expression for the temporal field in the form of frequential (polychromatic) wave packets, as defined in Chapter 3. Accordingly, each spatiotemporal field $\vec{E}(\vec{\rho}, t)$ or $\vec{H}(\vec{\rho}, t)$ is replaced by its Fourier reconstruction integral.

First, we examine the surface flux density $\frac{d\Phi}{dS}$:

$$\frac{d\Phi}{dS}(t) = \vec{P}.\vec{n} = \left\{\left[\int_f \vec{\mathcal{E}}(\vec{\rho}, f)e^{-2i\pi ft}\, df\right] \wedge \left[\int_{f'} \vec{\mathcal{H}}(\vec{\rho}, f')e^{-2i\pi f't}\, df'\right]\right\}.\vec{n}$$

$$= \int_f \int_{f'} e^{-2i\pi(f+f')t}\left[\vec{\mathcal{E}}(\vec{\rho}, f) \wedge \vec{\mathcal{H}}(\vec{\rho}, f')\right].\vec{n}\, df\, df'. \tag{5.13}$$

At this stage, the flux is a function of time and we need to take account of beats between the harmonic waves at the different frequencies $f$ and $f'$.

However, in the visible region the temporal oscillation frequency $f$ of the field is extremely high, on the order of several hundreds of terahertz:

$$f = \frac{c}{\lambda_0} \quad \rightarrow \quad f \simeq 600 \times 10^{12} \text{ Hz for } \lambda_0 = 0.5 \,\mu\text{m} \,,$$

corresponding to a temporal period $1/f$ no greater than $2 \times 10^{-15}$ s.

Given that the rise time of a standard detector is in the order of a nanosecond, it would be unable to follow the temporal flux oscillation. To take account of this constraint, we will consider that the detector behaves like a linear filter. Such filter connects the input (the flux $\Phi$ of the Poynting vector through the detector area $A_d$) to the output (the detector signal $v$) of the system in the form of a convolution product, that is,

$$v(t) = [h \star \Phi]_t = A_d \left[ h \star \frac{d\Phi}{dS} \right]_t = A_d \int_{t'} h(t') \frac{d\Phi}{dS}(t - t') \, dt' \,, \tag{5.14}$$

where $h(t)$ is given for the **receiver transfer function**.

Under this assumption, we can combine relations (5.13) and (5.14) and invert the integrals on time and frequencies:

$$v(t) = A_d \int_f \int_{f'} \int_{t'} h(t') e^{-2i\pi(f+f')(t-t')} \, dt' \left[ \vec{\mathcal{E}}(\vec{\rho}, f) \wedge \vec{\mathcal{H}}(\vec{\rho}, f') \right] \cdot \vec{n} \, df \,.$$

We can see that the Fourier transform of $h$, appears, taken at the frequency $-(f + f')$. So the previous equation becomes

$$v(t) = A_d \int_f \int_{f'} \hat{h}^*(f + f') e^{-2i\pi(f+f')t} \left[ \vec{\mathcal{E}}(\vec{\rho}, f) \wedge \vec{\mathcal{H}}(\vec{\rho}, f') \right] \cdot \vec{n} \, df \, df' \,, \tag{5.15}$$

where we used the Hermitian symmetry of the receiver transfer function.

At this step we notice that the transfer function is a low-pass filter and that its Fourier transform will tend toward a Dirac distribution when its cutoff frequency approaches 0. To be convinced (though this is not a general proof), let us consider the specific case where the $h(t)$ function performs a temporal flux integration over a period $T$, around the instant $\tau$. We have

$$h(t) = \text{rect} \left[ \frac{t - \tau}{T} \right] \quad \Rightarrow \quad \hat{h}(f) = T \, e^{-2i\pi f \tau} \, \text{Sinc}(\pi f T) \,, \tag{5.16}$$

with the property

$$T \, \text{Sinc}(\pi f T) \rightarrow \delta(f) \quad \text{for } T \rightarrow \infty \,. \tag{5.17}$$

As a consequence, we have

$$\widehat{h}^*(f + f')e^{-2i\pi(f+f')t} \quad \rightarrow \quad e^{-2i\pi(f+f')(t-\tau)}\,\delta(f + f') = \delta(f + f')\,. \quad (5.18)$$

Using the properties of the Dirac distribution, (5.15) can be written

$$v(t) = A_d \int_f \left[ \vec{\mathcal{E}}(\vec{\rho}, -f) \wedge \vec{\mathcal{H}}(\vec{\rho}, f) \right] . \vec{n}\ df = A_d \langle \frac{d\Phi}{dS} \rangle_T\,. \quad (5.19)$$

Making use of the Hermitian symmetry of the field, we hence have

$$\langle \frac{d\Phi}{dS} \rangle_T = \int_f \left[ \vec{\mathcal{E}}^*(\vec{\rho}, f) \wedge \vec{\mathcal{H}}(\vec{\rho}, f) \right] . \vec{n}\,df\,, \quad (5.20)$$

or, restricting our summation to positive frequencies only,

$$\langle \frac{d\Phi}{dS} \rangle_T = 2 \int_0^\infty \Re \left\{ [\vec{\mathcal{E}}^*(\vec{\rho}, f) \wedge \vec{\mathcal{H}}(\vec{\rho}, f)]. \vec{n} \right\} df\,. \quad (5.21)$$

Hence under the integral sign we find a term corresponding to the Poynting vector in the harmonic regime $\vec{\Pi} = \frac{1}{2}(\vec{\mathcal{E}}^* \wedge \vec{\mathcal{H}})$ introduced in Chapter 2. As a result, we can write

$$\langle \frac{d\Phi}{dS} \rangle_T = 4 \int_0^\infty \Re[\vec{\Pi}.\vec{n}]\,df\,, \quad (5.22)$$

or, introducing the surface flux density per unit temporal frequency, we get the **monochromatic surface flux density**:

$$\langle \frac{d^2\Phi}{df\,dS} \rangle_T = 4\Re[\vec{\Pi}.\vec{n}]\,. \quad (5.23)$$

This result extends to temporal frequencies a property demonstrated in Chapter 4 for spatial frequencies, namely that the flux transported by a wave packet is equal to the sum of the fluxes transported by each of its elementary components. As expressed in equation (5.23), it should also be pointed out that the harmonic Poynting flux corresponds to a temporal mean of the monochromatic flux.

The reader may verify that the same approach can be used to calculate the loss and power volume densities defined in Section 5.1.2. Hence, we obtain, for the **monochromatic volume power density**:

$$\langle \frac{d^2F}{df\,dV} \rangle_T = 2\Re \left[ \vec{\mathcal{M}}^*(\vec{\rho}, f).\vec{\mathcal{H}}(\vec{\rho}, f) - \vec{\mathcal{J}}(\vec{\rho}, f).\vec{\mathcal{E}}^*(\vec{\rho}, f) \right]\,, \quad (5.24)$$

and finally, for the **monochromatic volume absorption density** (related to the inertia of the material):

$$\langle \frac{d^2A}{df\,dV} \rangle_T = 2\Re \left[ -i\omega \vec{\mathcal{H}}^*(\vec{\rho}, f).\vec{\mathcal{B}}(\vec{\rho}, f) - i\omega \vec{\mathcal{E}}^*(\vec{\rho}, f).\vec{\mathcal{D}}(\vec{\rho}, f) \right]\,, \quad (5.25)$$

or

$$\langle\frac{d^2 A}{df\,dV}\rangle_T = 2\omega\Re\left[i\tilde{\mu}^*(f)|\vec{\mathcal{H}}(\vec{\rho},f)|^2 - i\tilde{\epsilon}(f)|\vec{\mathcal{E}}(\vec{\rho},f)|^2\right]$$
$$= 2\omega\left[\tilde{\mu}''(f)|\vec{\mathcal{H}}(\vec{\rho},f)|^2 + \tilde{\epsilon}''(f)|\vec{\mathcal{E}}(\vec{\rho},f)|^2\right], \tag{5.26}$$

where $\tilde{\mu}''(f)$ and $\tilde{\epsilon}''(f)$ are the imaginary parts of the magnetic permeability and electric permittivity of the medium respectively. On this occasion, note that these parameters are positive for passive materials.

Finally, observe that, from the point of view of an *aide memoire* (omitting the $\langle\ \rangle_T$ notation) the energy balance can be obtained by integration of the surface and volume complex spectral densities defined by

$$\frac{1}{4}\frac{d^2 F}{df\,dV} = \frac{1}{2}\left[\vec{\mathcal{M}}^*(\vec{\rho},f).\vec{\mathcal{H}}(\vec{\rho},f) - \vec{\mathcal{J}}(\vec{\rho},f).\vec{\mathcal{E}}^*(\vec{\rho},f)\right]$$

$$\frac{1}{4}\frac{d^2\Phi}{df\,dS} = \vec{\Pi}.\vec{n} \tag{5.27}$$

$$\frac{1}{4}\frac{d^2 A}{df\,dV} = \frac{i\omega}{2}\left[\tilde{\mu}^*(f)|\vec{\mathcal{H}}(\vec{\rho},f)|^2 - \tilde{\epsilon}(f)|\vec{\mathcal{E}}(\vec{\rho},f)|^2\right].$$

Only the real part of these variables will be considered in the energy balance. For instance, in the case of the third equation, we have

$$\Re\left[\frac{1}{4}\frac{d^2 A}{df\,dV}\right] = \frac{\omega}{2}\left[\tilde{\mu}''(f)|\vec{\mathcal{H}}(\vec{\rho},f)|^2 + \tilde{\epsilon}''(f)|\vec{\mathcal{E}}(\vec{\rho},f)|^2\right]. \tag{5.28}$$

Introduction of the factor $\frac{1}{4}$ provides a traditional normalization with respect to the flux transported by a unidirectional, monochromatic progressive wave. For such a plane wave, the temporal dependence of the Poynting vector $\vec{\Pi}$ varies as $\cos^2\omega t$, and its temporal mean is therefore $\frac{1}{2}$. The result is that the flux transported by this wave is indeed equal to $\Re[\vec{\Pi}.\vec{n}]$, which justifies, a posteriori, the introduction of the factor $\frac{1}{2}$ in the definition given in Chapter 2 of the harmonic Poynting vector.

In conclusion, this section provides justification that broad-spectrum light can be analyzed by considering all its wavelengths in turn, and that a separate energy balance for each wavelength can be established. Keep in mind, however, that it is the integrator aspect of the detector that makes it possible to eliminate beat (or interference) phenomena between wavelengths within the same beam. A fresh look might be taken at these results in the microwave domain, where detectors can *follow* the temporal variations in the electromagnetic signal [that is, $h(t) = h_0\delta(t)$] on account of the greater length of the temporal periods involved. Also, detectors may follow the optical signal in the case of beats between very close frequencies, which exhibit large temporal periods.

## 5.2 Kramers–Kronig Relations

At the end of Chapter 1 we emphasized the importance of the principle of causality, which states that *no effect can precede its cause*. For this reason, the susceptibility function (whether electric or magnetic), which relates the polarization of matter to the excitation that causes it, must satisfy the property

$$\chi(t) = 0 \quad \forall t < 0\,. \tag{5.29}$$

This condition can be put in the following equivalent form:

$$\chi(t) = \mathrm{Sgn}(t)\,\chi(t)\,, \tag{5.30}$$

where $\mathrm{Sgn}(t)$ is the Sign function defined in Section 1.7.3. If $t < 0$, $\mathrm{Sgn}(t) = -1$, meaning that the function $\chi(t)$ is equal to its opposite and hence identically zero. Conversely, if $t > 0$, $\mathrm{Sgn}(t) = +1$ and (5.30) is clearly satisfied and does not in any way constrain the susceptibility function.

If we now apply a Fourier transform to the two terms of equation (5.30), we immediately get

$$\tilde{\chi}(f) = \int_{-\infty}^{+\infty} \chi(t) e^{2i\pi f t} dt = [\widetilde{\mathrm{Sgn}} \star \tilde{\chi}](f)\,. \tag{5.31}$$

The Fourier transform of the Sign function is here given by the expression

$$\widetilde{\mathrm{Sgn}}(f) = -\frac{1}{i\pi} \mathrm{PV}\left(\frac{1}{f}\right)\,, \tag{5.32}$$

where the letters PV denote the Cauchy Principal Value distribution. It follows that

$$\tilde{\chi}(f) = -\frac{1}{i\pi} \mathrm{PV}\left(\frac{1}{f}\right) \star \tilde{\chi}(f)\,, \tag{5.33}$$

or, expanding the convolution product,

$$\tilde{\chi}(f) = -\frac{1}{i\pi} \int_{-\infty}^{+\infty} \frac{\tilde{\chi}(u)}{f - u} du\,. \tag{5.34}$$

Consequently, if we now separate the real and imaginary parts $\tilde{\chi}'(f)$ and $\tilde{\chi}''(f)$ of the function $\tilde{\chi}(f)$, we can write, for the electric susceptibility:

$$
\begin{cases}
\tilde{\chi}'(f) = \tilde{\epsilon}_r'(f) - 1 = -\dfrac{1}{\pi} \displaystyle\int_{-\infty}^{+\infty} \dfrac{\tilde{\chi}''(u)}{f - u}\, du \\[3mm]
\tilde{\chi}''(f) = \tilde{\epsilon}_r''(f) = \dfrac{1}{\pi} \displaystyle\int_{-\infty}^{+\infty} \dfrac{\tilde{\chi}'(u)}{f - u}\, du
\end{cases}
. \qquad (5.35)
$$

These last two expressions are the **Kramers–Kronig relations**. We also say that the real and imaginary parts of the frequential component of susceptibility are related through a **Hilbert transformation**.

Relations (5.35) show that if all the values of the function $\tilde{\chi}''$ (and $\tilde{\chi}'$ respectively) are known, then the function $\tilde{\chi}'$ (and $\tilde{\chi}''$ respectively) is completely determined. Note that the integrations appearing in (5.35) can be reduced to positive frequencies only. Given that susceptibility has Hermitian symmetry, the real and imaginary parts of its Fourier transform are respectively even and odd, i.e.,

$$
\tilde{\chi}'(-f) = \tilde{\chi}'(f) \quad \text{et} \quad \tilde{\chi}''(-f) = -\tilde{\chi}''(f). \qquad (5.36)
$$

These Kramers–Kronig relations are very useful in analyzing frequency dispersion phenomena. Observe that if the medium is nonabsorbing [$\tilde{\chi}'' = 0$ for all frequencies], we immediately obtain $\tilde{\chi}' = 0$, so its permittivity reduces to that of vacuum ($\tilde{\epsilon} = \epsilon_v$), i.e., the medium that is *perfect*.

This remark emphasizes the intrinsic connection between absorption and dispersion, both resulting from inertia of materials. As already stressed in Chapter 2, the case of metals is slightly different in the sense that there is an additional dispersion resulting from Ohm law, which is not necessarily connected with inertia. Hence the dispersion of the equivalent relative permittivity must be corrected for metals as

$$
\epsilon_{r,eq} = 1 + \chi + i\frac{\tilde{\gamma}}{\omega\epsilon_v}, \qquad (5.37)
$$

with $\tilde{\gamma}$ the Fourier transform of the metal conductivity.

Frequency variations of this conductivity take account of the metal inertia, but the equivalent permittivity dispersion holds for constant conductivity (no more inertia). Finally, for nondispersive metals it is sufficient to replace $\chi''$ by $\chi'' + \frac{\tilde{\gamma}}{\omega\epsilon_v}$.

We can go further by investigating the derivative of the real part of the susceptibility. We get

$$\frac{\partial \tilde{\chi}'}{\partial f} = \frac{1}{\pi} \int_{-\infty}^{+\infty} \frac{\tilde{\chi}''(u)}{(f-u)^2} \, du$$

$$= \frac{1}{\pi} \int_{-\infty}^{0} \frac{\tilde{\chi}''(u)}{(f-u)^2} \, du + \frac{1}{\pi} \int_{0}^{+\infty} \frac{\tilde{\chi}''(u)}{(f-u)^2} \, du \,. \qquad (5.38)$$

Given the odd symmetry of $\tilde{\chi}''(f)$, the first integral can be transformed as follows:

$$\int_{-\infty}^{0} \frac{\tilde{\chi}''(u)}{(f-u)^2} \, du = -\int_{0}^{+\infty} \frac{\tilde{\chi}''(u)}{(f+u)^2} \, du \,; \qquad (5.39)$$

so we can write

$$\pi \frac{\partial \tilde{\chi}'}{\partial f} = \int_{0}^{+\infty} \tilde{\chi}''(u) \left[ \frac{1}{(f-u)^2} - \frac{1}{(f+u)^2} \right] du$$

$$= 4 \int_{0}^{+\infty} \frac{u \tilde{\chi}''(u)}{(f^2 - u^2)^2} \, du \,. \qquad (5.40)$$

Given that $\tilde{\chi}'' = \epsilon_r''$ and that the imaginary part of the relative electric permittivity $\epsilon_r''$ is positive, equation (5.40) shows that the derivative $\partial \tilde{\chi}'/\partial f = \partial \tilde{\epsilon}_r'/\partial f$ is positive.

Hence in a transparent zone in a nonmagnetic medium ($\tilde{\mu}_r = 1$), the refractive index is real and given by $n = \sqrt{\tilde{\epsilon}_r}$, so it also increases with frequency; or equivalently, that it decreases with increasing wavelength. We used this property in Chapter 3 to justify a change of variable that was useful in calculating the group velocity. Notice in (5.35) that a transparent zone means $\tilde{\chi}''(f) = 0$ in this zone, which does not contradict the fact that the integral is not 0 (the transparent zone may be narrow with respect to the complete range of frequencies).

Similarly, the reader can easily verify that, for the imaginary part of susceptibility, we obtain

$$\pi \frac{\partial \tilde{\chi}''}{\partial f} = 2 \int_{0}^{+\infty} \frac{f^2 + u^2}{(f^2 - u^2)^2} \tilde{\chi}'(u) \, du \,. \qquad (5.41)$$

## 5.3 Coherence

We saw in Section 5.1.3 that the description of a detector functioning as a linear filter in the time domain was able to simplify the expression for the Poynting vector flux transported by polychromatic light. Such considerations lead naturally to the notion of coherence, this notion being closely related to the nature of detection.

Generally speaking, the idea of coherence describes how fields emitted by sources can interfere with one another when they are spatially and/or temporally superposed. However, it is useful to note at this stage that, since the fields have been described by spatiotemporal wave packets with no approximation, a simple algebraic sum will represent their interference perfectly. In other words, such sources and fields as we have described are perfectly coherent and interfere fully; this means that, with the (exact) formalism of wave packets, the notion of coherence cannot be addressed just by referring to Maxwell's equations.

Ideas of coherence (temporal and spatial) are related to the conditions of detection, most frequently through the phenomena of temporal and/or spatial integration; this approach can be generalized to effects involving the polarization of light. We shall show succinctly in the sections that follow how these various notions can be introduced and used.

### 5.3.1 Temporal Coherence

Consider a two-wave interferometry experiment, as shown diagrammatically in Figure 5.2.

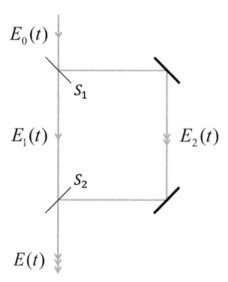

Figure 5.2 Temporal overlap of a plane wave $\vec{E}(\vec{\rho}, t)$ obtained using a two-beam Mach–Zehnder interferometer.

This arrangement splits an incident field $\vec{E}_0(t)$ into two distinct fields $\vec{E}_1(t)$ and $\vec{E}_2(t)$, using a beam splitter $S_1$ that is assumed balanced and lossless ($R = T = 1/2$ in energy terms); the two beams are then recombined through a second beam splitter $S_2$ with the same characteristics as the first. The mirrors used for the path of beam 2 are assumed to be perfect (energy reflection coefficient equal to 1, with no phase shift on reflection). In the interests of simplicity, we shall assume that the initial wave is TE polarized so that we can consider an algebraic, rather than vectorial, sum.

Taking account of the difference in length of the two arms of the interferometer, the field $E(t)$ resulting from the superposition of the two fields $E_1$ and $E_2$ can be written

$$E(t) = \frac{1}{\sqrt{2}}[E_1(t) + E_2(t)] \quad \text{avec} \quad E_i(t) = \frac{1}{\sqrt{2}}E_0(t - \tau_i). \quad (5.42)$$

The quantities $\tau_i$ represent the temporal delays introduced by each arm of the interferometer. Note that since we are working in the temporal domain with real fields, expression (5.42) is valid only because we have assumed that there is no frequential dispersion in the properties of the beam splitters and mirrors.

In Chapter 3 we discussed a similar interferometric experiment, where the temporal dependence of the incident field was perfectly known (an ultrashort pulse delivered by a femtosecond laser). In contrast, we are dealing here with sources whose fluctuations are so rapid and complex that their temporal behavior cannot be described in terms of a simple mathematical expression: a stochastic approach is more suited to such a description. This will be true for so-called incoherent sources such as halogen lamps, but also for sources usually considered to be coherent, such as quasi-monochromatic lasers that deliver an almost constant average power over the course of time.

In this very general case, we shall now include the effect of temporal filtering by a **quadratic** photodetector. Rigorously, we should again use the detectors transfer function $h(t)$ (see Section 5.1.3), but in the interests of simplicity we shall assume that this function behaves like an **ideal low-pass filter**. The convolution operation is then replaced by an integration, and we shall allow the existence of introduced quantities.

To a constant of proportionality, the signal $S$ recorded during this experiment is therefore defined by

$$S = \frac{1}{2}\langle[E_1(t) + E_2(t)]^2\rangle = \frac{1}{4}\langle E_0^2(t - \tau_1)\rangle + \frac{1}{4}\langle E_0^2(t - \tau_2)\rangle$$
$$+ \frac{1}{2}\langle E_0(t - \tau_1)E_0(t - \tau_2)\rangle, \quad (5.43)$$

where the $\langle\rangle$ symbols denote a temporal mean.

Using the assumption of **stationarity**, this expression is equivalent to

$$S = \frac{1}{2}\langle E_0^2(t)\rangle + \frac{1}{2}\langle E_0(t)E_0(t-\tau)\rangle\,. \tag{5.44}$$

At this stage, we introduce the autocorrelation function for field $E_0(t)$, defined by

$$\Gamma(\tau) = \langle E_0(t)E_0(t-\tau)\rangle\,, \tag{5.45}$$

so that equation (5.44) can be written in the form

$$2S = \Gamma(0) + \Gamma(\tau)\,. \tag{5.46}$$

Expression (5.46) therefore shows that the level of the interference signal is driven by the first-order temporal autocorrelation function for the field. It is customary to normalize this autocorrelation function, i.e.,

$$g^{(1)}(\tau) = \frac{\Gamma(\tau)}{\Gamma(0)}\,, \tag{5.47}$$

in order to put the expression for the signal in the form

$$S(\tau) = \Gamma(0)\left[1 + g^{(1)}(\tau)\right] \quad \text{where} \quad |g^{(1)}(\tau)| \leqslant 1\,. \tag{5.48}$$

Using the Parseval's theorem, this function $g^{(1)}(\tau)$ can be expressed in an alternative form, i.e.,

$$g^{(1)}(\tau) = \frac{\displaystyle\int_f |\mathcal{E}_0(f)|^2\, e^{-2i\pi f\tau}\, df}{\displaystyle\int_f |\mathcal{E}_0(f)|^2\, df}\,, \tag{5.49}$$

which shows that it can be extracted directly from a measurement of the energy spectrum of the source via a Fourier transform.

From equation (5.49) we get the notion of coherence time using the relationship given in Chapter 3 between the supports of a function and its Fourier transform, i.e., here

$$\Delta\tau.\Delta f \geqslant 1\,, \tag{5.50}$$

where $\Delta f$ defines a measure of the frequential range of $|\mathcal{E}_0(f)|^2$. This relation leads to the introduction of a coherence time $t_c = 1/\Delta f$, and a corresponding **coherence length** $l_c = c/\Delta f$. This quantity defines a difference in length between the two arms of the interferometer, less than which interference can be seen. Above this value, the function $g^{(1)}(\tau)$ in (5.48) becomes negligible, so the signal $S$ no longer depends on the delay $\tau$, and hence on the difference in path length.

For a beam of low spectral width $\Delta f$ and central wavelength $\lambda_0$, we get

$$l_c = \frac{c}{\Delta f} = \frac{\lambda_0^2}{\Delta \lambda}, \tag{5.51}$$

which shows that the coherence length increases as the spectral line width of the source decreases. Hence a source emitting at 600 nm with a spectral line width of 1 picometer will have a typical coherence length of 36 centimeters.

For (ideal) monochromatic light, note that the coherence length would be infinite, and the function $g^{(1)}(\tau)$ would behave periodically as $\cos(2\pi f_0 \tau)$. We also understand that it is always possible to restore temporal coherence to the light by filtering it, i.e., by reducing the frequency range of its spectrum.

We remark that the coherence length $l_c$ was introduced here in the temporal regime, but it is also possible to obtain this variable directly from the harmonic regime; loss of visibility corresponds simply to a blurring of the interference fringes as the spectral width of the source increases.

In conclusion, we observe that the notion of temporal coherence extends directly to each mode of light polarization. However, since these polarization modes are defined in a plane perpendicular to the direction of propagation (see Chapter 2, Section 2.9), this means that the beam being considered is polychromatic, but supposedly unidirectional. A spatial wave packet would therefore require other developments.

### 5.3.2 Light Depolarization

The notion of depolarization follows on naturally from that of temporal coherence. The underlying ideas are not very different, since now it is not a case of superposing the same scalar field at two different instants but of superposing at the same instant (or, in the interests of generality, at two different instants) the two previously aligned polarization modes of the same field. In other words, what we are interested in here are the results of **polarimetric interference**, a technique widely used in ellipsometry in the case of monochromatic (and hence polarized) light.

At the input to the interferometer the unidirectional field can be decomposed into two polarization modes (commonly denoted S and P), perpendicular to the direction $z$ of propagation, i.e.,

$$\vec{E}_0(t) = E_S(t)\vec{x} + E_P(t)\vec{y}. \tag{5.52}$$

The two resultant modes are superposed at the output, and are written

$$\vec{E}(t) = E_S(t - \tau_1)\vec{x} + E_P(t - \tau_2)\vec{y}. \tag{5.53}$$

These two fields do not interfere at this stage because they are perpendicular. Hence, we need to introduce a component into the system that will align these two modes, either by rotation (using a birefringent half-wave plate or a Faraday-effect component) or by projection (using an analyzer). Once this is done (e.g., with a polarizer oriented at 45 degrees to the directions of the two modes), the two modes can then interfere because their superposition leads to the algebraic sum

$$E(t) = \frac{1}{\sqrt{2}} E_S(t - \tau_1) + \frac{1}{\sqrt{2}} E_P(t - \tau_1) \,. \tag{5.54}$$

With the same stationarity assumption as before, the signal $S$ output by a quadratic photodetector can be expressed in the following form (give or take a constant of proportionality):

$$S(\tau) = \frac{1}{2} \left[ \Gamma_S(0) + \Gamma_P(0) + 2\Gamma_{SP}(\tau) \right], \tag{5.55}$$

where

$$\tau = \tau_2 - \tau_1 \quad ; \quad \Gamma_{SP}(\tau) = \langle E_S(t) E_P(t - \tau) \rangle \,, \tag{5.56}$$

and

$$\Gamma_S(0) = \langle E_S^2(t) \rangle \quad ; \quad \Gamma_P(0) = \langle E_P^2(t) \rangle \,. \tag{5.57}$$

The more common situation is where the two arms of the interferometer are the same length ($\tau = 0$), and we shall assume that this is the case from now on. After normalization, (5.55) can be written

$$S = \frac{1}{2} \left[ \Gamma_S + \Gamma_P + 2\gamma_{SP} \sqrt{\Gamma_S \Gamma_P} \right] \quad \text{where} \quad \gamma_{SP} = \frac{\Gamma_{SP}}{\sqrt{\Gamma_S \Gamma_P}} \,. \tag{5.58}$$

Hence it can be seen that the magnitude of the interferometric signal is now driven by the mutual coherence coefficient $\gamma_{SP}$.

It might be tempting at this stage to use the extreme values (i.e., 0 and 1) of this coefficient to define the notions of polarized and depolarized light; however, this will not work. Taking account of the Cauchy–Schwarz inequality, a unit value of $\gamma_{SP}$ would imply that the two modes of polarization were proportional; this would require a constant field direction, which is characteristic of light that is **linearly polarized**. This then excludes the possibility that elliptically polarized light can be considered totally polarized, though it is precisely this situation that might be encountered in the case of a monochromatic wave (see Section 2.9). Furthermore, a zero value of the coefficient $\gamma_{SP}$ would not require the two modes to be independent, a condition we might wish to attribute to depolarized light in order for its field to be uniformly distributed in direction, with in this case $\Gamma_S = \Gamma_P$.

Classically, we can correctly introduce the idea of partial polarization using the concept of a complex *analytical field* $E^a(t)$. Recall that this analytical field is defined by its spectrum $E^a(f)$, which is a restricted version of the real field $E(t)$ at positive frequencies only, i.e.,

$$\mathcal{E}^a(f) = H(f)\mathcal{E}(f), \tag{5.59}$$

where $H$ is the Heaviside unit step function. Coming back to the temporal domain, we get

$$E^a(t) = \left[\left\{\delta - \frac{i}{\pi}\mathrm{VP}\left(\frac{1}{t}\right)\right\} \star E\right](t), \tag{5.60}$$

where $\delta$ is the Dirac distribution and PV is the Cauchy principal value. This expression shows that the real field is just the real part of the analytical field, i.e.,

$$E(t) = \Re[E^a(t)]. \tag{5.61}$$

Note that this analytical field will be easy to manipulate only for quasi-monochromatic light, since this condition will guarantee that the analytical field is much the same as a classical complex expression of the real field, despite its frequency range. Consider, therefore, the two polarization modes of the real field of quasi-monochromatic light, i.e.,

$$\begin{aligned} E_S(t) &= A_S(t)\cos[2\pi f_0 t - \phi_S(t)] \\ E_P(t) &= A_P(t)\cos[2\pi f_0 t - \phi_P(t)], \end{aligned} \tag{5.62}$$

where the amplitudes $A_S(t)$ and $A_P(t)$ vary slowly with respect to the reciprocal of the central frequency $f_0$. Note also the temporal dependence of the phase terms $\phi_S$ and $\phi_P$ which, apart from the variable amplitudes, constitutes an essential difference from monochromatic light. These phase variations lead to a spectral width $\Delta f$ around the central frequency $f_0$, and they will be assumed sufficiently slow that the quasi-monochromatic regime condition is satisfied, i.e.,

$$\Delta f / f_0 \ll 1. \tag{5.63}$$

Under these conditions, it can be shown that the analytical fields associated with these two polarization modes can be expressed as

$$E_S^a(t) = A_S^a(t)\,e^{i[2\pi f_0 t - \phi_S(t)]} \quad \text{and} \quad E_P^a(t) = A_P^a(t)\,e^{i[2\pi f_0 t - \phi_P(t)]}. \tag{5.64}$$

With these new variables, depolarized light can now be defined using two special properties. The first characterizes light that is said to be *balanced*, in the sense that it transports the same flux for each mode of polarization, written as

$$\Gamma_S^a = \Gamma_P^a \quad \text{where} \quad \Gamma_u^a = \langle |E_u^a(t)|^2 \rangle = \langle [A_u^a(t)]^2 \rangle \text{ and } u = S, P . \quad (5.65)$$

The second property represents the *independence* of the analytical modes of polarization, and is written as

$$\gamma_{SP}^a = \frac{\Gamma_{SP}^a}{\sqrt{\Gamma_S^a \Gamma_P^a}} = 0 , \quad (5.66)$$

where

$$\Gamma_{SP}^a = \langle E_S^a(t) E_P^{a*}(t) \rangle \approx A_S^a(t) A_P^a(t) \langle e^{i[\phi_S(t) - \phi_P(t)]} \rangle , \quad (5.67)$$

and where it is assumed that the amplitudes barely change over the integration time of the detector.

Finally, these two conditions (5.65) and (5.66) show that any variations in the direction of the real field are sufficiently random that they are uniformly distributed in direction, hence defining a light that is said to be *depolarized*.

As for polarized light, this is defined by the other extreme value of $\gamma_{SP}^a$, i.e.,

$$|\gamma_{SP}^a| = \left| \frac{\langle E_S^a(t) E_P^a(t) \rangle}{\sqrt{\Gamma_S^a \Gamma_P^a}} \right| = 1 . \quad (5.68)$$

The Cauchy–Schwarz inequality requires that the two analytical modes of polarization are proportional. Since this time the coefficient of proportionality is complex (by comparison with $\gamma_{SP} = 1$, where the coefficient was real), this means that a constant phase shift $\psi$ between the two modes is introduced, in addition to a modulus modification. This is written as

$$E_P^a(t) = \alpha \, E_S^a(t) \text{ where } \alpha = |\alpha| \, e^{i\psi}$$
$$\Rightarrow \quad E_P(t) = \Re[E_P^a(t)] = |\alpha| \, A_S(t) \cos[2\pi f_0 t - \phi_S(t) + \psi], \quad (5.69)$$

and corresponds to a (generally elliptical) polarized light (see Chapter 2).

In summary, we need to remember that, with the quasi-monochromatic light defined in (5.62) and (5.63), the polarization or depolarization of the light is controlled first and foremost by the temporal variations in phase shift $\phi_P(t) - \phi_S(t)$. In all cases, the end of the vector $\vec{E}$ describes an ellipse (or a circle, in the case of depolarized light). On the other hand, whether this ellipse is traced out deterministically or stochastically is dependent on the polarized or depolarized nature of the light.

To be convinced of this, it is sufficient to recalculate the speed of rotation $d\eta/dt$ of the field, as we did in Chapter 2 for monochromatic light. We get

$$\frac{d\eta}{dt} = \frac{E_P(t)\frac{dE_S}{dt}(t) - E_S(t)\frac{dE_P}{dt}(t)}{E_P^2(t) + E_S^2(t)}, \tag{5.70}$$

and, in the case of very slowly varying amplitudes,

$$\frac{d\eta}{dt} = \frac{A_S A_P/2}{A_P^2 \cos^2[\omega_0 t - \phi_P] + A_S^2 \cos^2[\omega_0 t - \phi_S]}$$
$$\times \left\{ \left(\frac{d\phi_S}{dt} - \frac{d\phi_P}{dt}\right) \sin(2\omega_0 t - \phi_S - \phi_P) \right.$$
$$\left. + \left(\frac{d\phi_S}{dt} + \frac{d\phi_P}{dt}\right) \sin(\phi_P - \phi_S) - 2\omega_0 \sin(\phi_P - \phi_S) \right\}, \tag{5.71}$$

where, for clarity, the time dependence of the phases $\phi_S$ and $\phi_P$ has not been spelled out.

In light of the complexity of this formula, it is easy to see that, over time, speed of rotation can be subject to huge variations in modulus and sign. In the extreme case for which equation (5.66) is satisfied (depolarized light), the variations in phase are independent and sufficiently complex that the field direction is seen by the detector as random and uniformly distributed. Conversely, for polarized light the phase shift is constant $[\phi_P(t) - \phi_S(t) = \psi]$ and (5.71) reduces to

$$\frac{d\eta}{dt} = -\frac{A_S/A_P}{|\alpha|^2 \cos^2[\omega_0 t + \psi] + \cos^2 \omega_0 t} \left(\omega_0 - \frac{d\phi_S}{dt}\right) \sin \psi. \tag{5.72}$$

This result shows that the definition (5.69) of polarized light is broader than that for monochromatic light, for which we would have $\frac{d\phi_S}{dt} = 0$. The direction of rotation can change depending on the sign taken at each instant by the quantity $\omega_0 - d\phi_S/dt$.

To conclude this section, while polarized and depolarized light are from now defined by relations (5.65–5.66) and (5.68) respectively, it can be shown that a quasi-monochromatic light can be viewed as the superposition of these two sorts of light. We speak of **partially polarized light**, leading to the definition of a degree of polarization (DOP) as the ratio of polarized to total flux. This degree of polarization is calculated by diagonalizing an Hermitian matrix $\mathbf{M}$ and making use of the invariance properties of the trace and determinant of this matrix:

$$\mathbf{M} = \begin{bmatrix} \Gamma_S^a & \Gamma_{SP}^a \\ \Gamma_{SP}^{a*} & \Gamma_P^a \end{bmatrix} \quad \Rightarrow \quad \text{DOP} = \sqrt{1 - \frac{4\beta}{(1+\beta)^2}(1 - |\gamma_{SP}^a|^2)}. \tag{5.73}$$

Here $\gamma_{SP}^a$ describes the mutual coherence between the modes of the analytical field, and the level of polarization $\beta$ is defined as

$$\beta = \frac{\Gamma_P^a}{\Gamma_S^a}. \tag{5.74}$$

Note the particularity of the matrices describing polarized or depolarized light (zero determinant and zero off-diagonal terms respectively).

Finally, note that while the analytical fields led to a definition of the partial polarization of light, they must nevertheless allow the initial interference as described by (5.58) to be analyzed. Rigorously, we should verify that the same equation can be described by replacing the real fields by analytical fields. It is shown that this is indeed the case, namely that the intercorrelation functions of the real field can be deduced from the real part of the complex intercorrelation functions of the analytical field, defined as

$$\Gamma_{uv}^a(\tau) = \langle E_u^a(t) E_v^{a*}(t - \tau) \rangle \quad \Rightarrow \quad \Re[\Gamma_{uv}^a] = 2\Gamma_{uv}. \tag{5.75}$$

Consequently, the coefficient $\gamma_{SP}^a$ is involved in the definition of interference contrast, at the same time describing the complexity of the real field.

### 5.3.3 Spatial Coherence

The same ideas as those introduced in Section 5.3.1 to define temporal coherence govern the notion of spatial coherence, except that here the spatial coordinates $(x, y)$ replace time $t$.

Assume that the regime is harmonic and that the same field $\vec{\mathcal{E}}(\vec{r}, z, f)$ has been split into two secondary fields $\vec{\mathcal{E}}_1(\vec{r} - \vec{r}_1, z, f)$ and $\vec{\mathcal{E}}_2(\vec{r} - \vec{r}_2, z, f)$; the result of the superposition is recorded with a quadratic detector in the plane $z = 0$, as shown diagrammatically in Figure 5.3 in the case of a weakly divergent beam.

Question: is there a distance $\Delta\vec{r} = \vec{r}_2 - \vec{r}_1$ for which, if exceeded, interference between the two duplicate beams will no longer be visible?

As shown in Chapter 4, each field can be written in the form of a spatial wave packet:

$$\vec{\mathcal{E}}_i(\vec{r}, z, f) = \int_{\vec{\nu}} \vec{\mathbb{A}}_i(\vec{\nu}, f) \, e^{i[2\pi\vec{\nu}.\vec{r} + \alpha(\nu, f)z]} \, d^2\vec{\nu}, \tag{5.76}$$

where $\alpha^2(\nu, f) = k^2(f) - 4\pi^2\nu^2$, i.e., at $z = 0$ (assuming that the duplication was balanced),

$$\vec{\mathcal{E}}_i(\vec{r}, 0, f) = C\vec{\mathcal{E}}(\vec{r} - \vec{r}_i, 0, f) = C\int_{\vec{\nu}} \vec{\mathbb{A}}(\vec{\nu}, f) \, e^{2i\pi\vec{\nu}.(\vec{r} - \vec{r}_i)} \, d^2\vec{\nu}. \tag{5.77}$$

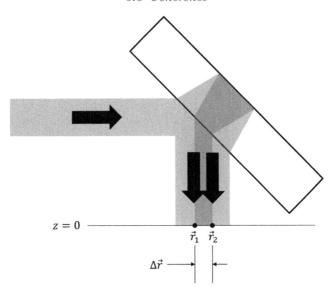

Figure 5.3 Spatial overlap of a monochromatic light beam $\vec{\mathcal{E}}(\vec{r}, 0, f)$ obtained using a shearing plate.

We deduce immediately that

$$\vec{\mathbb{A}}_i(\vec{\nu}, f) = C\vec{\mathbb{A}}(\vec{\nu}, f)\, e^{-2i\pi\vec{\nu}.\vec{r}_i}\,, \qquad (5.78)$$

where $\vec{\mathbb{A}}(\vec{\nu}, f)$ is the spatial Fourier transform of $\vec{\mathcal{E}}(\vec{r}, 0, f)$.

Now consider the Poynting vector flux transported by these two beams through the plane $z = 0$. As in Chapter 4, provided the receiving surface of the detector intercepts all the light, the signal $S$ that it provides (give or take a factor of proportionality) is given by

$$S = \frac{1}{2\omega\tilde{\mu}} \int_{\vec{\nu}} \alpha(\vec{\nu}, f)\, |\vec{\mathbb{A}}_1(\vec{\nu}, f) + \vec{\mathbb{A}}_1(\vec{\nu}, f)|^2\, d^2\vec{\nu} \qquad (5.79)$$

and

$$S = \frac{|C|^2}{\omega\tilde{\mu}} \int_{\vec{\nu}} \alpha(\vec{\nu}, f)\, |\vec{\mathbb{A}}(\vec{\nu}, f)|^2 \left\{1 + \cos[2\pi\vec{\nu}.(\vec{r}_2 - \vec{r}_1)]\right\}\, d^2\vec{\nu}. \qquad (5.80)$$

We recognize here a trigonometric function that will define the fringe contrast obtained by varying $\Delta r = |\vec{r}_2 - \vec{r}_1|$. However, in order for these fringes to be visible, the argument of the cosine cannot vary too much over the frequency range $\Delta\nu$ of $|\vec{A}(\vec{\nu}, f)|^2$, i.e.,

$$\Delta\nu\,\Delta r \leqslant 1 \quad \Rightarrow \quad \Delta r \leqslant l_s \text{ avec } l_s = \frac{\lambda}{n\sin\theta_{\max}}. \qquad (5.81)$$

and where $\theta_{\max}$ is the angular aperture of the beam. Hence, we see that the spatial coherence length $l_s$ decreases rapidly with increasing divergence of the beam.

# 6

# Thin-Film Filters

## 6.1 Introduction

We have investigated the characteristics of a light wave and highlighted the key governing parameters such as wavelength, refractive index, polarization state, amplitude, phase, pulse duration, and line width. On the other hand, we have not yet attempted to modify the properties of this wave, i.e., to *manipulate* or *control* the light. This control is effected generally via so-called *electromagnetic* objects, i.e., components for which at least one of the dimensions is of the same order of magnitude as the wavelength of the incident light. At the heart of what is known as **electromagnetic optics** is the recurrent problem of synthesizing an object (e.g., an antireflective coating, polarizer, dichroic beam-splitter, narrow band filter, broad band absorber, pulse compressor) that is capable of conferring the desired properties onto light.

These components can be extremely varied in nature, though the aim here is not to make an exhaustive list so much as to study those that are the most commonly used in order to gain a better understanding of their operation and the specific possibilities that they provide. Among these, we have chosen to focus attention on **optical interference filters** (or stacked layers of optical thin films); for optical filtering in free space, these give rise to a multitude of components found in applications for the public at large (photography, spectacles, lighting, automobiles, building, etc.) and also in areas of high technology (Earth observation, astronomy, free space telecommunications, fiber optic communications, defense, health, microscopy, etc.). It is rare these days for an optical surface to be used without some form of surface coating. To be convinced that this is so, note that in the absence of any antireflective coating a glass surface reflects approximately 4% of the incident visible light; in transmission, losses are therefore on the order of

8% for a single substrate, rapidly approaching 100% if many substrates are involved, as is the case for any even slightly complex optical system. Use of these antireflection layers is even more critical in the case of mid-infrared, where surfaces reflect nearly 40% of the incident flux! For all these reasons, it can be understood that, *wherever photons have to be moved, we will need optical thin films*, making this domain a major strategic issue.

From the manufacturing point of view, and regardless of the useable spectral domain (visible, ultraviolet, or infrared), an interference filter is obtained by depositing a stack of dielectric or metal layers onto a surface under vacuum; the elementary thicknesses lie typically between a few nanometers and several hundred nanometers. The partial reflections at the interfaces separating adjacent media having different optical properties create a multiplicity of secondary waves that interfere with one another. The correct choice of thickness and refractive index for the different layers that make up the stack have a direct influence not only on the number and relative amplitude of the secondary waves involved in the interference but also on the relative differences in phase between all these waves. In a planar component with an overall thickness of a few microns (or even several tens of microns), this allows a very rapid variation in the state of overall interference with the wavelength of the propagating light wave: this is the mechanism that constitutes *optical interference filtering*.

We shall also see how multilayer components can make up for the relatively small number of refractive indices found spontaneously in nature. To give an example, and as far as reflection is concerned, these components behave like artificial materials with arbitrary complex values of refractive index, the dispersion of which can be engineered.

However, before tackling these problems of synthesizing filters or materials, it is clearly essential to establish the mathematical framework on which these complex phenomena can be described. This we shall now do.

## 6.2 Principle of Spatiotemporal Control

There is often a tendency to reduce the incident wave interacting with an optical component to a simple plane wave, but in fact it is a beam of light rather than a plane wave that normally illuminates the component. Hence, we need to break this beam of light down into elementary components, namely **unidirectional monochromatic waves**, and then, using the principle of superposition, take advantage of the linearity of Maxwell's equations to reconstruct the overall response of the component from the individual response of each of these plane waves.

This procedure is applicable to most electromagnetic calculations, and consists of

1. Finding a basis of functions onto which the incident field can be decomposed down into elementary components
2. Studying the interaction of matter with each elementary component
3. Using the principle of superposition to reconstruct the total field

Clearly, this procedure has merit only if the chosen function basis is consistent with the geometry of the problem (and in particular, the locations of any associated discontinuities) so as to simplify the investigation of the elementary interaction when writing down the continuity equations. Most often used for concentric or radial geometries is a decomposition into spherical harmonics, or into complex exponentials for a planar geometry. It is the latter, which brings the spatial Fourier transformation into play, which is of interest here.

### 6.2.1 Elementary Components in Free Space

We have already used the temporal and spatial Fourier transforms of the field, which quickly allow these elementary components to be obtained. The temporal Fourier transformation uses a basis of exponentials in $e^{-2i\pi ft}$ (or $e^{-i\omega t}$):

$$\vec{E}(\vec{\rho}, t) = \int_f \vec{\mathcal{E}}(\vec{\rho}, f) \, e^{-2i\pi ft} df , \qquad (6.1)$$

where $f$ is the temporal frequency, while the spatial Fourier transformation uses an analogous basis in $e^{2i\pi \vec{\nu}.\vec{r}}$ (or $e^{i\vec{\sigma}.\vec{r}}$), i.e.,

$$\vec{\mathcal{E}}(\vec{\rho}, f) = \int_{\vec{\nu}} \vec{\mathbb{E}}(\vec{\nu}, z, f) \, e^{2i\pi \vec{\nu}.\vec{r}} \, d\vec{\nu} = \int_{\vec{\nu}} \vec{\mathbb{A}}(\vec{\nu}, f) \, e^{i\alpha(\vec{\nu},f)z} e^{2i\pi \vec{\nu}.\vec{r}} \, d\vec{\nu} , \qquad (6.2)$$

where $\vec{\nu}$ is the spatial frequency and where $\alpha(\vec{\nu}, f)$ is defined by

$$\alpha^2(\vec{\nu}, f) = k^2(f) - \sigma^2 \quad \text{with} \quad k = \frac{2\pi n}{\lambda} \quad \text{and} \quad \sigma = |\vec{\sigma}| = 2\pi|\vec{\nu}|. \qquad (6.3)$$

The space variable itself is written as $\vec{\rho} = (\vec{r}, z)$.
Finally, combining these two bases, we get

$$\vec{E}(\vec{\rho}, t) = \vec{E}(\vec{r}, z, t) = \int_f \int_{\vec{\nu}} \vec{\mathbb{A}}(\vec{\nu}, f) \, e^{i\alpha(\vec{\nu},f)z} e^{2i\pi \vec{\nu}.\vec{r}} e^{-2i\pi ft} \, d\vec{\nu} \, df . \qquad (6.4)$$

This equation shows that here the elementary progressive component has the form

$$\vec{\mathbb{A}}(\vec{\nu}, f)\, e^{i\alpha(\vec{\nu}, f)z}\, e^{2i\pi\vec{\nu}\cdot\vec{r}}\, e^{-2i\pi ft} = \vec{\mathbb{A}}(\vec{\nu}, f)\, e^{i\vec{\beta}(\vec{\nu}, f)\cdot\vec{\rho}}\, e^{-2i\pi ft}, \qquad (6.5)$$

and we find the **unidirectional monochromatic wave** that has already been covered in the previous chapter, and which identifies itself as a plane wave for low spatial frequencies ($\sigma < k$) in a transparent medium ($k$ real).

In the case of the planar components of interest here, namely optical interference filters, we have the benefit of invariance along the $y$-axis, such that we can work in the plane of incidence and restrict ourselves to the 2D case; the elementary component is then expressed as

$$\vec{\mathbb{A}}(\nu, f)\, e^{2i\pi\nu x}\, e^{i\alpha(\nu, f)z}\, e^{-2i\pi ft}. \qquad (6.6)$$

Note at this point that the decomposition basis is well chosen, since the discontinuities occur at the planes of equation $z = \mathrm{const}$, where the transverse dependencies are all the same. Note also that in the harmonic regime, temporal dependence simply appears as the multiplicative factor $e^{-2i\pi ft}$ which can be temporarily omitted.

### 6.2.2 Transfer Function of a Multilayer System

Now consider the reflection of a progressive electromagnetic field on a planar component as shown schematically in Figure 6.1.

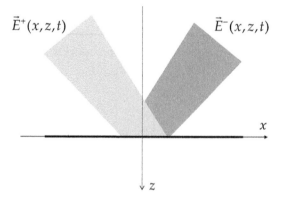

Figure 6.1 Reflection of a progressive field on a plane multilayer component.

To avoid any confusion, we shall add a + superscript to the progressive elementary components of the incident field (propagating in the positive $z$

direction), and a $-$ superscript to the retrograde elementary components of the reflected field (propagating in the negative $z$ direction), and the index 0 to the wave parameter $\alpha$ in the incident medium. In the interests of simplification, we shall also assume in this section that the electric field is perpendicular to the plane in Figure 6.1 (Transverse Electric linear polarization).

Omitting their time dependence, the elementary algebraic components of the incident field can be expressed as

$$\mathbb{A}^+(\nu, f)\, e^{i(2\pi\nu x + \alpha_0 z)}\,, \tag{6.7}$$

while those for the reflected field are written as

$$\mathbb{A}^-(\nu, f)\, e^{i(2\pi\nu x - \alpha_0 z)} = r(\nu, f)\mathbb{A}^+(\nu, f)\, e^{i(2\pi\nu x - \alpha_0 z)}\,, \tag{6.8}$$

where $r = A^-/A^+$ is the reflection coefficient of the multilayer for a wave of temporal frequency $f$ (defining the wavelength) and spatial frequency $\nu$ (defining the angle of incidence).

Using the principle of superposition, the reflected field can be expressed in the form

$$E^-(x, z, t) = \iint_{f,\nu} r(\nu, f)\mathbb{A}^+(\nu, f)\, e^{-i\alpha_0(\nu, f)z}\, e^{2i\pi(\nu x - ft)}\, d\nu\, df\,. \tag{6.9}$$

This can also be written compactly as

$$E^-(x, z, t) = \mathrm{FT}^{-1}\left[r(\nu, f)\mathbb{A}^+(\nu, f)\, e^{-i\alpha_0(\nu, f)z}\right]\,, \tag{6.10}$$

where the inverse Fourier transform ($\mathrm{FT}^{-1}$) is here taken simultaneously with respect to the temporal and spatial frequencies. At $z = 0$ we get

$$E^-(x, 0, t) = \mathrm{FT}^{-1}\left[r(\nu, f)\mathbb{A}^+(\nu, f)\right]\,. \tag{6.11}$$

We know that the Fourier transform of a simple product of two functions is equal to the convolution of their Fourier transforms, written as

$$E^-(x, 0, t) = [\check{r} \underset{x,t}{\star} A^+](x, t)\,, \tag{6.12}$$

with

$$\check{r}(x, t) = \iint_{f,\nu} r(\nu, f)\, e^{2i\pi(\nu x - ft)}\, d\nu\, df\,, \tag{6.13}$$

and

$$\vec{A}^+(x, t) = \iint_{f,\nu} \vec{\mathbb{A}}^+(\nu, f)\, e^{2i\pi(\nu x - ft)}\, d\nu\, df\,. \tag{6.14}$$

This last integral is simply the spatiotemporal reconstruction of the incident field on the $z = 0$ abscissa, leading to the final relationship

$$\vec{E}^-(x,0,t) = \check{r}(x,t) \underset{x,t}{\star} \vec{E}^+(x,0,t). \tag{6.15}$$

This equation shows that the field reflected by a multilayer system is equal to the convolution over space and time of the incident field with the inverse double Fourier transform of the reflection factor with respect to temporal and spatial frequencies. The variable $\check{r}(x,t)$ can therefore be considered as the transfer function of the multilayer system. This is the response that would be obtained in the ideal case, from a spatial and temporal point of view, of an infinitely *narrow* excitation $\delta(x,t)$. This transfer function also lies at the origin of the spatial and temporal resolution of multilayers.

Observe that (6.15) is given for the reflected field at $z = 0$. To take account of propagation at arbitrary $z$ planes, an additional convolution by the inverse Fourier transform of $\exp[-i\alpha_0(\nu, f)z]$ is required.

Note also that equation (6.15) lies at the heart of work on synthesizing components, since it demonstrates the possibility of the spatiotemporal control of light; however, it should be capable of determining those characteristics of the multilayer system (refractive indices, thicknesses, sequences or layouts, etc.) that will bring about the desired result.

## 6.3 Response of Stratified Media

In what follows, due to the fact that the incident field is monochromatic, we will omit the temporal frequency $(f)$ dependence in the notation for the fields.

### 6.3.1 Introduction

Planar components with a stratified structure are presented in the form of stacked layers of varying refractive index, whose elementary thicknesses are of the same order of magnitude as a wavelength. A schematic representation of such a component is given in Figure 6.2, where the stack being considered contains $p$ elementary layers deposited on a substrate of refractive index $n_{p+1} = n_s$. The refractive index of the incident medium (or superstrate) is denoted $n_0$. All materials are assumed to be linear, isotropic, homogeneous, and free of sources.

This component with a stratified structure is illuminated by a progressive plane wave of known characteristics: wavelength $\lambda$ and angle of incidence $\theta_0$.

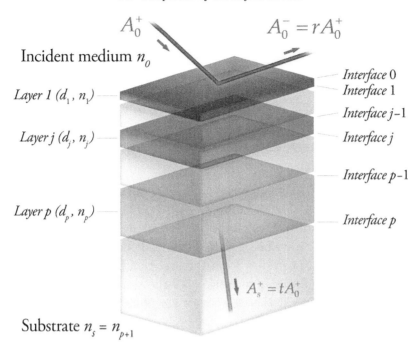

Figure 6.2 General structure of a multilayer stack.

The couple $(f, \nu)$ is therefore an input for the problem, since the temporal frequency $f$ is defined by the only wavelength $[f = c/\lambda]$, while the spatial frequency $\nu$ is defined by the wavelength and the angle of incidence in the incident medium $[\nu = n_0 \sin \theta_0 / \lambda]$. The $z$-dependence of the elementary component of this incident wave is then described as

$$\vec{\mathbb{E}}_0(\vec{\nu}, z) = \vec{\mathbb{A}}_0^+ e^{i\alpha_0 z}, \tag{6.16}$$

with $\alpha_0 = \sqrt{k_0^2 - \sigma^2}$, $\sigma = k_0 \sin \theta_0$ and $k_0 = \frac{2\pi n_0}{\lambda}$ and where the amplitude $\vec{\mathbb{A}}_0^+$ is also known, since it is a characteristic of the incident flux. It is important here to stress the fact that the angles are introduced to describe a far field illumination on the device, but that the formalism remains valid for near field illumination.

We immediately observe that in the substrate there exists a unique progressive wave, whose $z$-dependence is written

$$\vec{\mathbb{E}}_s(\vec{\nu}, z) = \vec{\mathbb{A}}_s^+ e^{i\alpha_s z} \quad \text{with} \quad \alpha_s^2 = k_s^2 - \sigma^2. \tag{6.17}$$

On the other hand, a stationary wave, resulting from the combination of a progressive and a retrograde wave, develops in the layer of rank $j$ ($1 \leqslant j \leqslant p$):

$$\vec{E}_j(\vec{v}, z) = \vec{A}_j^+ e^{i\alpha_j z} + \vec{A}_j^- e^{-i\alpha_j z} \quad \text{with} \quad \alpha_j^2 = k_j^2 - \sigma^2. \tag{6.18}$$

In contrast with the previous case, the modulus of this stationary wave varies with height $z$ within the thin layer.

Note also that the spatial pulsation $\sigma$ takes the same form in all the layers. Indeed it is a real variable, proportional to the Fourier conjugate of the spatial variable $x$, and that characterizes identical transverse variations in all media. As a consequence, if we consider transparent media ($n_j$ real) and introduce angles $\theta_j$, we can write

$$\sigma = k_j \sin\theta_j = \text{const} \quad \Rightarrow \quad n_j \sin\theta_j = n_0 \sin\theta_0, \tag{6.19}$$

which is simply the **Snell–Descartes invariant**.

Hence the pertinent physical variable is indeed the spatial frequency, the angles being defined a posteriori from this frequency and for each medium. In what follows we will prefer to work with the spatial frequency, in order to keep all results valid whether the waves are plane or evanescent, and the media transparent or absorbing.

Now that this notation has been established, we will address three configurations of increasing complexity, as shown in Figure 6.3.

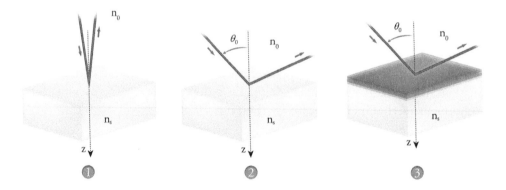

Figure 6.3 Illustration of the procedure for three configurations of increasing complexity.

1. Reflection at zero incidence ($\theta_0 = 0$) of an elementary wave on a plane surface separating two media of differing refractive index, the parameter $r$ being defined by a relation involving the values of these refractive indices:

$$r = \frac{n_0 - n_s}{n_0 + n_s}. \tag{6.20}$$

2. Reflection at nonzero incidence ($\theta_0 \neq 0$) of this same wave on the same plane surface; the parameter $r$ is identical to the preceding, except that the refractive indices are replaced by **effective refractive indices** $\tilde{n}_0$ and $\tilde{n}_s$ that depend on the angle of incidence and the state of polarization:

$$r = \frac{\tilde{n}_0 - \tilde{n}_s}{\tilde{n}_0 + \tilde{n}_s} . \qquad (6.21)$$

3. The most general case of reflection from this same surface when the latter is coated with a stack of thin layers; the effective refractive index of the substrate is then replaced by a quantity that takes account of the optical and geometrical characteristics of the stack deposited on the surface, namely its **complex admittance $Y_0$**:

$$r = \frac{\tilde{n}_0 - Y_0}{\tilde{n}_0 + Y_0} . \qquad (6.22)$$

These different results are demonstrated in the sections that follow, where all waves are monochromatic with a unique wave vector.

### *6.3.2 The Concept of Effective Refractive Index*

Consider a progressive wave propagating in the $xOz$ plane of a homogeneous medium. The $z$-dependence of the electric field is given by the spatial Fourier transform expressed as

$$\vec{\mathbb{E}}^+(\vec{\nu}, z) = \vec{\mathbb{A}}^+(\vec{\nu}) \, e^{i\alpha z} , \qquad (6.23)$$

while its 2D spatial dependence is defined by

$$\vec{\mathcal{E}}^+(x, z) = \vec{\mathbb{A}}^+(\vec{\nu}) \, e^{i(2\pi\nu x + \alpha z)} . \qquad (6.24)$$

In the same way, we have, for the magnetic field,

$$\vec{\mathcal{H}}^+(x, z) = \vec{\mathbb{B}}^+(\vec{\nu}) \, e^{i(2\pi\nu x + \alpha z)} . \qquad (6.25)$$

The mutual orthogonality rule introduced in Section 2.8 allows us to write

$$\begin{cases} \vec{\mathbb{B}}^+ = \dfrac{1}{\omega\tilde{\mu}} \, \vec{\beta}^+ \wedge \vec{\mathbb{A}}^+ \\[2mm] \vec{\mathbb{A}}^+ = -\dfrac{1}{\omega\tilde{\epsilon}} \, \vec{\beta}^+ \wedge \vec{\mathbb{B}}^+ \end{cases} \quad \text{with} \quad \vec{\beta}^+ = \sigma\vec{x} + \alpha\vec{z} = \begin{vmatrix} \sigma = 2\pi\nu \\ 0 \\ \alpha \end{vmatrix} . \qquad (6.26)$$

We need at this stage to introduce the polarization modes of light. In Chapter 2 we saw in transparent media that the electric field lies in a plane perpendicular to the wave vector, so that it can be described as the sum of two perpendicular fields in this plane, called the **polarization modes**.

In what follows we shall work with the two classical (TE and TM) specific modes which, as a result of the boundary conditions, provide the simplest calculations when a planar device is analyzed. This will allow us to address the interaction of the device with each polarization mode, before the field is reconstructed by mode superposition. The definition of the polarization modes is related to the definition of an incidence plane given by the normal to the sample and the incident wave vector (that is, here the plane $y = 0$).

## *TE Polarization*

First, we consider a TE (Transverse Electric) linearly polarized plane wave, i.e., one for which the electric field vector is perpendicular to the plane of incidence $xOz$. This polarization state is also denoted by the letter S, being the initial letter of the German word *senkrecht*, meaning perpendicular. For this particular polarization state we can write

$$\vec{\mathbb{A}}^+ = \mathbb{A}^+ \, \vec{y},$$

which leads to the following expression for $\vec{\mathbb{B}}^+$:

$$\vec{\mathbb{B}}^+ = \frac{1}{\omega \tilde{\mu}} \vec{\beta}^+ \wedge \vec{\mathbb{A}}^+ = \frac{1}{\omega \tilde{\mu}} \left( \sigma \vec{x} + \alpha \vec{z} \right) \wedge \mathbb{A}^+ \, \vec{y} = \frac{\sigma \mathbb{A}^+}{\omega \tilde{\mu}} \vec{z} + \frac{\alpha}{\omega \tilde{\mu}} \vec{z} \wedge \vec{\mathbb{A}}^+ , \quad (6.27)$$

and hence to the following relationship between the tangential components (i.e., within the $xOy$ plane) of the magnetic induction and the electric field:

$$\vec{\mathbb{B}}_{\text{tg}}^+ = \frac{\alpha}{\omega \tilde{\mu}} \vec{z} \wedge \vec{\mathbb{A}}^+ = \frac{\alpha}{\omega \tilde{\mu}} \vec{z} \wedge \vec{\mathbb{A}}_{\text{tg}}^+ . \quad (6.28)$$

The quantity $\tilde{n}$ is defined by

$$\tilde{n} = \frac{\alpha}{\omega \tilde{\mu}} , \quad (6.29)$$

and is known as the effective refractive index, so the relationship between tangential components can be written in the following compact form:

$$\vec{\mathbb{B}}_{\text{tg}}^+ = \tilde{n} \left[ \vec{z} \wedge \vec{\mathbb{A}}_{\text{tg}}^+ \right] . \quad (6.30)$$

It remains to calculate the relationship between the customary refractive index $n$ and the effective index $\tilde{n}$ in the case of this TE polarization. We can write

$$\tilde{n} = \frac{\alpha}{\omega \tilde{\mu}} = \frac{n\alpha}{kc\tilde{\mu}} , \quad (6.31)$$

with $\tilde{\mu} = \tilde{\mu}_r \mu_v$ and $\epsilon_v \mu_v c^2 = 1$. As a result we get

$$\tilde{n} = \frac{n\alpha}{k} \sqrt{\epsilon_v \mu_v} \frac{1}{\tilde{\mu}_r \mu_v} = \frac{n\alpha}{k} \sqrt{\frac{\epsilon_v}{\mu_v}} \frac{1}{\tilde{\mu}_r} = \frac{1}{\eta_v \tilde{\mu}_r} \frac{n\alpha}{k} , \quad (6.32)$$

with $\eta_v = \sqrt{\mu_v/\epsilon_v}$ (the quantity $\eta_v$ is known as the vacuum impedance).

Furthermore, in the low-frequency case (propagative components) we can replace $\alpha$ by expressing it in terms of $\theta$, i.e.,

$$\alpha = \sqrt{k^2 - \sigma^2} = \sqrt{k^2 - k^2 \sin^2\theta} = k\cos\theta, \tag{6.33}$$

which leads to the following particular form for the effective index $\tilde{n}$:

$$\tilde{n} = \frac{1}{\eta_v\tilde{\mu}_r}\frac{n\alpha}{k} = \frac{1}{\eta_v\tilde{\mu}_r}n\cos\theta. \tag{6.34}$$

### TM Polarization

Now consider a TM (Transverse Magnetic) linearly polarized progressive wave, where, as its name implies, this time it is the magnetic induction vector that is tangential, and hence perpendicular to the plane of incidence ($\vec{\mathbb{B}}^+ = \mathbb{B}^+\vec{y}$). This polarization state is also denoted by the letter P, the initial letter of the word parallel, a reminder that the electric field vector is this time parallel to the plane of incidence.

The second of the two orthogonality relations (6.26) allows us to write in this case

$$\vec{\mathbb{A}}^+ = -\frac{1}{\omega\tilde{\epsilon}}\vec{\beta}^+\wedge\vec{\mathbb{B}}^+ = -\frac{1}{\omega\tilde{\epsilon}}(\sigma\vec{x}+\alpha\vec{z})\wedge\mathbb{B}^+\vec{y} = -\frac{\sigma\mathbb{B}^+}{\omega\tilde{\epsilon}}\vec{z} - \frac{\alpha}{\omega\tilde{\epsilon}}\vec{z}\wedge\vec{\mathbb{B}}^+. \tag{6.35}$$

Consequently,

$$\vec{z}\wedge\vec{\mathbb{A}}^+ = -\frac{\alpha}{\omega\tilde{\epsilon}}\vec{z}\wedge(\vec{z}\wedge\vec{\mathbb{B}}^+) = -\frac{\alpha}{\omega\tilde{\epsilon}}[(\vec{z}.\mathbb{B}^+\vec{y})\vec{z} - (\vec{z}.\vec{z})\vec{\mathbb{B}}^+] = \frac{\alpha}{\omega\tilde{\epsilon}}\vec{\mathbb{B}}^+, \tag{6.36}$$

and hence finally

$$\vec{\mathbb{B}}_{tg}^+ = \frac{\omega\tilde{\epsilon}}{\alpha}\vec{z}\wedge\vec{\mathbb{A}}_{tg}^+ = \tilde{n}\,\vec{z}\wedge\vec{\mathbb{A}}_{tg}^+. \tag{6.37}$$

Using the general relationship between propagation velocity, permittivity, and permeability, $\tilde{\epsilon}\tilde{\mu}v^2 = 1$, we can transform the term $\omega\tilde{\epsilon}/\alpha$ to reveal again the vacuum impedance, i.e.,

$$\frac{\omega\tilde{\epsilon}}{\alpha} = \frac{kv}{\alpha}\frac{1}{\tilde{\mu}v^2} = \frac{nk}{\alpha}\frac{1}{c\tilde{\mu}} = \frac{1}{\eta_v\tilde{\mu}_r}\frac{nk}{\alpha}. \tag{6.38}$$

As before in the low-frequency case, we can replace $\alpha$ by its equivalent expression in $\theta$, and hence express the effective index $\tilde{n}$ in the following particular form:

$$\tilde{n} = \frac{1}{\eta_v\tilde{\mu}_r}\frac{nk}{\alpha} = \frac{1}{\eta_v\tilde{\mu}_r}\frac{n}{\cos\theta}. \tag{6.39}$$

*Conclusion*

The relationships established in the previous sections lead us to conclude that, in the case of the elementary component of a progressive wave, regardless of the polarization state, there exists a scalar $\tilde{n}$ such that

$$\vec{\mathbb{B}}_{tg}^{+} = \tilde{n}\ \vec{z} \wedge \vec{\mathbb{A}}_{tg}^{+}, \tag{6.40}$$

with

$$\tilde{n} = \begin{cases} \alpha/\omega\tilde{\mu} & \text{in TE polarization} \\ \omega\tilde{\epsilon}/\alpha & \text{in TM polarization} \end{cases}, \tag{6.41}$$

or

$$\tilde{n} = \frac{1}{\eta_v \tilde{\mu}_r} \begin{cases} n\alpha/k & \text{in TE polarization} \\ nk/\alpha & \text{in TM polarization} \end{cases}. \tag{6.42}$$

This relation remains true for evanescent waves (high spatial frequencies) or dissociated waves (absorbing medium). For low spatial frequencies in a transparent medium, i.e., for plane waves, we get

$$\tilde{n} = \frac{1}{\eta_v \tilde{\mu}_r} \begin{cases} n\cos\theta & \text{in TE polarization} \\ n/\cos\theta & \text{in TM polarization} \end{cases}. \tag{6.43}$$

The procedure is exactly the same in the case of a retrograde wave, except that now the vector $\vec{\beta}$ is expressed as $\vec{\beta}^{-} = \sigma\vec{x} - \alpha\vec{z}$ which equates to reversing the sign of $\alpha$, and hence that of $\tilde{n}$, in the final expressions. The generic relationship between tangential components is therefore written in this case:

$$\vec{\mathbb{B}}_{tg}^{-} = -\tilde{n}\ \vec{z} \wedge \vec{\mathbb{A}}_{tg}^{-}. \tag{6.44}$$

Note that the idea of effective index applies only to progressive or retrograde waves. For stationary waves (that is, the sum of progressive and retrograde waves), we will have to turn to the concept of admittance.

### 6.3.3 Tangential and Normal Components

*Tangential Components, Poynting Flux*

In the previous sections our analysis was limited to the tangential components of the field. These components are the only ones involved in calculating the flux through a $z$ plane, and will consequently be sufficient to establish the far field energy balance.

This general result can be proved immediately. We saw in Chapter 4 that the surface flux density of the Poynting vector could be written, for one spatial frequency, as

$$\frac{d\Phi}{dS} = \frac{1}{2}\Re\left[(\vec{\mathbb{E}}^* \wedge \vec{\mathbb{H}}).\vec{z}\right]. \tag{6.45}$$

Decomposing the field into its tangential and normal components

$$(\vec{\mathbb{E}}^* \wedge \vec{\mathbb{H}}).\vec{z} = [(\vec{\mathbb{E}}_{tg}^* + \vec{\mathbb{E}}_N^*) \wedge (\vec{\mathbb{H}}_{tg} + \vec{\mathbb{H}}_N)].\vec{z}, \tag{6.46}$$

and using the properties of vector products, we get

$$(\vec{\mathbb{E}}^* \wedge \vec{\mathbb{H}}).\vec{z} = (\vec{\mathbb{E}}_{tg}^* \wedge \vec{\mathbb{H}}_{tg}).\vec{z}. \tag{6.47}$$

This expression only involves the tangential field components. Calculation of the flux can therefore be finished off for plane waves in transparent media as follows:

$$\vec{\mathbb{B}}_{tg} = \tilde{n}\left[\vec{z} \wedge \vec{\mathbb{A}}_{tg}\right] \quad \Rightarrow \quad \frac{d\Phi}{dS} = \frac{1}{2}\Re[\tilde{n}]\,|\vec{\mathbb{A}}_{tg}|^2. \tag{6.48}$$

### Normal Components

In some cases, such as the computation of absorbance, it is useful to know not only the tangential components of the fields, but also their normal (sometimes called longitudinal) components.

Indeed, in accordance with general results established in Chapter 5, the monochromatic linear absorption density of a homogeneous medium, invariant in $x$ and $y$, is given by

$$\frac{d^2 A}{df\,dz} = \frac{1}{2}\omega\left[\tilde{\epsilon}''(f)|\vec{\mathbb{E}}(\vec{\nu}, z, f)|^2 + \tilde{\mu}''(f)|\vec{\mathbb{H}}(\vec{\nu}, z, f)|^2\right], \tag{6.49}$$

with, for instance

$$\vec{\mathbb{E}}(\vec{\nu}, z, f) = \vec{\mathbb{E}}_{tg}(\vec{\nu}, z, f) + \vec{\mathbb{E}}_N(\vec{\nu}, z, f). \tag{6.50}$$

For TE polarization, the electric field is purely tangential [$\vec{\mathbb{A}}_{tg} = \mathbb{A}^+\vec{y}$ and $\vec{\mathbb{A}}_z = \vec{0}$], whereas the magnetic field $\vec{\mathbb{B}}$ is described by two components, tangential and normal, respectively defined for a progressive wave by (see equation 6.27)

$$\vec{\mathbb{B}}_{tg}^+ = \tilde{n}[\vec{z} \wedge \vec{\mathbb{A}}_{tg}^+] = -\frac{\alpha}{\omega\tilde{\mu}}\mathbb{A}^+\vec{x} \quad \text{and} \quad \vec{\mathbb{B}}_N^+ = \frac{\sigma}{\omega\tilde{\mu}}\mathbb{A}^+\vec{z}. \tag{6.51}$$

Conversely, for TM polarization, it is the magnetic field that is purely tangential [$\vec{\mathbb{B}}_{tg} = \mathbb{B}^+\vec{y}$ and $\vec{\mathbb{B}}_z = \vec{0}$], whereas the tangential and normal components of the electric field are respectively given for a progressive wave by (see equation 6.35)

$$\vec{\mathbb{A}}_{tg}^+ = \frac{\alpha}{\omega\tilde{\epsilon}}\mathbb{B}^+\vec{x} \quad \text{and} \quad \vec{\mathbb{A}}_N^+ = -\frac{\sigma}{\omega\tilde{\epsilon}}\mathbb{B}^+\vec{z}. \tag{6.52}$$

Note that, by combining these two last relations, we can write for TM polarization

$$\sigma\,\vec{x}\cdot\vec{\mathbb{A}}_{\mathrm{tg}}^{+} + \alpha\,\vec{z}\cdot\vec{\mathbb{A}}_{\mathrm{N}}^{+} = \sigma\mathbb{A}_{\mathrm{tg}}^{+} + \alpha\mathbb{A}_{\mathrm{N}} = 0\,. \tag{6.53}$$

This last relation is simply the consequence of the third Maxwell's equation for an elementary progressive wave in a homogeneous medium. Indeed,

$$\mathrm{div}[\tilde{\epsilon}\vec{\mathbb{A}}^{+}(\vec{\nu},f)e^{i(\sigma x+\alpha z)}] = 0 \quad \Rightarrow \quad \sigma\vec{x}\cdot\vec{\mathbb{A}}^{+} + \alpha\vec{z}\cdot\vec{\mathbb{A}}^{+} = 0\,. \tag{6.54}$$

In the case of a retrograde wave the same results hold provided $\alpha$ is replaced by $-\alpha$. Note that all these relations are no longer valid in the case of a stationary wave. We will analyze this situation in Section 6.3.4.

### 6.3.4 Application to a Plane Interface

To familiarize ourselves with the idea of effective index and its use, we shall first consider a simple plane interface in the configuration illustrated in Figure 6.4.

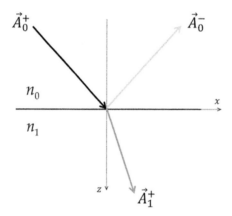

Figure 6.4 Plane interface.

In both media, $\vec{\mathbb{E}}(\vec{\nu},z)$ and $\vec{\mathbb{H}}(\vec{\nu},z)$ are solutions of an homogeneous Helmholtz equation and can be expressed as

$$\begin{cases} \vec{\mathbb{E}}_0(\vec{\nu},z) = \vec{\mathbb{A}}_0^{+}e^{i\alpha_0 z} + \vec{\mathbb{A}}_0^{-}e^{-i\alpha_0 z} \\ \vec{\mathbb{E}}_1(\vec{\nu},z) = \vec{\mathbb{A}}_1^{+}e^{i\alpha_1 z} \end{cases}, \tag{6.55}$$

and

$$\begin{cases} \vec{\mathbb{H}}_0(\vec{\nu},z) = \vec{\mathbb{B}}_0^{+}e^{i\alpha_0 z} + \vec{\mathbb{B}}_0^{-}e^{-i\alpha_0 z} \\ \vec{\mathbb{H}}_1(\vec{\nu},z) = \vec{\mathbb{B}}_1^{+}e^{i\alpha_1 z} \end{cases}. \tag{6.56}$$

*Boundary Conditions for Tangential Components*

For each polarization mode, the coefficients of reflection $r$ and transmission $t$ (in amplitude) of this plane optical interface are defined using the following two relations:

$$\vec{A}_{0,tg}^{-} = r\,\vec{A}_{0,tg}^{+} \qquad \vec{A}_{1,tg}^{+} = t\,\vec{A}_{0,tg}^{+}\,. \tag{6.57}$$

Note here that these coefficients $r$ and $t$, often known as the *Fresnel coefficients*, are ratios of **tangential** electric field components and depend on light polarization.

The continuity of the tangential component of the electric field crossing the optical interface $(z = 0)$ allows us to write

$$\vec{E}_{0,tg}(\vec{\nu},0) = \vec{E}_{1,tg}(\vec{\nu},0) \quad \Rightarrow \quad \vec{A}_{0,tg}^{+} + \vec{A}_{0,tg}^{-} = \vec{A}_{1,tg}^{+}\,, \tag{6.58}$$

or, incorporating the coefficients $r$ and $t$ defined in (6.57):

$$\vec{A}_{0,tg}^{+} + r\vec{A}_{0,tg}^{+} = t\vec{A}_{0,tg}^{+}\,, \tag{6.59}$$

whence the classical relationship between the amplitudes of the reflected and incident fields:

$$1 + r = t\,. \tag{6.60}$$

It should not be forgotten here that $r$ and $t$ depend on polarization and are complex numbers. They should not be confused with the energy coefficients of reflection $R$ and transmission $T$ that we shall be introducing later.

The continuity of the tangential component of magnetic induction crossing the plane optical interface leads similarly to

$$\vec{H}_{0,tg}(\vec{\nu},0) = \vec{H}_{1,tg}(\vec{\nu},0) \quad \Rightarrow \quad \vec{B}_{0,tg}^{+} + \vec{B}_{0,tg}^{-} = \vec{B}_{1,tg}^{+}\,. \tag{6.61}$$

Introducing at this stage the effective indices of the two media and taking account of the progressive or retrograde character of the waves being considered, this equation can be put in the form

$$\tilde{n}_0[\vec{z} \wedge \vec{A}_{0,tg}^{+}] - \tilde{n}_0[\vec{z} \wedge \vec{A}_{0,tg}^{-}] = \tilde{n}_1[\vec{z} \wedge \vec{A}_{1,tg}^{+}]\,. \tag{6.62}$$

It only remains to express all quantities as a function of the value of the incident electric field, i.e.,

$$\tilde{n}_0[\vec{z} \wedge \vec{A}_{0,tg}^{+}] - \tilde{n}_0 r[\vec{z} \wedge \vec{A}_{0,tg}^{+}] = \tilde{n}_1 t[\vec{z} \wedge \vec{A}_{0,tg}^{+}]\,, \tag{6.63}$$

or

$$\tilde{n}_0(1 - r) = \tilde{n}_1 t\,, \tag{6.64}$$

and hence, combined with (6.60)

$$r = \frac{\tilde{n}_0 - \tilde{n}_1}{\tilde{n}_0 + \tilde{n}_1} \quad \text{and} \quad t = \frac{2\tilde{n}_0}{\tilde{n}_0 + \tilde{n}_1}. \tag{6.65}$$

Clearly, these coefficients depend on the refractive indices of the media being considered, but also on the angle of incidence and the polarization state of the light. Note the usefulness of the concept of effective index, not only because it simplifies the calculations, but also because it allows the coefficients $r$ and $t$ to be expressed in the same way whatever the polarization and the incidence: in other words, this means that at oblique incidence, the material of refractive index $n$ behaves like a material of index $\tilde{n}$ at normal incidence.

### Angle Dependence

At normal incidence, the TE and TM effective indices are identical and given by

$$\tilde{n}_{\text{TE}} = \tilde{n}_{\text{TM}} = \frac{n}{\eta_v \tilde{\mu}_r}, \tag{6.66}$$

where $\tilde{\mu}_r$ can be taken as 1 for all common (nonmagnetic) optical materials. As a result,

$$r = \frac{\tilde{n}_0 - \tilde{n}_1}{\tilde{n}_0 + \tilde{n}_1} = \frac{n_0 - n_1}{n_0 + n_1}, \tag{6.67}$$

so that, in the case of an air–glass interface illuminated by visible light ($n_0 = 1$ and $n_1 \approx 1.5$), we have

$$r = \frac{n_0 - n_1}{n_0 + n_1} = \frac{1 - 1,5}{1 + 1,5} = -\frac{0,5}{2,5} = -\frac{1}{5}.$$

In amplitude, the reflection coefficient of this interface is therefore $-20\%$, the $-$ sign indicating a phase shift of $\pi$ on reflection. In energy terms, the reflection coefficient $R$ (to be defined later as $R = |r|^2$) is therefore 4%.

Equation (6.65) can also be used to find the conditions under which the reflection coefficient at the boundary between two transparent media vanishes, here satisfied by $\tilde{n}_0 = \tilde{n}_1$. This can occur only in the case of TM polarization, and leads to the relation

$$\frac{n_0}{\cos\theta_0} = \frac{n_1}{\cos\theta_1}, \tag{6.68}$$

which, combined with the law of refraction $n_0 \sin\theta_0 = n_1 \sin\theta_1$, enables us to determine the angle of incidence $\theta_0$ for which this condition is satisfied:

$$n_0 \cos\theta_1 = n_1 \cos\theta_0 \quad \Rightarrow \quad n_1^2 \cos^2\theta_0 = n_0^2 \left(1 - \frac{n_0^2}{n_1^2} \sin^2\theta_0\right). \tag{6.69}$$

Dividing both sides of this second equation by $\cos^2\theta_0$, we get

$$n_1^2 = \frac{n_0^2}{\cos^2\theta_0} - \frac{n_0^4}{n_1^2}\tan^2\theta_0 = n_0^2(1 + \tan^2\theta_0) - \frac{n_0^4}{n_1^2}\tan^2\theta_0, \qquad (6.70)$$

or

$$n_1^2 - n_0^2 = \frac{n_0^2}{n_1^2}(n_1^2 - n_0^2)\tan^2\theta_0, \qquad (6.71)$$

and hence

$$\tan\theta_0 = \frac{n_1}{n_0}. \qquad (6.72)$$

This particular angle is known as the Brewster angle $\theta_B$ and is often used to linearly polarize a beam of unpolarized light through simple reflection at an air–glass interface. For the two TE and TM polarization states, Figure 6.5 shows how the reflection coefficient $r$ varies versus the angle of incidence at the air–glass interface ($n_0 = 1$, $n_1 \sim 1.5$).

Notice that the refraction angle is given by

$$\tan\theta_1 = \frac{1}{\tan\theta_0}, \qquad (6.73)$$

and that the corresponding spatial pulsation is

$$\sigma_B = \frac{k_0 k_1}{\sqrt{k_0^2 + k_1^2}}. \qquad (6.74)$$

Finally, we observe that the same Brewster effect occurs regardless of the illumination medium.

### Boundary Conditions for Normal Components

We have seen that the Poynting flux through a $z$ plane can be calculated directly using the tangential coefficients $r$ and $t$. However in some situations such as polarization analysis, it is necessary to determine the direction of the reflected (or transmitted) field, and for that, it is useful, in TM polarization, to know the reflection (or transmission) coefficient of the normal components.

For TM polarization, we already stated the following relation between the elementary components of the electric and magnetic fields

$$\mathbb{A}_{j,\mathrm{tg}}^{\pm} = \pm\frac{\alpha_j}{\omega\tilde{\epsilon}_j}\mathbb{B}_j^{\pm} = \pm\frac{1}{\tilde{n}_j}\mathbb{B}_j^{\pm} \quad \text{and} \quad \mathbb{A}_{j,\mathrm{N}}^{\pm} = -\frac{\sigma}{\omega\tilde{\epsilon}_j}\mathbb{B}_j^{\pm}. \qquad (6.75)$$

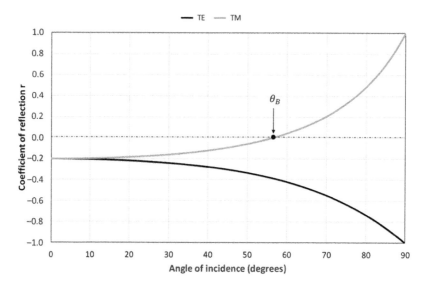

Figure 6.5 Variation of the reflection coefficient $r$ at an air–glass inter-face versus the angle of incidence $\theta_0$ [dark: TE polarization – gray: TM polarization].

So we can write, using the subscript N to identify the reflection and trans-mission coefficients associated with the normal components

$$r_{\mathrm{N}} = \frac{\mathbb{A}_{0,\mathrm{N}}^-}{\mathbb{A}_{0,\mathrm{N}}^+} = \frac{\mathbb{B}_0^-}{\mathbb{B}_0^+} = \frac{\mathbb{A}_{0,\mathrm{tg}}^-}{\mathbb{A}_{0,\mathrm{tg}}^+} = -r\,, \tag{6.76}$$

$$t_{\mathrm{N}} = \frac{\mathbb{A}_{1,\mathrm{N}}^+}{\mathbb{A}_{0,\mathrm{N}}^+} = \frac{\mathbb{B}_1^+}{\tilde{\epsilon}_1}\frac{\tilde{\epsilon}_0}{\mathbb{B}_0^+} = \frac{\mathbb{A}_{1,\mathrm{tg}}^+}{\alpha_1}\frac{\alpha_0}{\mathbb{A}_{0,\mathrm{tg}}^+} = \frac{\alpha_0}{\alpha_1}\,t\,. \tag{6.77}$$

### 6.3.5 The Concept of Complex Admittance

We propose to generalize to stationary waves the idea of effective refractive index that we introduced for progressive or retrograde waves. We must again find the relationship between the tangential components of the electric and magnetic fields. To this end, recall that a stationary wave is the sum of a progressive wave and a retrograde wave, and that the modulus of the wave varies with height $z$.

The stationary wave within layer $j$ of the stack shown in Figure 6.2 is therefore described by the equation

$$\vec{\mathcal{E}}_{j,\mathrm{tg}}(x, z, f) = \vec{\mathbb{E}}_{j,\mathrm{tg}}(\vec{\nu}, z, f)\,e^{2i\pi\nu x} = (\vec{\mathbb{A}}_{j,\mathrm{tg}}^+ e^{i\alpha_j z} + \vec{\mathbb{A}}_{j,\mathrm{tg}}^- e^{-i\alpha_j z})e^{i\sigma x}\,, \tag{6.78}$$

with $\alpha_j^2 = k_j^2 - \sigma^2$ and $k_j = \frac{2\pi n_j}{\lambda}$.

To fix our ideas, and following a procedure analogous to that adopted in Section 6.3.2, it is easy to check that the tangential components of the two quantities $\vec{\mathbb{H}}_j$ and $\vec{z} \wedge \vec{\mathbb{E}}_j$ are parallel. Therefore there exists a complex scalar $Y$ such that we can write

$$\vec{\mathbb{H}}_{j,\mathrm{tg}}(\vec{\nu}, z, f) = Y_j(\vec{\nu}, z, f) \left[ \vec{z} \wedge \vec{\mathbb{E}}_{j,\mathrm{tg}}(\vec{\nu}, z, f) \right]. \tag{6.79}$$

This quantity $Y_j(\vec{\nu}, z, f)$ is simply a factor of proportionality between the tangential components of these two fields. For $z = z_j$, we will use the contracted notation $Y_j(\vec{\nu}, z_j, f) = Y_j$, and this last quantity is known as the complex admittance at the interface at which $\vec{\mathbb{E}}_{\mathrm{tg}}$ and $\vec{\mathbb{H}}_{\mathrm{tg}}$ are being considered. For a stack of thin layers, the complex admittance $Y_j(z)$ plays the same role as the effective refractive index in the case of a unique homogeneous medium. However, and in contrast with the effective index that is constant in each layer, this parameter varies with height $z$. Furthermore, since the tangential fields are continuous within the stack, the complex admittance $Y(z)$ is a continuous function of height $z$ throughout the stack, and $Y_j(z)$ is its restriction to medium $j$.

For each state of polarization, in order to deduce the amplitude reflection coefficient $r$ of the stack shown in Figure 6.2, we follow a procedure similar to that given in section 6.3.4 for a single plane interface, namely

• At the upper 0 interface, we use the definition of the complex admittance

$$\vec{\mathbb{H}}_{0,\mathrm{tg}}(\vec{\nu}, 0) = Y_0 \left[ \vec{z} \wedge \vec{\mathbb{E}}_{0,\mathrm{tg}}(\vec{\nu}, 0) \right], \tag{6.80}$$

which leads to the following relation:

$$\vec{\mathbb{B}}_{0,\mathrm{tg}}^+ + \vec{\mathbb{B}}_{0,\mathrm{tg}}^- = Y_0 \left[ \vec{z} \wedge \vec{\mathbb{A}}_{0,\mathrm{tg}}^+ + \vec{z} \wedge \vec{\mathbb{A}}_{0,\mathrm{tg}}^- \right]. \tag{6.81}$$

• The reflection coefficient $r$ is defined by

$$\vec{\mathbb{A}}_{0,\mathrm{tg}}^- = r \vec{\mathbb{A}}_{0,\mathrm{tg}}^+. \tag{6.82}$$

• Moreover, we have in the incident medium

$$\vec{\mathbb{B}}_{0,\mathrm{tg}}^+ = \tilde{n}_0 \left[ \vec{z} \wedge \vec{\mathbb{A}}_{0,\mathrm{tg}}^+ \right] \quad ; \quad \vec{\mathbb{B}}_{0,\mathrm{tg}}^- = -\tilde{n}_0 \left[ \vec{z} \wedge \vec{\mathbb{A}}_{0,\mathrm{tg}}^- \right]. \tag{6.83}$$

By combining (6.81), (6.82), and (6.83), we obtain

$$\tilde{n}_0 \left[ \vec{z} \wedge \vec{\mathbb{A}}_{0,\mathrm{tg}}^+ \right] - \tilde{n}_0 r \left[ \vec{z} \wedge \vec{\mathbb{A}}_{0,\mathrm{tg}}^+ \right] = Y_0 (1 + r) \left[ \vec{z} \wedge \vec{\mathbb{A}}_{0,\mathrm{tg}}^+ \right], \tag{6.84}$$

and hence

$$r = \frac{\tilde{n}_0 - Y_0}{\tilde{n}_0 + Y_0}. \tag{6.85}$$

The quantity $Y_0$, the complex admittance or inverse impedance of the component at the upper interface of the stack, depends not only on the angle of incidence and on the polarization state but also on all the optical and geometrical characteristics of the multilayer. Equation (6.85) thus demonstrates that, for the calculation of reflection, the multilayer component behaves as an artificial substrate of refractive index $Y_0$. This result forms the basis of the first calculation in analytical synthesis, insofar as the admittance $Y$ can take arbitrary values in the complex plane. For example, we can obtain a nonreflective surface ($r = 0$) by choosing a stack giving an admittance $Y_0$ equal to $\tilde{n}_0$. In the same way, perfect mirrors can be obtained with zero admittance ($r = 1$) or infinite admittance ($r = -1$). Keep in mind, however, that the equivalence between refractive index and admittance is valid only for reflection calculations; indeed we shall see that transmission in the multilayer takes a form different from that of the single interface ($t \neq \frac{2\tilde{n}_0}{\tilde{n}_0 + Y_0}$).

It now remains to define how to calculate this complex admittance in practice.

### 6.3.6 Transfer Matrix

Consider inside the stack shown in Figure 6.6 the layer $j$ of thickness $d_j$ and refractive index $n_j$ located between the $(j-1)$th and $j$th interfaces. The origin of the $z$-axis is taken at interface 0.

As previously indicated, we are interested only in the tangential field components that are continuous throughout the whole stack. At some arbitrary point on the $z$ abscissa within the thickness of the layer $j$, we can write

$$\begin{cases} \vec{\mathbb{E}}_{j,\text{tg}}(z) = \vec{\mathbb{A}}_{j,\text{tg}}^+ e^{i\alpha_j z} + \vec{\mathbb{A}}_{j,\text{tg}}^- e^{-i\alpha_j z} \\ \vec{\mathbb{H}}_{j,\text{tg}}(z) = \vec{\mathbb{B}}_{j,\text{tg}}^+ e^{i\alpha_j z} + \vec{\mathbb{B}}_{j,\text{tg}}^- e^{-i\alpha_j z} \end{cases} \quad \text{for} \quad z_{j-1} \leqslant z \leqslant z_j, \qquad (6.86)$$

where we omitted to recall the common $\vec{\nu}$-dependence of the two elementary components $\vec{\mathbb{E}}_{j,\text{tg}}(\vec{\nu}, z)$ and $\vec{\mathbb{H}}_{j,\text{tg}}(\vec{\nu}, z)$.

This leads immediately to the following two groups of equations, corresponding respectively to the particular abscissae $z = z_{j-1}$ and $z = z_j = z_{j-1} + d_j$, and which describe the values of the tangential field components at the two interfaces $j - 1$ and $j$:

$$\begin{cases} \vec{\mathbb{E}}_{j,\text{tg}}(z_{j-1}) = \vec{\mathbb{A}}_{j,\text{tg}}^+ e^{i\alpha_j z_{j-1}} + \vec{\mathbb{A}}_{j,\text{tg}}^- e^{-i\alpha_j z_{j-1}} \\ \vec{\mathbb{H}}_{j,\text{tg}}(z_{j-1}) = \vec{\mathbb{B}}_{j,\text{tg}}^+ e^{i\alpha_j z_{j-1}} + \vec{\mathbb{B}}_{j,\text{tg}}^- e^{-i\alpha_j z_{j-1}} \end{cases}, \qquad (6.87)$$

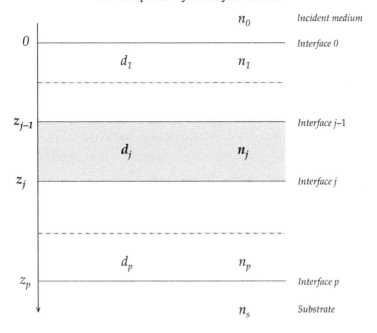

Figure 6.6 Layer $j$ embedded in a stack of $p$ layers.

$$\begin{cases} \vec{\mathbb{E}}_{j,\text{tg}}(z_j) = \vec{\mathbb{A}}_{j,\text{tg}}^{+} e^{i\alpha_j z_j} + \vec{\mathbb{A}}_{j,\text{tg}}^{-} e^{-i\alpha_j z_j} \\ \vec{\mathbb{H}}_{j,\text{tg}}(z_j) = \vec{\mathbb{B}}_{j,\text{tg}}^{+} e^{i\alpha_j z_j} + \vec{\mathbb{B}}_{j,\text{tg}}^{-} e^{-i\alpha_j z_j} \end{cases} . \tag{6.88}$$

In accordance with the conclusions of Section 6.3.2, we also have

$$\vec{\mathbb{B}}_{j,\text{tg}}^{\pm} = \pm \tilde{n}_j \, \vec{z} \wedge \vec{\mathbb{A}}_{j,\text{tg}}^{\pm} . \tag{6.89}$$

Forming the quantities $\tilde{n}_j \, \vec{z} \wedge \vec{\mathbb{E}}_{j,\text{tg}}(z_{j-1})$ and $\tilde{n}_j \, \vec{z} \wedge \vec{\mathbb{E}}_{j,\text{tg}}(z_j)$, and using equations (6.89), we immediately get

$$\begin{cases} \tilde{n}_j \, \vec{z} \wedge \vec{\mathbb{E}}_{j,\text{tg}}(z_{j-1}) = \vec{\mathbb{B}}_{j,\text{tg}}^{+} e^{i\alpha_j z_{j-1}} - \vec{\mathbb{B}}_{j,\text{tg}}^{-} e^{-i\alpha_j z_{j-1}} \\ \tilde{n}_j \, \vec{z} \wedge \vec{\mathbb{E}}_{j,\text{tg}}(z_j) = \vec{\mathbb{B}}_{j,\text{tg}}^{+} e^{i\alpha_j z_j} - \vec{\mathbb{B}}_{j,\text{tg}}^{-} e^{-i\alpha_j z_j} \end{cases} . \tag{6.90}$$

If we now add to this system the two following relations, not yet used, namely

$$\begin{cases} \vec{\mathbb{H}}_{j,\text{tg}}(z_{j-1}) = \vec{\mathbb{B}}_{j,\text{tg}}^{+} e^{i\alpha_j z_{j-1}} + \vec{\mathbb{B}}_{j,\text{tg}}^{-} e^{-i\alpha_j z_{j-1}} \\ \vec{\mathbb{H}}_{j,\text{tg}}(z_j) = \vec{\mathbb{B}}_{j,\text{tg}}^{+} e^{i\alpha_j z_j} + \vec{\mathbb{B}}_{j,\text{tg}}^{-} e^{-i\alpha_j z_j} \end{cases} , \tag{6.91}$$

we deduce, through a linear combination of the first equations of systems (6.90) and (6.91), that

$$
\begin{cases}
\vec{\mathbb{B}}_{j,\mathrm{tg}}^{+} = \dfrac{1}{2}\left(\tilde{n}_j\,\vec{z}\wedge\vec{\mathbb{E}}_{j,\mathrm{tg}}(z_{j-1}) + \vec{\mathbb{H}}_{j,\mathrm{tg}}(z_{j-1})\right)e^{-i\alpha_j z_{j-1}} \\[2mm]
\vec{\mathbb{B}}_{j,\mathrm{tg}}^{-} = -\dfrac{1}{2}\left(\tilde{n}_j\,\vec{z}\wedge\vec{\mathbb{E}}_{j,\mathrm{tg}}(z_{j-1}) - \vec{\mathbb{H}}_{j,\mathrm{tg}}(z_{j-1})\right)e^{i\alpha_j z_{j-1}}
\end{cases}. \tag{6.92}
$$

It only remains to substitute these expressions into the second equations of systems (6.90) and (6.91) to obtain the following linear relations:

$$
\begin{cases}
\tilde{n}_j\,\vec{z}\wedge\vec{\mathbb{E}}_{j,\mathrm{tg}}(z_j) = \cos(\alpha_j d_j)\left[\tilde{n}_j\,\vec{z}\wedge\vec{\mathbb{E}}_{j,\mathrm{tg}}(z_{j-1}\right] + i\sin(\alpha_j d_j)\vec{\mathbb{H}}_{j,\mathrm{tg}}(z_{j-1}) \\[2mm]
\vec{\mathbb{H}}_{j,\mathrm{tg}}(z_j) = i\sin(\alpha_j d_j)\left[\tilde{n}_j\,\vec{z}\wedge\vec{\mathcal{E}}_{j,\mathrm{tg}}(z_{j-1})\right] + \cos(\alpha_j d_j)\vec{\mathbb{H}}_{j,\mathrm{tg}}(z_{j-1})
\end{cases}.
$$

Moreover, the tangential components of the fields are continuous at the interface $j-1$, i.e.,

$$
\vec{\mathbb{E}}_{j,\mathrm{tg}}(z_{j-1}) = \vec{\mathbb{E}}_{j-1,\mathrm{tg}}(z_{j-1}) \quad\text{and}\quad \vec{\mathbb{H}}_{j,\mathrm{tg}}(z_{j-1}) = \vec{\mathbb{H}}_{j-1,\mathrm{tg}}(z_{j-1}). \tag{6.93}
$$

Finally, using a matrix representation and denoting the quantity $\alpha_j d_j$ by $\delta_j$, we obtain

$$
\begin{bmatrix} \vec{z}\wedge\vec{\mathbb{E}}_{j,\mathrm{tg}}(z_j) \\[2mm] \vec{\mathbb{H}}_{j,\mathrm{tg}}(z_j) \end{bmatrix} = \begin{bmatrix} \cos\delta_j & \dfrac{i}{\tilde{n}_j}\sin\delta_j \\[3mm] i\tilde{n}_j\sin\delta_j & \cos\delta_j \end{bmatrix}\begin{bmatrix} \vec{z}\wedge\vec{\mathbb{E}}_{j-1,\mathrm{tg}}(z_{j-1}) \\[2mm] \vec{\mathbb{H}}_{j-1,\mathrm{tg}}(z_{j-1}) \end{bmatrix}, \tag{6.94}
$$

more commonly written as

$$
\begin{bmatrix} \vec{z}\wedge\vec{\mathbb{E}}_{j-1,\mathrm{tg}}(z_{j-1}) \\[2mm] \vec{\mathbb{H}}_{j-1,\mathrm{tg}}(z_{j-1}) \end{bmatrix} = M_j\begin{bmatrix} \vec{z}\wedge\vec{\mathbb{E}}_{j,\mathrm{tg}}(z_j) \\[2mm] \vec{\mathbb{H}}_{j,\mathrm{tg}}(z_j) \end{bmatrix}, \tag{6.95}
$$

with

$$
M_j = \begin{bmatrix} \cos\delta_j & -\dfrac{i}{\tilde{n}_j}\sin\delta_j \\[3mm] -i\tilde{n}_j\sin\delta_j & \cos\delta_j \end{bmatrix}, \tag{6.96}
$$

The matrix $M_j$ is known as the **transfer matrix** associated with layer $j$, and enables the quantities $\vec{z}\wedge\vec{\mathbb{E}}_{\mathrm{tg}}$ and $\vec{\mathbb{H}}_{\mathrm{tg}}$ to be calculated at interface $j-1$, given the values of these same quantities at interface $j$. It is important here to emphasize that the use of this transfer matrix allows us to *work our way up* inside the stack, starting from the substrate and ending at the superstrate.

Note also that the phase term $\delta_j$ is dimensionless, and is given at low frequencies by $\delta_j = \frac{2\pi}{\lambda_v}n_j d_j\cos\theta_j$.

Finally, it is worth pointing out that defined in the literature are other transfer matrices that relate variables at the interfaces. In our case, this transfer matrix relates the tangential components of the stationary fields $\vec{\mathbb{E}}$

and $\vec{\mathbb{H}}$ but among other authors, these can connect the progressive or retrograde fields, or electric or magnetic fields depending on their polarization state.

### 6.3.7 Recurrence Relations Between Complex Admittances

By definition, the complex admittances of the interfaces $j - 1$ and $j$ are defined by the following relations (see Section 6.3.5):

$$
\begin{aligned}
\vec{\mathbb{H}}_{j-1,\text{tg}}(z_{j-1}) &= Y_{j-1} \left[ \vec{z} \wedge \vec{\mathbb{E}}_{j-1,\text{tg}}(z_{j-1}) \right] \\
\vec{\mathbb{H}}_{j,\text{tg}}(z_j) &= Y_j \left[ \vec{z} \wedge \vec{\mathbb{E}}_{j,\text{tg}}(z_j) \right] .
\end{aligned}
\tag{6.97}
$$

As a result, inserting these definitions into the matrix equation (6.95) and then constructing the ratio of the two equations thus obtained, we get

$$
Y_{j-1} = \frac{Y_j \cos \delta_j - i\tilde{n}_j \sin \delta_j}{\cos \delta_j - i(Y_j/\tilde{n}_j) \sin \delta_j} .
\tag{6.98}
$$

This equation enables us to calculate the admittance of interface $j-1$, given that we know that of interface $j$. By inserting a virtual interface at altitude $z$ within layer $j$, we can also compute the complex admittance at any height $z$, namely

$$
Y_j(z) = \frac{Y_j \cos[\alpha_j(z_j - z)] - i\tilde{n}_j \sin[\alpha_j(z_j - z)]}{\cos[\alpha_j(z_j - z)] - i(Y_j/\tilde{n}_j) \sin[\alpha_j(z_j - z)]} .
\tag{6.99}
$$

Moreover, relation (6.98) can also be written as

$$
Y_j = \frac{Y_{j-1} \cos \delta_j + i\tilde{n}_j \sin \delta_j}{\cos \delta_j + i(Y_{j-1}/\tilde{n}_j) \sin \delta_j} .
\tag{6.100}
$$

It now remains to initialize these recurrence relations; a known admittance is required to do this.

For an incident wave that is progressive in nature, this known admittance is given by the substrate, which is the sole medium where the wave is not stationary (it is progressive). The admittance of the substrate is therefore equal to its effective refractive index ($Y_p = \tilde{n}_s$). This enables the recurrence relation (6.98) to be initialized, and hence progress up to the value of admittance at the first interface of the multilayer, i.e., $Y_0$, and then to calculate the value of the reflection coefficient using equation (6.85), i.e.,

$$
r = \frac{\tilde{n}_0 - Y_0}{\tilde{n}_0 + Y_0} .
$$

In some situations it is also useful to know the optical properties of a stack when the illumination comes from the substrate. In this case where the incident wave is retrograde in nature, the only admittance known a priori is that given by the incident medium, where the wave is not stationary but retrograde. The admittance of this medium is therefore equal to its effective refractive index $(Y_0' = -\tilde{n}_0)$; in order to avoid any confusion with the admittance $Y_0$ previously calculated for the progressive case, this admittance is distinguished from $Y_0$ by a prime symbol.

The recurrence relations remain unchanged for this illumination from the substrate. Indeed, by reversing the direction of propagation of light, $\alpha_j$ is replaced by $-\alpha_j$ and $\tilde{n}_j$ by $-\tilde{n}_j$, leaving $\cos \delta_j$, $\tilde{n}_j \sin \delta_j$ and $[\sin \delta_j]/\tilde{n}_j$ invariant.

Then applying the recurrence relation (6.100) allows a progression up to the value of admittance at the first interface encountered by the retrograde wave, i.e., $Y_s'$, and then calculation of the associated reflection coefficient, i.e.,

$$r' = \frac{-\tilde{n}_s - Y_s'}{-\tilde{n}_s + Y_s'}, \tag{6.101}$$

where the superscript $'$ is again used for the reflection coefficient to avoid any confusion.

In all the following sections, we will consider only the case where the incident wave is progressive in nature. The transposition to the case where this incident wave is retrograde in nature (illumination coming from the substrate) is quite straightforward.

### 6.3.8 Field Components at the Interfaces

#### Tangential Components

Once the admittances are known at the various optical interfaces, the matrix equation (6.95) immediately relates the tangential components of the fields at the interfaces. For the electric field we get

$$\vec{z} \wedge \vec{\mathbb{E}}_{j-1,\text{tg}}(z_{j-1}) = \cos \delta_j \, (\vec{z} \wedge \vec{\mathbb{E}}_{j,\text{tg}})(z_j) - \frac{i}{\tilde{n}_j} \sin \delta_j \vec{\mathbb{H}}_{j,\text{tg}}(z_j). \tag{6.102}$$

Then

$$\vec{z} \wedge \vec{\mathbb{E}}_{j-1,\text{tg}}(z_{j-1}) = \cos \delta_j \, [\vec{z} \wedge \vec{\mathbb{E}}_{j,\text{tg}}](z_j) - i\frac{Y_j}{\tilde{n}_j} \sin \delta_j \, [\vec{z} \wedge \vec{\mathbb{E}}_{j,\text{tg}}](z_j), \tag{6.103}$$

and finally

$$\vec{\mathbb{E}}_{j-1,\text{tg}}(z_{j-1}) = \left[ \cos \delta_j - i\frac{Y_j}{\tilde{n}_j} \sin \delta_j \right] \vec{\mathbb{E}}_{j,\text{tg}}(z_j), \tag{6.104}$$

or

$$\vec{\mathbb{E}}_{j,\text{tg}}(z_j) = \frac{\vec{\mathbb{E}}_{j-1,\text{tg}}(z_{j-1})}{\cos \delta_j - i\frac{Y_j}{\tilde{n}_j} \sin \delta_j}. \tag{6.105}$$

This recurrence relation can be initialized at the upper interface, where we have

$$\vec{\mathbb{E}}_{0,\text{tg}}(z_0) = \vec{\mathbb{A}}_{0,\text{tg}}^+ + \vec{\mathbb{A}}_{0,\text{tg}}^- = (1+r)\vec{\mathbb{A}}_{0,\text{tg}}^+. \tag{6.106}$$

Observe that the incident field $\vec{\mathbb{A}}_{0,\text{tg}}^+$ will vanish after normalization by the incident flux (see Section 6.3.11).

Note that, using the properties of the transfer matrix, equation (6.105) can also be easily transformed into

$$\vec{\mathbb{E}}_{j,\text{tg}}(z_j) = \left[\cos \delta_j + i\frac{Y_{j-1}}{\tilde{n}_j} \sin \delta_j\right] \vec{\mathbb{E}}_{j-1,\text{tg}}(z_{j-1}). \tag{6.107}$$

### Normal Components

For TM polarization, the normal component of the electric field can be computed using (6.75)

$$\begin{aligned}
\vec{\mathbb{E}}_{j,\text{N}}(z_j) &= \vec{\mathbb{A}}_{j,\text{N}}^+ e^{i\alpha_j z_j} + \vec{\mathbb{A}}_{j,\text{N}}^- e^{-i\alpha_j z_j} \\
&= -\frac{\sigma}{\omega\tilde{\epsilon}_j}[\mathbb{B}_j^+ e^{i\alpha_j z_j} + \mathbb{B}_j^- e^{-i\alpha_j z_j}]\vec{z} = -\frac{\sigma}{\omega\tilde{\epsilon}_j}[\vec{x} \wedge \vec{\mathbb{H}}_{j,\text{tg}}(z_j)]. \tag{6.108}
\end{aligned}$$

By using the complex admittance $Y_j$ of interface $j$, we can replace $\vec{\mathbb{H}}_{j,\text{tg}}(z_j)$ by $Y_j[\vec{z} \wedge \vec{\mathbb{E}}_{j,\text{tg}}(z_j)]$ and obtain finally

$$\mathbb{E}_{j,\text{N}}(z_j) = -\frac{\sigma}{\omega\tilde{\epsilon}_j}Y_j\mathbb{E}_{j,\text{tg}}(z_j), \tag{6.109}$$

which connects the normal and tangential components of the TM stationary electric field at interface $j$.

### Remark

Note that in this section, all relationships given for stationary waves remain valid for progressive or retrograde waves (the admittance is then identified with the effective index).

### 6.3.9 Transmission Coefficient of the Stack

We have seen that the reflection coefficient $r$ is deduced from the admittance $Y_0$ by

$$r = \frac{\tilde{n}_0 - Y_0}{\tilde{n}_0 + Y_0}.$$

Hence for reflection, the formula is analogous to that of a single surface with a substrate of complex index $Y_0$. As already said, this is not the case for the transmission coefficient $t$.

This coefficient $t$ can be determined from equation (6.104) that relates the tangential components of the fields:

$$\vec{\mathbb{E}}_{j-1,\mathrm{tg}}(z_{j-1}) = \left[\cos\delta_j - i\frac{Y_j}{\tilde{n}_j}\sin\delta_j\right]\vec{\mathbb{E}}_{j,\mathrm{tg}}(z_j),$$

We immediately deduce a relationship between the variables $\vec{\mathbb{E}}_{0,\mathrm{tg}}(0)$ and $\vec{\mathbb{E}}_{p,\mathrm{tg}}(z_p)$, namely

$$\vec{\mathbb{E}}_{0,\mathrm{tg}}(0) = \prod_{j=1}^{p}\left[\cos\delta_j - i\frac{Y_j}{\tilde{n}_j}\sin\delta_j\right]\vec{\mathbb{E}}_{p,\mathrm{tg}}(z_p),\qquad(6.110)$$

with

$$\vec{\mathbb{E}}_{p,\mathrm{tg}}(z_p) = \vec{\mathbb{E}}_{s,\mathrm{tg}}(z_p) = \vec{\mathbb{A}}_{s,\mathrm{tg}}^{+}e^{i\alpha_s z_p}.\qquad(6.111)$$

We will define the transmission coefficient $t$ as

$$\vec{\mathbb{E}}_{s,\mathrm{tg}}(z_p) = t\vec{\mathbb{A}}_{0,\mathrm{tg}}^{+}.\qquad(6.112)$$

By using this definition and equation (6.106), we get

$$(1+r)\vec{\mathbb{A}}_{0,\mathrm{tg}}^{+} = \prod_{j=1}^{p}\left[\cos\delta_j - i\frac{Y_j}{\tilde{n}_j}\sin\delta_j\right]t\vec{\mathbb{A}}_{0,\mathrm{tg}}^{+}.\qquad(6.113)$$

That is,

$$t = \frac{1+r}{\prod_{j=1}^{p}\left[\cos\delta_j - i\frac{Y_j}{\tilde{n}_j}\sin\delta_j\right]}.\qquad(6.114)$$

As already stressed in Section 6.3.5, this transmission formula is rather complicated and differs greatly from that of a single interface (given by $t=1+r$), even though all admittances are involved in (6.114). This is a key difference with the reflection calculation [see equation (6.85)], where the admittance of the top interface plays the role of an effective index.

### 6.3.10 Field Components Within the Layered Media

Complementary relationships are given in this section to connect the stationary (normal or tangential) field within a layer to the same stationary field at the boundaries of the layer. These can be useful for faster calculation for instance, due to the fact that most often fields and admittances are first calculated at the only interfaces. Under these conditions, the field distribution at each altitude can be retrieved without knowing the admittance distribution within the whole stack [in other words, equation (6.99)]. For completeness, we also give the field distribution in the incident medium and the substrate.

### Tangential Components in the Bulk of the Layers

To determine the value of the tangential component of the field at depth $z$ within layer $j$, we first go back to the definition that appears in equation (6.86) for electric fields, i.e.,

$$\vec{\mathbb{E}}_{j,\text{tg}}(z) = \vec{\mathbb{A}}_{j,\text{tg}}^{+}e^{i\alpha_j z} + \vec{\mathbb{A}}_{j,\text{tg}}^{-}e^{-i\alpha_j z}, \tag{6.115}$$

and then consider the following additional relation at $z = z_{j-1}$:

$$\vec{\mathbb{E}}_{j,\text{tg}}(z_{j-1}) = \vec{\mathbb{E}}_{j-1,\text{tg}}(z_{j-1}),$$

so as to calculate the progressive and retrograde quantities $\vec{\mathbb{A}}_{j,\text{tg}}^{+}$ and $\vec{\mathbb{A}}_{j,\text{tg}}^{-}$ as a function of these stationary components of the electric field at the interfaces, i.e.,

$$\begin{cases} \vec{\mathbb{A}}_{j,\text{tg}}^{+} = \dfrac{\vec{\mathbb{E}}_{j,\text{tg}}(z_j) - e^{-i\alpha_j d_j}\vec{\mathbb{E}}_{j-1,\text{tg}}(z_{j-1})}{2i\sin\alpha_j d_j}e^{-i\alpha_j z_{j-1}} \\[3ex] \vec{\mathbb{A}}_{j,\text{tg}}^{-} = -\dfrac{\vec{\mathbb{E}}_{j,\text{tg}}(z_j) - e^{i\alpha_j d_j}\vec{\mathbb{E}}_{j-1,\text{tg}}(z_{j-1})}{2i\sin\alpha_j d_j}e^{i\alpha_j z_{j-1}} \end{cases}. \tag{6.116}$$

We finally insert these expressions into the expression for $\vec{\mathbb{E}}_{j,\text{tg}}(z)$:

$$\begin{aligned} \vec{\mathbb{E}}_{j,\text{tg}}(z) = \frac{1}{\sin\alpha_j d_j}\Big[&\sin\alpha_j(z - z_{j-1}) \times \vec{\mathbb{E}}_{j,\text{tg}}(z_j) \\ &- \sin\alpha_j(z - z_j) \times \vec{\mathbb{E}}_{j-1}(z_{j-1})\Big]. \end{aligned} \tag{6.117}$$

### Tangential Component within the Incident Medium

In the incident medium, the $z$-dependence of the tangential component of the electric field is defined by, for $-\infty < z \leqslant 0$

$$\mathbb{E}_{0,\text{tg}}(z) = \mathbb{A}_{0,\text{tg}}^{+}e^{i\alpha_0 z} + \mathbb{A}_{0,\text{tg}}^{-}e^{-i\alpha_0 z} = [e^{i\alpha_0 z} + r\,e^{-i\alpha_0 z}]\mathbb{A}_{0,\text{tg}}^{+}. \tag{6.118}$$

### Tangential Component Within the Substrate

In the substrate, the $z$-dependence of the tangential component of the electric field is defined by

$$\mathbb{E}_{s,\text{tg}}(z) = t\mathbb{A}_{0,\text{tg}}^{+}e^{i\alpha_s(z-z_p)} \quad \text{for } z_p \leqslant z < +\infty, \tag{6.119}$$

which shows that the phase accumulated during the crossing of the stack is, by definition, **included** in the transmission coefficient $t$.

### Normal Components in the Bulk of the Layers

In TM polarization, the $z$-dependence of the normal component of the electric field within the layer $j$ is defined by

$$\mathbb{E}_{j,\mathrm{N}}(z) = -\frac{\sigma}{\omega\tilde{\epsilon}_j}[\mathbb{B}_j^+ e^{i\alpha_j z} + \mathbb{B}_j^- e^{-i\alpha_j z}], \tag{6.120}$$

which is equivalent to

$$\tilde{\epsilon}_j\mathbb{E}_{j,\mathrm{N}}(z) = -\frac{\sigma}{\omega}[\mathbb{B}_j^+ e^{i\alpha_j z} + \mathbb{B}_j^- e^{-i\alpha_j z}]. \tag{6.121}$$

Moreover, the continuity of the normal component of the electric displacement $\vec{\mathbb{D}} = \tilde{\epsilon}\vec{\mathbb{E}}$ at the interface $j-1$ allows us to write

$$\tilde{\epsilon}_j\mathbb{E}_{j,\mathrm{N}}(z_{j-1}) = \tilde{\epsilon}_{j-1}\mathbb{E}_{j-1,\mathrm{N}}(z_{j-1}). \tag{6.122}$$

By using these two preceding equations, we can compute the quantities $\mathbb{B}_{j,\mathrm{N}}^+$ and $\mathbb{B}_{j,\mathrm{N}}^-$, i.e.,

$$\begin{cases} \mathbb{B}_j^+ = -\dfrac{e^{-i\alpha_j z_{j-1}}}{2i\sin\alpha_j d_j}\dfrac{\omega}{\sigma}\left\{\tilde{\epsilon}_j\mathbb{E}_{j,\mathrm{N}}(z_j) - e^{-i\alpha_j d_j}\tilde{\epsilon}_{j-1}\mathbb{E}_{j-1,\mathrm{N}}(z_{j-1})\right\} \\[2ex] \mathbb{B}_j^- = \dfrac{e^{i\alpha_j z_{j-1}}}{2i\sin\alpha_j d_j}\dfrac{\omega}{\sigma}\left\{\tilde{\epsilon}_j\mathbb{E}_{j,\mathrm{N}}(z_j) - e^{i\alpha_j d_j}\tilde{\epsilon}_{j-1}\mathbb{E}_{j-1,\mathrm{N}}(z_{j-1})\right\} \end{cases}, \tag{6.123}$$

and insert these results into the expression for $\tilde{\epsilon}_j\mathbb{E}_{j,\mathrm{N}}(z)$:

$$\begin{aligned} \tilde{\epsilon}_j\mathbb{E}_{j,\mathrm{N}}(z) = \frac{1}{\sin\alpha_j d_j}\{&\sin\alpha_j(z - z_{j-1}) \times \tilde{\epsilon}_j\mathbb{E}_{j,\mathrm{N}}(z_j) \\ &- \sin\alpha_j(z - z_j) \times \tilde{\epsilon}_{j-1}\mathbb{E}_{j-1,\mathrm{N}}(z_{j-1})\}. \end{aligned} \tag{6.124}$$

Finally, we obtain, after dividing by $\tilde{\epsilon}_j$

$$\begin{aligned} \mathbb{E}_{j,\mathrm{N}}(z) = \frac{1}{\sin\alpha_j d_j}\Big\{&\sin\alpha_j(z - z_{j-1}) \times \mathbb{E}_{j,\mathrm{N}}(z_j) \\ &- \sin\alpha_j(z - z_j) \times \frac{\tilde{\epsilon}_{j-1}}{\tilde{\epsilon}_j}\mathbb{E}_{j-1,\mathrm{N}}(z_{j-1})\Big\}. \end{aligned} \tag{6.125}$$

This relation is similar to (6.117), but for the normal components of the stationary field.

*Normal Component Within the Substrate*

In the substrate, we only have a progressive wave. So, in accordance with the third Maxwell's equation, we can write

$$\alpha_s \mathbb{A}^+_{s,\mathrm{N}} + \sigma \mathbb{A}^+_{s,\mathrm{tg}} = 0 \,.$$

As a consequence

$$\mathbb{E}_{s,\mathrm{N}}(z) = -\frac{\sigma}{\alpha_s} \mathbb{A}^+_{s,\mathrm{tg}} e^{i\alpha_s(z-z_p)} = -\frac{\sigma}{\alpha_s} t \mathbb{A}^+_{0,\mathrm{tg}} e^{i\alpha_s(z-z_p)} \,. \tag{6.126}$$

Again making use of the third Maxwell's equation for the incoming elementary wave, namely $\alpha_0 \mathbb{A}^+_{0,\mathrm{N}} + \sigma \mathbb{A}^+_{0,\mathrm{tg}} = 0$, we finally obtain

$$\mathbb{E}_{s,\mathrm{N}}(z) = \frac{\alpha_0}{\alpha_s} t \mathbb{A}^+_{0,\mathrm{N}} e^{i\alpha_s(z-z_p)} = t_\mathrm{N} \mathbb{A}^+_{0,\mathrm{N}} e^{i\alpha_s(z-z_p)} \quad \text{for } z_p \leqslant z < +\infty \,. \tag{6.127}$$

where $t_\mathrm{N}$ is the transmission coefficient of the stack for the normal components of the electric field.

*Normal Component Within the Incident Medium*

In the incident medium, the $z$-dependence of the normal component of the electric field is defined by

$$\mathbb{E}_{0,\mathrm{N}}(z) = \mathbb{A}^+_{0,\mathrm{N}} e^{i\alpha_0 z} + \mathbb{A}^-_{0,\mathrm{N}} e^{-i\alpha_0 z} \,. \tag{6.128}$$

At the top interface of the stack, we have

$$\mathbb{E}_{0,\mathrm{N}}(0) = \mathbb{A}^+_{0,\mathrm{N}} + \mathbb{A}^-_{0,\mathrm{N}} = -\frac{\sigma}{\omega \tilde{\epsilon}_0} Y_0 \mathbb{E}_{0,\mathrm{tg}}$$

$$= -\frac{\sigma}{\omega \tilde{\epsilon}_0} Y_0 (1+r) \mathbb{A}^+_{0,\mathrm{tg}} = \frac{\alpha_0}{\omega \tilde{\epsilon}_0} Y_0 (1+r) \mathbb{A}^+_{0,\mathrm{N}} \,. \tag{6.129}$$

By dividing both members by $\mathbb{A}^+_{0,\mathrm{N}}$ and using the definition of effective index in TM polarization ($\tilde{n}_0 = \omega \tilde{\epsilon}_0 / \alpha_0$), we obtain

$$1 + r_\mathrm{N} = \frac{Y_0}{\tilde{n}_0}(1+r) \;\Rightarrow\; 1 + r_\mathrm{N} = \frac{2 Y_0}{\tilde{n}_0 + Y_0}$$

$$\Rightarrow \; r_\mathrm{N} = -\frac{\tilde{n}_0 - Y_0}{\tilde{n}_0 + Y_0} = -r \,. \tag{6.130}$$

In the incident medium, the $z$-dependence of the normal component of the electric field is hence defined by

$$\mathbb{E}_{0,\mathrm{N}}(z) = [e^{i\alpha_0 z} + r_\mathrm{N} e^{-i\alpha_0 z}] \mathbb{A}^+_{0,\mathrm{N}} \quad \text{for } -\infty < z \leqslant 0 \,. \tag{6.131}$$

### 6.3.11 Energy-Related Variables

As always, energy is balanced relative to a closed surface $\Sigma$ restricting a domain containing the component, as shown in Figure 6.7. Given the geometry of our problem, this surface is where the planes of heights $z_0 = 0$ and $z_s = \sum_{j=1}^{p} d_j$ meet, located in the incident medium and substrate respectively. Note that because the normal has to be outwardly directed, it merges with $-\vec{z}$ and $+\vec{z}$ in the superstrate (surface $\Sigma_1$) and substrate (surface $\Sigma_2$), respectively.

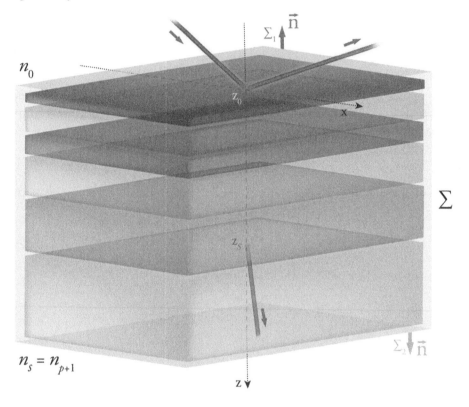

Figure 6.7 Geometry of the problem.

Note here that the side faces are not considered; indeed they are at infinity while the beam size is limited (as predicted by spatial wave packets), so they take no part in the energy balance. This is correct as long as all interfaces are planar and all sources are in the far field. In other situations (including light scattering or diffraction, or luminescent microcavities), guided modes may propagate within and parallel to the stack and reach these side surfaces in the absence of absorption (see Chapters 7, 8, and 9).

*Reflection and Transmission*

To compute the energy reflectance and transmittance of the stack, we have first to calculate the fluxes crossing the two $\Sigma_i$-surfaces. We consider an energy balance carried out in the far field, as it is usually the case experimentally. This will allow not needing to take into account the superposition of the incident and reflected fields in the immediate vicinity of the component.

Under these conditions, there are just three fluxes (two progressive and one retrograde) to be calculated in the superstrate and substrate; the spectral densities of these fluxes have already been given in Chapter 4 dealing with spatial wave packets. With a unique spatial frequency we have

$$\text{Incident flux: } \Phi_0^+ = \frac{1}{2\omega\tilde{\mu}_0}\Re[\alpha_0]\,|\vec{A}_0^+|^2$$

$$\text{Reflected flux: } \Phi_0^- = \frac{1}{2\omega\tilde{\mu}_0}\Re[\alpha_0]\,|\vec{A}_0^-|^2 \qquad (6.132)$$

$$\text{Transmitted flux: } \Phi_s^+ = \frac{1}{2\omega\tilde{\mu}_s}\Re[\alpha_s]\,|\vec{A}_s^+|^2\,.$$

These expressions, which involve the total amplitude of the field, are valid for both polarizations. To remain consistent with the method so far, which makes use only of the tangential field components, we can also rewrite the fluxes as indicated in Section 6.3.3:

$$\text{Incident flux: } \Phi_0^+ = \frac{1}{2}\Re[\tilde{n}_0]\,|\vec{A}_{0,\text{tg}}^+|^2$$

$$\text{Reflected flux: } \Phi_0^- = \frac{1}{2}\Re[\tilde{n}_0]\,|\vec{A}_{0,\text{tg}}^-|^2 \qquad (6.133)$$

$$\text{Transmitted flux: } \Phi_s^+ = \frac{1}{2}\Re[\tilde{n}_s]\,|\vec{A}_{s,\text{tg}}^+|^2\,.$$

It should be noted that these flux expressions are given at the top and bottom surfaces of the stack. In the case of transparent extreme media (superstrate and substrate), they hold at any $z < 0$ or $z > z_s$ planes, so that the definitions of $R$ and $T$ are not ambiguous. Otherwise we would have to consider the decrease of transmission within the substrate; since this substrate is here semi-infinite, the transmitted light would be absorbed. Note also in the formalism that, since the incident beam originates from the far field in the superstrate, this superstrate would be transparent; otherwise we would have to consider a near field source and its attenuation before it meets the component.

From now on we can define the energy reflection coefficient or reflectance as the ratio of reflected flux to incident flux:

$$R = \frac{\Phi_0^-}{\Phi_0^+} = \frac{\Re[\tilde{n}_0]\,|\vec{\mathbb{A}}_{0,\text{tg}}^-|^2}{\Re[\tilde{n}_0]\,|\vec{\mathbb{A}}_{0,\text{tg}}^+|^2} = |r|^2, \tag{6.134}$$

and the energy transmission coefficient or transmittance as the ratio of transmitted flux to incident flux:

$$T = \frac{\Phi_s^+}{\Phi_0^+} = \frac{\Re[\tilde{n}_s]\,|\vec{\mathbb{A}}_{s,\text{tg}}^+|^2}{\Re[\tilde{n}_0]\,|\vec{\mathbb{A}}_{0,\text{tg}}^+|^2} = \frac{\Re[\tilde{n}_s]}{\Re[\tilde{n}_0]}|t|^2. \tag{6.135}$$

These coefficients $R$ and $T$ are those that are normally measured in spectrophotometry, where we are essentially interested in measurements of illumination. In the absence of absorption, obviously we would have $R + T = 1$, an expression not to be confused with the amplitude relation $(1 + r = t)$ for the plane optical interface.

### Absorbance

We can directly calculate the overall normalized absorption (or absorbance) $A$ of a layered component using the principle of energy conservation ($A = 1 - R - T$), but we can also examine in detail the absorption in each layer of the stack in two different ways, hereafter detailed.

In the first approach, we consider a closed surface surrounding layer $j$ within the stack and we compute the flux through this closed surface gathering the planes $z = z_{j-1}$ and $z = z_j$. In the absence of sources, the result is the opposite of the layer absorption, that we can write as

$$\Phi(z_{j-1}) + \Phi(z_j) = -A_j \Phi_0^+. \tag{6.136}$$

We first observe from (6.136) that the absolute value of the flux through a $z$ plane is constant within a transparent multilayer (the sign is given by the normal of the closed surface at interfaces $j - 1$ and $j$.

Then following a procedure similar to that of Section 6.3.3, it is easy to check that relation (6.48) can be generalized to stationary waves, provided that the effective index is replaced by the admittance. This gives

$$\Phi(z_j) = \frac{1}{2}\Re[Y_j]|\vec{\mathbb{E}}_{j,\text{tg}}(z_j)|^2 \;;\; \Phi(z_{j-1}) = -\frac{1}{2}\Re[Y_{j-1}]|\vec{\mathbb{E}}_{j-1,\text{tg}}(z_{j-1})|^2, \tag{6.137}$$

with $\Phi_0^+ = \frac{1}{2}\Re[\tilde{n}_0]|\vec{\mathbb{A}}_{0,\text{tg}}|^2$. Consequently, making use of (6.136),

$$A_j = \frac{\Re[Y_{j-1}]|\vec{\mathbb{E}}_{j-1,\text{tg}}(z_{j-1})|^2 - \Re[Y_j]|\vec{\mathbb{E}}_{j,\text{tg}}(z_j)|^2}{\Re[\tilde{n}_0]|\vec{\mathbb{A}}_{0,\text{tg}}|^2}. \tag{6.138}$$

Such a relation allows the computation of absorbance in layer $j$ versus the stationary tangential fields at the two interfaces. Note that only the real parts of the admittances are involved in this calculation.

In the second approach, we compute this absorption directly by applying relation (6.49). We obtain for nonmagnetic materials ($\tilde{\mu}_j'' = 0$)

$$d\Phi_j(z) = \Phi_0^+ dA_j(z) = \frac{\omega}{2}\tilde{\epsilon}_j''|\vec{\mathbb{E}}_j(z)|^2 dz \,, \tag{6.139}$$

and, after integration over the thickness of layer $j$,

$$A_j\Phi_0^+ = \frac{\omega}{2}\tilde{\epsilon}_j'' \int_{z_{j-1}}^{z_j} |\vec{\mathbb{E}}_j(z)|^2 dz \,. \tag{6.140}$$

Consequently, the absorbance of layer $j$ is given by

$$A_j = \omega\frac{\tilde{\epsilon}_j''}{\tilde{n}_0} \int_{z_{j-1}}^{z_j} \frac{|\vec{\mathbb{E}}_{j,tg}(z)|^2 + |\vec{\mathbb{E}}_{j,N}(z)|^2}{|\vec{\mathbb{A}}_{0,tg}|^2} dz \,. \tag{6.141}$$

Finally, the overall normalized absorption of the multilayer is given by summing the absorbance of all the layers, i.e.,

$$A = \sum_{j=1}^{p} A_j \,. \tag{6.142}$$

To conclude, knowing the absorption within each layer may help in designing a coating for minimal losses; to that end, materials with the higher imaginary index would be located (when possible) in the regions of lower electric field.

## 6.4 Optical Properties of Some Basic Thin-Film Coatings

### 6.4.1 Single Layer

Now consider the particular case of a single layer of thickness $d_1$ and refractive index $n_1$ deposited onto the surface of a semi-infinite substrate of index $n_s$, as shown in Figure 6.8. The refractive index of the incident medium is $n_0$.

Using all the ideas introduced in Section 6.3, we are in a position to calculate the following quantities:

• Complex admittance of interface 1:

$$Y_1 = \tilde{n}_s \,. \tag{6.143}$$

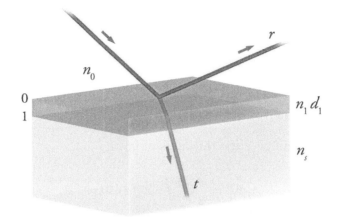

Figure 6.8 Single layer deposited onto the surface of a semi-infinite substrate.

- Complex admittance of interface 0:

$$Y_0 = \frac{Y_1 \cos \delta_1 - i \tilde{n}_1 \sin \delta_1}{\cos \delta_1 - i(Y_1/\tilde{n}_1) \sin \delta_1} = \frac{\tilde{n}_s \cos \delta_1 - i \tilde{n}_1 \sin \delta_1}{\cos \delta_1 - i(\tilde{n}_s/\tilde{n}_1) \sin \delta_1}, \qquad (6.144)$$

  with $\delta_1 = \alpha_1 d_1$.
- Reflection coefficient $r$:

$$r = \frac{\tilde{n}_0 - Y_0}{\tilde{n}_0 + Y_0} = \frac{(\tilde{n}_0 - \tilde{n}_s) \cos \delta_1 - i(\tilde{n}_0 \tilde{n}_s/\tilde{n}_1 - \tilde{n}_1) \sin \delta_1}{(\tilde{n}_0 + \tilde{n}_s) \cos \delta_1 - i(\tilde{n}_0 \tilde{n}_s/\tilde{n}_1 + \tilde{n}_1) \sin \delta_1}. \qquad (6.145)$$

- Transmission coefficient $t$:

$$t = \frac{1 + r}{\cos \delta_1 - i(Y_1/\tilde{n}_1) \sin \delta_1}$$

$$= \frac{2\tilde{n}_0}{(\tilde{n}_0 + \tilde{n}_s) \cos \delta_1 - i(\tilde{n}_0 \tilde{n}_s/\tilde{n}_1 + \tilde{n}_1) \sin \delta_1}, \qquad (6.146)$$

or

$$\frac{2\tilde{n}_0}{t} = (\tilde{n}_0 + \tilde{n}_s) \cos \delta_1 - i(\tilde{n}_0 \tilde{n}_s/\tilde{n}_1 + \tilde{n}_1) \sin \delta_1. \qquad (6.147)$$

### Half-Wave Layer

A layer whose optical thickness satisfies the condition

$$n_j d_j = q \frac{\lambda_0}{2}, \qquad (6.148)$$

where $q$ is a positive integer, is known as a *half-wave* layer (relative to the wavelength $\lambda_0$). Obviously, this condition makes sense only for transparent media.

Consequently, at zero incidence and for this wavelength $\lambda_0$, the phase difference $\delta_j$ is an integer multiple of $\pi$ ($\delta_j = q\pi$), so that, in the absence of absorption, the complex admittance of interface 0 takes the following particular form:

$$Y_0 = n_s, \tag{6.149}$$

i.e., which corresponds to the bare substrate. Everything happens as if the half-wave layer did not exist at that particular wavelength $\lambda_0$, for which reason these half-wave layers are often known as *absentee* layers.

Note that absentee layers also exist at oblique incidence. Such *matched* layers follow

$$n_j d_j \cos \theta_j = q\frac{\lambda_0}{2}. \tag{6.150}$$

### Quarter-Wave Layer

A transparent layer whose optical thickness satisfies the condition

$$n_j d_j = (2q+1)\frac{\lambda_0}{4} \tag{6.151}$$

is known as a *quarter-wave* layer (relative to the wavelength $\lambda_0$, as before), where $q$ is an integer that is positive or 0. At zero incidence and for wavelength $\lambda_0$ we immediately deduce that $\delta_j = (2q+1)\frac{\pi}{2}$, so the complex admittance at interface 0 here has the form

$$Y_0 = \frac{n_1^2}{n_s}. \tag{6.152}$$

If, for example, we assume a glass substrate ($n_s = 1.5$) onto which a quarter-wave layer of refractive index $n_1 = 2.15$ has been deposited (corresponding to tantalum pentoxide $Ta_2O_5$ at 600 nm), everything happens as if the substrate had an index of 3.08 at wavelength $\lambda_0$, giving rise to a significant increase in the associated reflection coefficient, which goes from 4% to more than 25% in energy terms. We shall return to this aspect in the section of this chapter dedicated to quarter-wave mirrors.

Notice that quarter-wave (matched) layers can also be defined for oblique incidence.

### Antireflective Layer

A layer whose properties of index and thickness allow it to satisfy the condition $r = 0$ is known as an antireflective layer at wavelength $\lambda_0$. This means that the complex quantity in the numerator in equation (6.145) must be identically 0, and hence that its real and imaginary parts both vanish at the same time, i.e., for a transparent layer

$$\begin{cases} (\tilde{n}_0 - \tilde{n}_s)\cos\delta_1 = 0 \\ (\tilde{n}_0\tilde{n}_s/\tilde{n}_1 - \tilde{n}_1)\sin\delta_1 = 0 \end{cases} . \tag{6.153}$$

Insofar as the incident medium and the substrate are different $(n_0 \neq n_s)$, the first condition requires the layer to be quarter-wave, while the second imposes a condition on its refractive index. At zero incidence, the properties of the resulting antireflective layer are as follows:

$$n_1 d_1 = (2q + 1)\frac{\lambda_0}{4} \quad \text{and} \quad n_1 = \sqrt{n_0 n_s}. \tag{6.154}$$

As illustrated in Figure 6.9, the energy reflection coefficient $R$ vanishes only

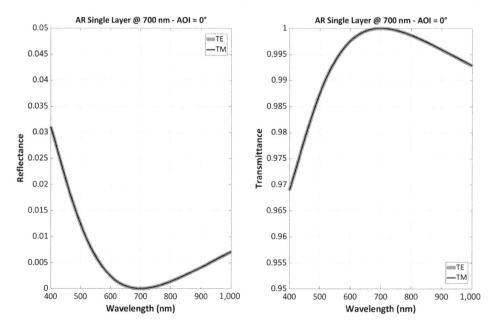

Figure 6.9 Spectral dependence of the reflectance and transmittance of a quarter-wave monolayer at 700 nm and optimal refractive index ($n_1 = \sqrt{n_0 n_s}$) at zero incidence.

at a single wavelength ($\lambda_0 = 700$ nm) and requires the use of a material with very low refractive index ($n_1 = 1.206$ if the substrate is silica) that cannot be found with common materials. This is why it is necessary to use at least two layers to achieve an antireflective coating that can be produced with common materials, for which reason we shall now be turning our attention to the optical properties of a double layer.

As previously, note that equation (6.154) can be directly *matched* for oblique incidence. To conclude, we stress the fact that antireflection layers are often known as low-index layers; this originates from relation (6.154), which gives a layer index lower than that of the substrate, provided that normal illumination is used with an air superstrate.

### 6.4.2 Double Layer

Figure 6.10 shows the configuration investigated.

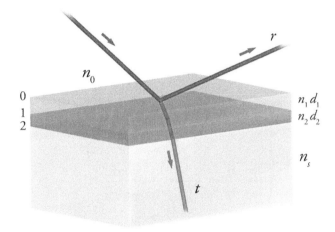

Figure 6.10 Double layer deposited onto the surface of a semi-infinite substrate.

We again use the ideas introduced in Section 6.3 to calculate the different characteristic variables for this particular transparent stacking, i.e.,

- Complex admittance of interface 2:

$$Y_2 = \tilde{n}_s. \tag{6.155}$$

- Complex admittance of interface 1:

$$Y_1 = \frac{\tilde{n}_s \cos \delta_2 - i \tilde{n}_2 \sin \delta_2}{\cos \delta_2 - i (\tilde{n}_s / \tilde{n}_2) \sin \delta_2} \quad \text{avec} \quad \delta_2 = \alpha_2 d_2. \tag{6.156}$$

- Complex admittance of interface 0:

$$Y_0 = \frac{[\tilde{n}_s \cos \delta_1 \cos \delta_2 - \tilde{n}_s \frac{\tilde{n}_1}{\tilde{n}_2} \sin \delta_1 \sin \delta_2] - i[\tilde{n}_1 \sin \delta_1 \cos \delta_2 + \tilde{n}_2 \sin \delta_2 \cos \delta_1]}{[\cos \delta_1 \cos \delta_2 - \frac{\tilde{n}_2}{\tilde{n}_1} \sin \delta_1 \sin \delta_2] - i[\frac{\tilde{n}_s}{\tilde{n}_1} \sin \delta_1 \cos \delta_2 + \frac{\tilde{n}_s}{\tilde{n}_2} \sin \delta_2 \cos \delta_1]}.$$

- transmission coefficient $t$:

$$\frac{2\tilde{n}_0}{t} = \left[ (\tilde{n}_0 + \tilde{n}_s) \cos \delta_1 \cos \delta_2 - \left( \frac{\tilde{n}_s \tilde{n}_1}{\tilde{n}_2} + \frac{\tilde{n}_0 \tilde{n}_2}{\tilde{n}_1} \right) \sin \delta_1 \sin \delta_2 \right]$$
$$- i \left[ \left( \tilde{n}_1 + \frac{\tilde{n}_0 \tilde{n}_s}{\tilde{n}_1} \right) \sin \delta_1 \cos \delta_2 + \left( \tilde{n}_2 + \frac{\tilde{n}_0 \tilde{n}_s}{\tilde{n}_2} \right) \sin \delta_2 \cos \delta_1 \right].$$

(6.157)

In order for this two-layer stack to be anti-reflective, the complex admittance $Y_0$ must be equal to the effective index of the incident medium $\tilde{n}_0$ and, separating the real and imaginary parts, the following two conditions must be satisfied:

$$\begin{cases} (\tilde{n}_s - \tilde{n}_0) \cos \delta_1 \cos \delta_2 - (\tilde{n}_s \dfrac{\tilde{n}_1}{\tilde{n}_2} - \tilde{n}_0 \dfrac{\tilde{n}_2}{\tilde{n}_1}) \sin \delta_1 \sin \delta_2 = 0 \\[2mm] (\tilde{n}_1 - \dfrac{\tilde{n}_0 \tilde{n}_s}{\tilde{n}_1}) \sin \delta_1 \cos \delta_2 + (\tilde{n}_2 - \dfrac{\tilde{n}_0 \tilde{n}_s}{\tilde{n}_2}) \sin \delta_2 \cos \delta_1 = 0 \end{cases}.$$

(6.158)

This gives

$$\begin{cases} \tan \delta_1 \tan \delta_2 = \dfrac{\tilde{n}_1 \tilde{n}_2 (\tilde{n}_s - \tilde{n}_0)}{\tilde{n}_s \tilde{n}_1^2 - \tilde{n}_0 \tilde{n}_2^2} \\[3mm] \dfrac{\tan \delta_1}{\tan \delta_2} = -\dfrac{\tilde{n}_1}{\tilde{n}_2} \cdot \dfrac{\tilde{n}_2^2 - \tilde{n}_0 \tilde{n}_s}{\tilde{n}_1^2 - \tilde{n}_0 \tilde{n}_s} \end{cases}.$$

(6.159)

and hence

$$\begin{cases} \tan^2 \delta_1 = \dfrac{\tilde{n}_1^2 (\tilde{n}_s - \tilde{n}_0)(\tilde{n}_2^2 - \tilde{n}_0 \tilde{n}_s)}{(\tilde{n}_0 \tilde{n}_2^2 - \tilde{n}_s \tilde{n}_1^2)(\tilde{n}_1^2 - \tilde{n}_0 \tilde{n}_s)} \\[3mm] \tan \delta_2 = \dfrac{\tilde{n}_1 \tilde{n}_2 (\tilde{n}_s - \tilde{n}_0)}{\tilde{n}_s \tilde{n}_1^2 - \tilde{n}_0 \tilde{n}_2^2} \cdot \dfrac{1}{\tan \delta_1} \end{cases}.$$

(6.160)

As opposed to the previous single layer antireflection coating, these relations now allow the two layer thicknesses versus the material indices to be adjusted. If the choice of the two constitutive materials of this stack is prescribed, namely tantalum pentoxide ($Ta_2O_5$, $n_2 = 2.130$ at $\lambda_0 = 700$ nm) and silicon dioxide ($SiO_2$, $n_1 = 1.489$ at $\lambda_0 = 700$ nm), there are two possible solutions that differ by the sign of $\tan \delta_1$, namely

- Solution 1: Air / 0.672L 1.645H / Silica
- Solution 2: Air / 1.328L 0.355H / Silica

The notation $bL$ in the formulas defining the two stack structures is given for a low refractive index layer (L) with optical thickness $n_L d_L = b\lambda_0/4$; similarly, the notation $hH$ is used for a high refractive index layer (H) with optical thickness $n_H d_H = h\lambda_0/4$.

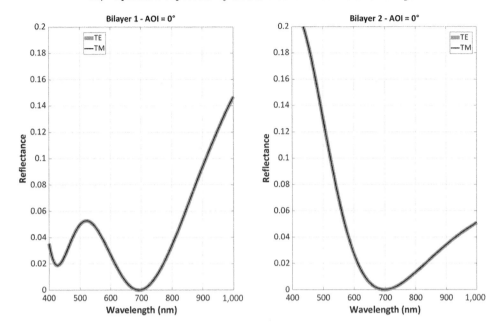

Figure 6.11 Spectral dependence of the reflection coefficient of the two anti-reflective double layers corresponding to Solutions 1 and 2 at zero angle of incidence (AOI).

We observe in Figure 6.11 that antireflection effectively occurs at a wavelength of 700 nm, but that the reflection coefficient rapidly increases either side of this particular wavelength and quickly exceeds the value of the reflection coefficient of the bare substrate, which is 3.5%.

This is why more than two layers (generally six to eight optimized layers) are required to maintain a low reflection coefficient over a broad spectral domain. More generally, design techniques must be used both to cover the widest spectral range and minimize reflection in this range.

### 6.4.3 Quarter-Wave Mirror

#### Definition and General Properties

Assume that a sequence of quarter-wave layers has been deposited onto a glass substrate, alternating from high to low index; the resultant stack can be described by the following compact formula:

$$\text{Air} \; / \; (\text{HL})^q \text{H} \; / \; \text{Substrate},$$

where, as introduced in the previous section, H and L denote layers of quarter-wave optical thickness of high and low refractive index. This quarter-wave stack hence comprises $p = 2q + 1$ layers.

Because of the particular value of phase term $\delta_j$ associated with these quarter-wave layers, it is possible to calculate analytically the reflection coefficient of such a stack at the central wavelength. Indeed the recurrence relation between complex admittances here takes the particular form $(\delta_j = \pi/2)$

$$Y_{j-1} = \frac{-i\tilde{n}_j}{-iY_j/\tilde{n}_j} \quad \Rightarrow \quad Y_{j-1}Y_j = \tilde{n}_j^2. \tag{6.161}$$

The complex admittance $Y_0$ can be expressed analytically as

$$Y_0 = \frac{\tilde{n}_H^2}{\tilde{n}_s}\left[\frac{\tilde{n}_H}{\tilde{n}_B}\right]^{2q}. \tag{6.162}$$

This shows that for both polarizations at normal incidence the modulus of $Y_0$ rapidly increases with $q$ if the ratio $n_H/n_L$ is large (we refer to the *index ratio*).

We can deduce the reflectance $R$ of this stack at wavelength $\lambda_0$ and observe how it changes as a function of the (odd) number $p$ of layers, i.e.,

$$R(\lambda_0) = \left|\frac{n_0 - Y_0}{n_0 + Y_0}\right|^2 \simeq 1 - 4\frac{n_0}{Y_0} = 1 - 4\frac{n_0 n_s}{n_H^2}\left[\frac{n_B}{n_H}\right]^{p-1}. \tag{6.163}$$

Table 6.1 summarizes the change in this coefficient $R(\lambda_0)$ as a function of the number $p$ of quarter-wave layers still with the two H and L materials (tantalum pentoxide and silicon dioxide) at the mirror central wavelength of 700 nm.

Table 6.1 Change in reflectance of a mirror $M_p$ with number $p$ of quarter-wave layers (H $\equiv$ Ta$_2$O$_5$, L $\equiv$ SiO$_2$)

| Number of layers $p$ | 3 | 5 | 7 | 9 | 11 | 13 | 15 |
|---|---|---|---|---|---|---|---|
| Reflectance $R$ (%) | 37.3 | 69.4 | 85.0 | 92.7 | 96.4 | 98.3 | 99.1 |

As the number of layers increases, the reflectance $R$ tends to 1, and values as high as 99.995% are almost routinely available these days. Such values, which cannot be obtained with metal mirrors, are essential for components such as the mirrors for ring laser gyroscopes (where the blind zone of rotation rate measurement is related to the losses), or to avoid degradation in catoptric systems under intense light (rise in temperature caused by absorption). Bear in mind, however, that as the number of layers increases, the reflection coefficient will be in practice decreased by absorption and scattering.

Figure 6.12 shows the variation of this energy reflection coefficient as a function of wavelength $\lambda$, where the number $p$ of layers is equal to 9. The modification at oblique incidence originates from the angular dependence of the effective indices. Matched layers and alternate designs can be introduced to improve this point.

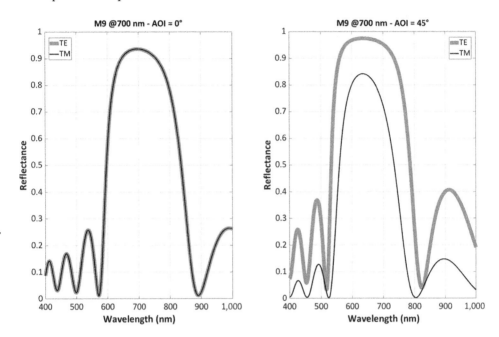

Figure 6.12 Spectral dependence of the reflectance of a quarter-wave mirror M9 centered at 700 nm for two different angles of incidence (0 and 45 degrees) and both states of polarization.

### Field Distribution Within the Stack

To calculate the tangential and normal components of the electric field within the depth of the stack corresponding to this M9 quarter-wave mirror, we shall here apply the relations of Section 6.3.10.

Figure 6.13 shows how the modulus squared of the total electric field normalized to that of the incident field varies with depth $z$ within the stack for both states of polarization and two different angles of incidence (0 and 45 degrees). These curves are drawn at the wavelength that corresponds to the maximum of the reflectance for each angle of incidence (i.e., 700 nm for normal incidence and 643 nm for 45 degrees).

Note that for TE polarization the field intensity is maximum at HL interfaces and minimum at LH interfaces. The highest values are in the upper

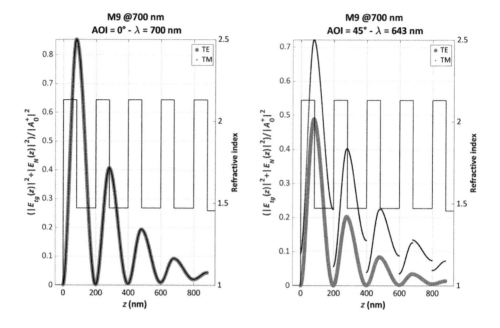

Figure 6.13 Variation with stack depth $z$ of the normalized modulus squared of the field value (TE, gray dots; TM, black dots; left axis) and of the refractive index (continuous black curve, right axis), for a M9 quarter-wave mirror centered at 700 nm, for two angles of incidence (0 and 45 degrees). The incident field arrives from the left.

stack and decrease toward the substrate (the reduction between two maxima is the square of the index ratio), highlighting a damped periodic behavior. For TM polarization the behavior is similar but at oblique incidence, discontinuities can be seen at the interfaces, which is due to the normal field component.

### 6.4.4 Fabry–Perot cavity

#### Definition and General Properties

A Fabry–Perot cavity is a passband filter consisting in two identical quarter-wave mirrors enclosing a half-wave layer at the central wavelength of the mirrors. One formula for such a filter is therefore

$$\text{Air} \, / \, (\text{HL})^q \text{H} \, 2\text{L} \, \text{H}(\text{LH})^q \, / \, \text{Substrate},$$

and the spectral dependence of the transmittance is shown in Figure 6.14 for nine-layer mirrors at two different angles of incidence (0 and 45 degrees).

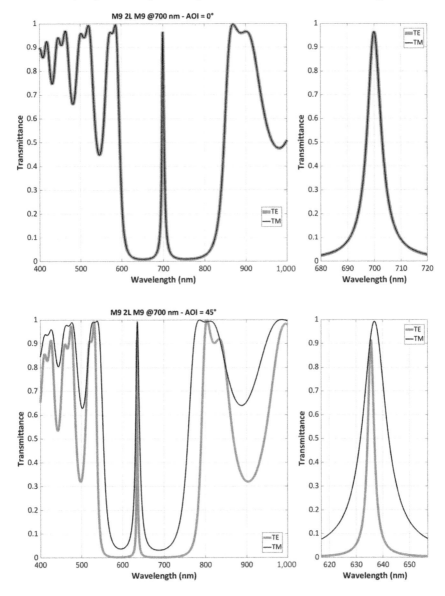

Figure 6.14 Spectral transmittance of a M9 2L M9 Fabry–Perot cavity centered at 700 nm for two different angles of incidence (0 and 45 degrees) and both states of polarization.

At zero incidence we observe a narrow transmission peak at the central wavelength $\lambda_0 = 700$ nm, bounded on either side by two deep rejection zones. Beyond these rejection zones, the transmission coefficient increases and exhibits an oscillatory structure characteristic of this type of component.

To eliminate these, it is necessary to complete the design of such a filter with blocking mirrors that reduce the transmission factor down to low values over the whole spectral range of use (here from 400 to 1,000 nm), obviously outside the domain of wavelengths that correspond to the filter passband.

If the angle of incidence is nonzero, here 45 degrees, the central wavelength of the filter shifts toward shorter wavelengths, while the transmission profiles become functions of the polarization state of the light.

Stacking several Fabry–Perot cavities allows a more square transmittance profile in the passband to be obtained, as illustrated in Figure 6.15 by a two-cavity configuration (synthetic formula Air / M9 2L M9 L M9 2L M9 / Substrate) at normal incidence.

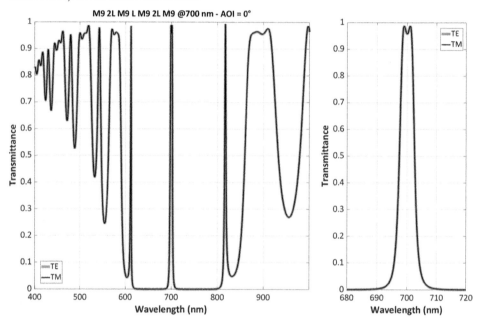

Figure 6.15 Spectral transmittance of a dual Fabry–Perot cavity (M9 2L M9 L M9 2L M9) centered at 700 nm (normal incidence).

### Field Distribution Within the Stack

Using the same approach as that described in Section 6.4.3 for a quarter-wave mirror, we can compute the $z$-dependence of the modulus squared of the total electric field normalized to that of the incident field (see Figure 6.16). Note the particularly high value of this ratio at the center of the 2L spacer [for TE polarization, roughly 25 (respectively 35) at 0 (respectively 45) degree(s) of incidence]. This field magnification reflects the phenomenon of resonance which occurs within this Fabry–Perot cavity at a

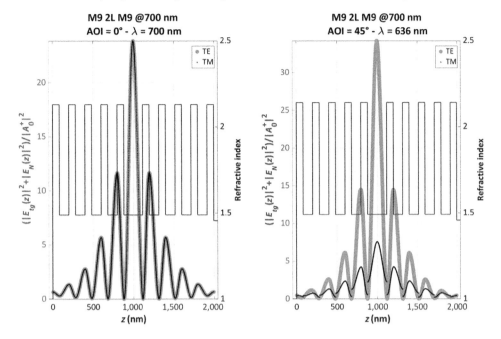

Figure 6.16 Variation with stack depth $z$ of the normalized modulus squared of the field value (TE, gray dots; TM, black dots; left axis) and of the refractive index (continuous black curve, right axis), for a M9 2L M9 Fabry–Perot cavity centered at 700 nm, for two angles of incidence (0 and 45 degrees).

specific wavelength, namely the central wavelength (700 nm) at normal incidence, and a slightly shifted wavelength (636 nm) at 45 degrees of incidence. We shall return to such resonances in greater detail in Chapter 7.

### *Absorbance Properties*

We showed in Section 6.3.11 how the absorbance of a stack can be computed by three different approaches, namely the direct application of the principle of conservation of energy, the difference of the Poynting vector flux through the top and the bottom layers of the stack, and finally the direct computation of the $z$-dependence of the modulus squared of the total electric field and its integration over the whole stack thickness.

Figure 6.17 shows the results of this computation, for a M9 2L M9 Fabry–Perot cavity, both states of polarization, and two angles of incidence (0 and 45 degrees); the spectral dependence of the real and imaginary parts of the complex refractive index of the materials that we used is shown in Figure 6.18.

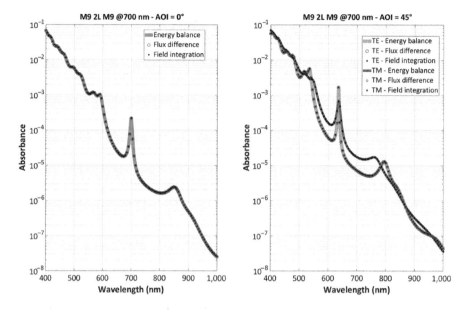

Figure 6.17 Spectral dependence of the absorbance for a M9 2L M9 Fabry–Perot cavity centered at 700 nm, for two angles of incidence (0 and 45 degrees) [computation achieved by three different independent approaches (energy balance, flux difference, and square field integration) for both states of polarization].

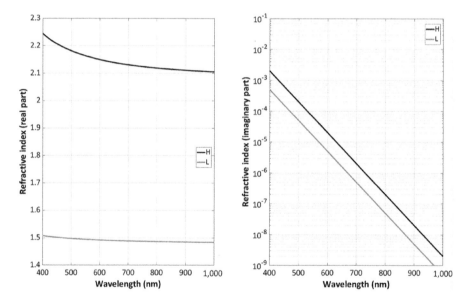

Figure 6.18 Spectral dependence of the real and imaginary parts of the refractive index of H (black continuous line) and L (gray continuous line) materials.

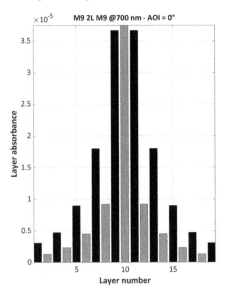

Figure 6.19 Absorbance in each layer of the M9 2L M9 Fabry–Perot at the resonance wavelength for normal incidence (H layers in black, L layers in gray).

The influence on absorption of the field magnification around the resonance wavelength is clearly seen. We also observe the perfect agreement between the values obtained using the three independent calculation approaches, despite the very low values of the corresponding absorbance.

Finally, Figure 6.19 shows the amount of absorbance in each layer of the stack at the resonance wavelength (H layers in black and L layers in gray).

### *6.4.5 Complex Interference Filters*

Since the end of the 1990s, the simultaneous availability of powerful optical filter design software and reliable thin-film deposition techniques using energetic processes such as ion beam assistance, ion beam sputtering, or magnetron sputtering have made possible the manufacture of high-performance optical interference coatings comprising a great number of layers, from one hundred up to a few thousand. During the past 20 years, a large number of high-performance filtering functions have also been developed for demanding applications such as observation of the Earth from space (stripe bandpass filters), wavelength multiplexing of high data rate optical telecommunications channels (narrow bandpass filters, gain flattening filters), study of the organization of living matter at the cellular scale (short-wave pass and long-wave pass filters), or interferometric detection of gravitational waves (ultra low loss mirrors).

The design, manufacture, and control of all these complex filtering functions are based on the theoretical background described throughout this chapter.

## 6.5  Admittance Locus

To better understand the construction of a filtering function and the consequences of growing a layer on the instantaneous optical features of a multilayer stack, it is useful to analyze the evolution of the complex admittance of the top interface of the stack during the deposition of the various layers. This is known as defining the admittance locus of this stack.

To illustrate this approach, consider a stack of $p$ transparent alternating high and low index layers deposited onto a semi-infinite substrate. Let $Y_{j-1}(d)$ be the complex admittance of this stack when a thickness $d$ of layer $j$ is deposited on the previous layers that constitute the substack.

We take the illumination incidence to be 0. Using the recurrence relation between complex admittances we can write

$$Y_{j-1}(\delta) = \frac{Y_j \cos\delta - in_j \sin\delta}{\cos\delta - i(Y_j/n_j)\sin\delta} \quad \text{with} \quad \delta = \frac{2\pi}{\lambda}n_j(\lambda)d, \qquad (6.164)$$

where we have replaced thickness $d$ by the effective variable $\delta$.

This function $Y_{j-1}(\delta)$ is a $\pi$ periodic function, and furthermore satisfies the initial condition: $Y_{j-1}(0) = Y_j$.

Let $(x, y)$ be the coordinates of the point representative of the admittance $Y_{j-1}(\delta)$ in the complex plane, and let $(x_j, y_j)$ be those for the starting admittance $Y_j$; then equation (6.164) can be written:

$$x + iy = n_j \frac{(x_j + iy_j)\cos\delta - in_j \sin\delta}{n_j \cos\delta - i(x_j + iy_j)\sin\delta}$$

$$= n_j \frac{x_j \cos\delta + i(y_j \cos\delta - n_j \sin\delta)}{(n_j \cos\delta + y_j \sin\delta) - ix_j \sin\delta}, \qquad (6.165)$$

or, separating the real and imaginary parts:

$$\begin{cases} x(n_j \cos\delta + y_j \sin\delta) + yx_j \sin\delta = n_j x_j \cos\delta \\ y(n_j \cos\delta + y_j \sin\delta) - xx_j \sin\delta = n_j(y_j \cos\delta - n_j \sin\delta) \end{cases}. \qquad (6.166)$$

It is sufficient to eliminate $\cos\delta$ and $\sin\delta$ to reveal the equation for the locus of points described by the complex admittance (Admittance locus) of the stack during the course of deposition of layer $j$, i.e.,

$$x^2 + y^2 - \frac{x}{x_j}(x_j^2 + y_j^2 + n_j^2) + n_j^2 = 0 \, . \tag{6.167}$$

Slightly rearranging this equation in the form

$$\left( x - \frac{x_j^2 + y_j^2 + n_j^2}{2x_j} \right)^2 + y^2 = \frac{(x_j^2 + y_j^2 + n_j^2)^2}{4x_j^2} - n_j^2 \, , \tag{6.168}$$

we see that this curve is a circle of center $C_j$ and radius $R_j$ defined by

$$C_j \left( \frac{x_j^2 + y_j^2 + n_j^2}{2x_j}, 0 \right) \quad ; \quad R_j = \sqrt{\frac{(x_j^2 + y_j^2 + n_j^2)^2}{4x_j^2} - n_j^2} \, . \tag{6.169}$$

Figure 6.20 shows this *admittance locus* in the particular case of a stack reduced to two layers, the first being of type 0.5H (black dots) and the second of type 1.5L (gray dots). Using the sign conventions adopted in the definitions of the refractive indices, the direction followed by these curves is anticlockwise.

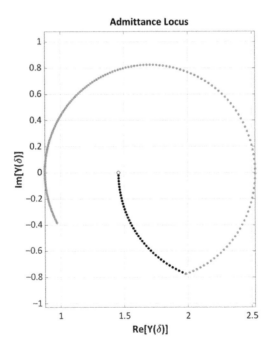

**Admittance Locus**

Figure 6.20 Locus of points taken by the complex admittance of a 0.5H 1.5L stack during the course of its deposition onto a silica substrate (silica substrate, black circle; H layer, black dots; L layer, gray dots).

Note that the density of points along these curves is not uniform (for example, the points are much closer together on the left-hand side of the gray circle than on the right-hand side), while two consecutive points are still separated by the deposition of one nanometer thickness of material, of high or low refractive index. This is why the circles are not traced out with a constant angular step.

## 6.6 Influence of the Rear Face of the Substrate

Until now, we have only considered the case where the two extreme media (incident medium and substrate) are semi-infinite. Clearly, however, this situation does not correspond with reality, and it is now necessary to examine how to tackle the case of a substrate of finite thickness $d_s$, as shown in Figure 6.21.

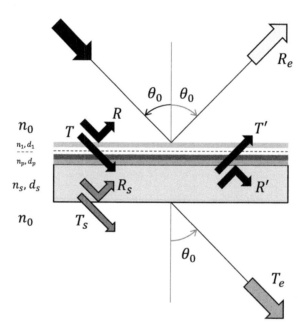

Figure 6.21 Monochromatic plane wave impinging at incidence $\theta_0$ on a stack of thin optical layers deposited onto the surface of a substrate of thickness $d_s$ with plane parallel faces.

The presence of the rear face of this substrate of finite thickness will essentially have two effects: on the one hand, it will generate the appearance of a retrograde wave in the substrate (energy reflection coefficient $R_s$) and on the other hand, attenuate the progressive wave (energy transmission coefficient $T_s$). In turn, the substrate will therefore harbor a stationary wave.

There are two possible approaches to address this problem, though they lead ultimately to identical results: the so-called **coherent** and **incoherent** approaches.

### 6.6.1 Coherent Approach

In the *coherent* approach, the substrate is regarded as a constitutive layer of the stack, and the final exterior medium of index $n_0$ is taken to be the effective substrate of the new configuration thus constituted. The stack therefore now contains $(p + 1)$ layers, with the last one having a thickness much greater than all the others. The consequence is that the phase term $\delta_{p+1}$ varies rapidly with wavelength. Under normal incidence, we can write

$$\delta_{p+1} = \alpha_{p+1} d_{p+1} = \frac{2\pi}{\lambda} n_s d_s = \delta_s . \tag{6.170}$$

To fix our ideas, if we consider a silica substrate ($n_s \approx 1.5$) of thickness $d_s = 1$ mm, the period of spectral modulation associated with the crossing of the substrate is $\Delta\lambda = \lambda^2/2\, n_s\, d_s = 0.12$ nm if the wavelength is taken to be 600 nm. How apparent this phenomenon of parasitic interference is will be essentially a function of the spectral resolution of the measuring device used. Now as far as calculation is concerned, recall that the approach developed throughout this chapter corresponds to the harmonic regime, in other words that the unidirectional waves considered are infinitely coherent (since they are monochromatic) and that any interference that occurs is always characterized by a 100% contrast, regardless of the path differences involved.

However, we obviously cannot guarantee that the two faces of the substrate are perfectly parallel. This residual error in parallelism $\alpha$ will have the effect of reducing the effective visibility of this interference phenomenon, unless the variation in path difference associated with traversing the substrate does not exceed $\lambda/10$ within the footprint of the beam. If $D$ is the transverse dimension of this beam, this condition is written

$$n_s D \alpha < \frac{\lambda}{10} , \tag{6.171}$$

i.e., a parallelism better than $4 \times 10^{-6}$ radians (0.8 arcsecond) if the diameter $D$ of the beam is taken to be 10 mm. The substrates we are using have errors in parallelism that are typically on the order of 1 arcminute, which is much larger; therefore the effective visibility of this coherent effect will be greatly reduced, regardless of the spectral resolution of the measuring equipment used.

We can describe these two phenomena by averaging the spectral response of the component over an equivalent $2\pi$ period, i.e.,

$$T_e(\lambda) = \frac{1}{2\pi} \int\limits_{\delta_s - \pi}^{\delta_s + \pi} T_{\mathrm{coh}}(\lambda, \delta) \, d\delta \quad \text{and} \quad R_e(\lambda) = \frac{1}{2\pi} \int\limits_{\delta_s - \pi}^{\delta_s + \pi} R_{\mathrm{coh}}(\lambda, \delta) \, d\delta.$$

$$(6.172)$$

It remains to calculate the coherent transmission and reflection coefficients $T_{\mathrm{coh}}$ and $R_{\mathrm{coh}}$ that appear in equation (6.172), which is achieved by using the following set of equations:

- Initialization of the recurrence relation between complex admittances:

$$Y_{p+1} = \tilde{n}_0. \qquad (6.173)$$

- Recursive calculation of the complex admittance of the stack:

$$Y_p = \frac{\tilde{n}_0 \cos \delta_s - i \tilde{n}_s \sin \delta_s}{\cos \delta_s - i(\tilde{n}_0/\tilde{n}_s) \sin \delta_s}$$

$$\cdots \qquad (6.174)$$

$$Y_0 = \frac{Y_1 \cos \delta_1 - i \tilde{n}_1 \sin \delta_1}{\cos \delta_1 - i(Y_1/\tilde{n}_1) \sin \delta_1},$$

- Calculation of reflection coefficients:

$$r = \frac{\tilde{n}_0 - Y_0}{\tilde{n}_0 + Y_0} \quad \text{and} \quad R_{\mathrm{coh}} = |r|^2. \qquad (6.175)$$

- Calculation of transmission coefficients:

$$t = \frac{1 + r}{\prod\limits_{j=1}^{p+1} \left( \cos \delta_j - i \frac{Y_j}{\tilde{n}_j} \sin \delta_j \right)} \quad \text{and} \quad T_{\mathrm{coh}} = |t|^2, \qquad (6.176)$$

since the extreme media are identical.

### 6.6.2 Incoherent Approach

In this second, so-called *incoherent*, approach we shall assume that the various waves generated by the reflection on the rear face of the substrate do not interfere with all the others because of their rapid phase variations and that it is therefore possible to sum their contributions in energy without considering phase terms.

As illustrated in Figure 6.21, we shall now denote as

- $R$ (respectively $T$), the reflection (respectively transmission) coefficient of a progressive wave at the stack initially considered (semi-infinite incident medium of index $n_0$, semi-infinite substrate of index $n_s$)

- $R'$ (respectively $T'$), the reflection (respectively transmission) coefficient of a retrograde wave at the stack initially considered (semi-infinite incident medium of index $n_s$, semi-infinite substrate of index $n_0$)

- $R_s$ (respectively $T_s$), the reflection (respectively transmission) coefficient of a progressive wave at the rear face of the substrate (semi-infinite incident medium of index $n_s$, semi-infinite substrate of index $n_0$)

Summing incoherently the different contributions to the reflected and transmitted fluxes, we get immediately

- The energy reflection coefficient of the ensemble comprising stack plus finite-thickness substrate:

$$
\begin{aligned}
R_e &= R + TR_sT' + TR_sR'R_sT' + \cdots \\
&= R + TR_sT'\{1 + R_sR' + \cdots\} \\
&= R + \frac{TT'R_s}{1 - R_sR'} .
\end{aligned}
\tag{6.177}
$$

- The energy transmission coefficient of this same ensemble:

$$
\begin{aligned}
T_e &= TT_s + TR_sR'T_s + \cdots \\
&= TT_s\{1 + R_sR' + \cdots\} \\
&= \frac{TT_s}{1 - R_sR'} .
\end{aligned}
\tag{6.178}
$$

The analytical expressions for the coefficients $R$ and $T$ have already been established in Sections 6.3.9 and 6.3.11. Those for the coefficients $R'$ and $T'$ can be deduced by reversing the direction of light propagation (the waves being considered are retrograde, as emphasized Section 6.3.7) and exchanging the roles of the media with indices $n_0$ and $n_s$ (the initially incident medium becomes the substrate and vice versa). Hence, we get

- Initialization of the recurrence relation between complex admittances:

$$
Y_0' = -\tilde{n}_0 .
\tag{6.179}
$$

- Recursive calculation of the complex admittance $Y_s' = Y_p'$ of the stack:

$$Y_1' = \frac{Y_0' \cos \delta_1 + i\tilde{n}_1 \sin \delta_1}{\cos \delta_1 + i(Y_0'/\tilde{n}_1) \sin \delta_1}$$

$$\cdots$$

$$Y_p' = \frac{Y_{p-1}' \cos \delta_p + i\tilde{n}_p \sin \delta_p}{\cos \delta_p + i(Y_{p-1}'/\tilde{n}_p) \sin \delta_p}$$

(6.180)

- Calculation of the reflection coefficients:

$$r' = \frac{(-\tilde{n}_s) - Y_p'}{(-\tilde{n}_s) + Y_p'} \quad \text{et} \quad R' = |r'|^2 .$$

(6.181)

- Calculation of the transmission coefficients:

$$t' = \frac{1 + r'}{\prod\limits_{j=1}^{p} \left(\cos \delta_j + i\frac{Y_{j-1}'}{\tilde{n}_j} \sin \delta_j\right)} \quad \text{et} \quad T' = \frac{\Re[\tilde{n}_0]}{\Re[\tilde{n}_s]} |t'|^2 .$$

(6.182)

It remains finally to establish analytical expressions for the reflection and transmission coefficients $R_s$ and $T_s$ of the rear face of the substrate, i.e.,

$$r_s = \frac{\tilde{n}_s - \tilde{n}_0}{\tilde{n}_s + \tilde{n}_0} \quad \Rightarrow \quad R_s = \left|\frac{\tilde{n}_s - \tilde{n}_0}{\tilde{n}_s + \tilde{n}_0}\right|^2 ,$$

(6.183)

and

$$t_s = 1 + r_s = \frac{2\tilde{n}_s}{\tilde{n}_s + \tilde{n}_0} \quad \Rightarrow \quad T_s = \frac{\Re[\tilde{n}_0]}{\Re[\tilde{n}_s]} \left|\frac{2\tilde{n}_s}{\tilde{n}_s + \tilde{n}_0}\right|^2 .$$

(6.184)

We now have all the relations necessary for calculating the spectral transmittance $T_e(\lambda)$ of a stack deposited on a substrate of finite thickness, adopting either the so-called coherent or incoherent approach. We shall now compare the results given by these two approaches for the particular case where the stack reduces to a single layer.

### 6.6.3 Comparison Between the Two Approaches

Consider a single layer of $Ta_2O_5$ ($n_1 = 2.15$ at 600 nm) 500 nm thick, deposited onto the surface of a silica substrate of 1 mm; we shall calculate the spectral transmittance $T(\lambda)$ of this ensemble at zero incidence.

Figure 6.22 shows the result obtained by adopting the coherent approach (left-hand graph, gray curve), as well as that obtained by adopting the incoherent approach (right-hand graph, black curve); the result obtained from the coherent approach after integration over a range of wavelengths corresponding to a phase variation of $2\pi$ is shown on the same right-hand graph (gray dots).

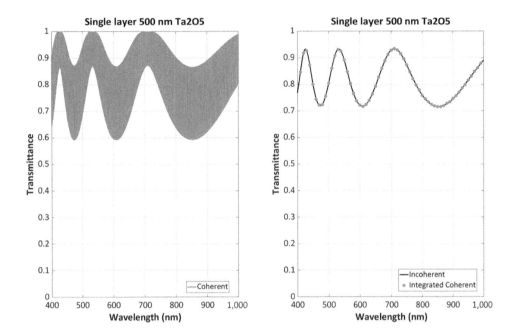

Figure 6.22 Spectral dependence of the transmission of a 500-nm layer of $Ta_2O_5$ deposited onto a 1-mm-thick silica substrate [left, gray curve, coherent approach; right, incoherent approach (black curve) and the result of integrating over $2\pi$ the result provided by the coherent approach (gray dots)].

We note on the left graph that the transmission coefficient obtained from the coherent approach exhibits extremely rapid variations with wavelength; the graphical representation cannot resolve these variations (this is directly related to the large ratio of silica substrate thickness to wavelength). Note that this transmission passes through maximum values of 100% at the wavelengths for which the $Ta_2O_5$ layer is *absent* and where the transmission of the Fabry–Perot formed by the silica substrate is equal to 100%. Indeed, the maximum transmission of a symmetrical Fabry–Perot filter is always 1.

If we now take account of the imperfect parallelism of this substrate and calculate the mean value defined by the integral in equation (6.172), the results (right-hand graph, gray dots) are significantly different, particularly in terms of the maximum values of transmission, but they are a perfect fit to those obtained by adopting the incoherent approach (right-hand graph, black curve).

For this reason this incoherent approach (which is simpler to implement) is most often used to describe the performance of a stack deposited onto a substrate of finite thickness. Notice finally that similar incoherent effects may result from the divergence of the incident beam.

## 6.7 Reciprocity Theorem

One important property of interference filters is that their transmittance does not depend on whether the illumination comes from the superstrate (progressive illumination) or from the substrate (retrograde illumination), that is, $T = T'$. Though this property classically results from the general reciprocity theorem, it is interesting to write below a proof that relies on the thin-film formalism.

To this end, recall first that the fields at the top and bottom surfaces of the device are connected as (see Section 6.3.6):

$$
\begin{bmatrix} \vec{z} \wedge \vec{\mathbb{E}}_{0,\text{tg}}(z_0) \\ \vec{\mathbb{H}}_{0,\text{tg}}(z_0) \end{bmatrix} = M \begin{bmatrix} \vec{z} \wedge \vec{\mathbb{E}}_{p,\text{tg}}(z_p) \\ \vec{\mathbb{H}}_{p,\text{tg}}(z_p) \end{bmatrix}, \tag{6.185}
$$

with

$$
M = \begin{bmatrix} a & b \\ c & d \end{bmatrix} = \prod_{j=1}^{p} M_j \quad \text{and} \quad M_j = \begin{bmatrix} \cos \delta_j & -\dfrac{i}{\tilde{n}_j} \sin \delta_j \\ -i\tilde{n}_j \sin \delta_j & \cos \delta_j \end{bmatrix}. \tag{6.186}
$$

Consequently, we have

$$
\det(M) = ad - bc = \det\left( \prod_{j=1}^{p} M_j \right) = \prod_{j=1}^{p} \det(M_j) = 1. \tag{6.187}
$$

Note that the matrix $M$ does not depend on the nature (progressive or retrograde) of the illumination.

This allows us to write, for each state of polarization:

$$
\vec{z} \wedge \vec{\mathbb{E}}_{0,\text{tg}}(z_0) = a[\vec{z} \wedge \vec{\mathbb{E}}_{p,\text{tg}}(z_p)] + b\vec{\mathbb{H}}_{p,\text{tg}}(z_p) = (a + bY_p)[\vec{z} \wedge \vec{\mathbb{E}}_{p,\text{tg}}(z_p)] \tag{6.188}
$$

and

$$
Y_0 = \frac{c + dY_p}{a + bY_p}. \tag{6.189}
$$

Now for a progressive illumination, we have

$$
\vec{z} \wedge \vec{\mathbb{E}}_{0,\text{tg}}(z_0) = (1 + r)\,[\vec{z} \wedge \vec{\mathbb{A}}_{0,\text{tg}}^{+}] \quad ; \quad \vec{z} \wedge \vec{\mathbb{E}}_{p,\text{tg}}(z_p) = t\,[\vec{z} \wedge \vec{\mathbb{A}}_{0,\text{tg}}^{+}], \tag{6.190}
$$

and

$$Y_p = \tilde{n}_s \quad ; \quad r = \frac{\tilde{n}_0 - Y_0}{\tilde{n}_0 + Y_0}, \tag{6.191}$$

so that relations (6.188) and (6.189) can be written as

$$1 + r = (a + b\tilde{n}_s)\, t \quad ; \quad Y_0 = \frac{c + d\tilde{n}_s}{a + b\tilde{n}_s}. \tag{6.192}$$

In a similar way, we get for the retrograde illumination:

$$\vec{z} \wedge \vec{\mathbb{E}}_{p,\text{tg}}(z_p) = (1 + r')\,[\vec{z} \wedge \vec{\mathbb{A}}^-_{p,\text{tg}}] \quad ; \quad \vec{z} \wedge \vec{\mathbb{E}}_{0,\text{tg}}(z_0) = t'\,[\vec{z} \wedge \vec{\mathbb{A}}^-_{p,\text{tg}}], \tag{6.193}$$

and

$$Y_0' = -\tilde{n}_0 \quad ; \quad r' = \frac{-\tilde{n}_s - Y_p'}{-\tilde{n}_s + Y_p'}, \tag{6.194}$$

so that relations (6.188) and (6.189) can be written as

$$t' = (a + bY_p')\,(1 + r') \quad ; \quad -\tilde{n}_0 = \frac{c + dY_p'}{a + bY_p'}. \tag{6.195}$$

Observe that in all these relations $Y_0$ and $Y_p'$ are respectively calculated from the substrate ($\tilde{n}_s$) and the air ($-\tilde{n}_0$) effective indices.

From (6.192) and (6.195), we write the transmission ratio $t/t'$ as

$$\frac{t}{t'} = \frac{1 + r}{1 + r'}\, \frac{1}{(a + b\tilde{n}_s)(a + bY_p')}. \tag{6.196}$$

Using the relationship between reflection and admittance, and equation (6.192), we get

$$1 + r = \frac{2\tilde{n}_0}{\tilde{n}_0 + Y_0} \quad \Rightarrow \quad 1 + r = \frac{2\tilde{n}_0(a + b\tilde{n}_s)}{(c + d\tilde{n}_s) + \tilde{n}_0(a + b\tilde{n}_s)}. \tag{6.197}$$

In a similar way, we have

$$1 + r' = \frac{-2\tilde{n}_s}{-\tilde{n}_s + Y_p'}. \tag{6.198}$$

Here $Y_p'$ must be extracted from (6.195), i.e.,

$$Y_p' = -\frac{c + a\tilde{n}_0}{d + b\tilde{n}_0}, \tag{6.199}$$

to express (6.198) as

$$1 + r' = \frac{2\tilde{n}_s(d + b\tilde{n}_0)}{(c + a\tilde{n}_0) + \tilde{n}_s(d + b\tilde{n}_0)}. \tag{6.200}$$

In relation (6.196), $Y_p'$ must also be introduced to write

$$a + bY_p' = \frac{ad - bc}{d + b\tilde{n}_0} = \frac{1}{d + b\tilde{n}_0} . \tag{6.201}$$

Finally, substituting (6.197), (6.200), and (6.201) in (6.196) we get

$$\frac{t}{t'} = \frac{\tilde{n}_0}{\tilde{n}_s} , \tag{6.202}$$

so that the transmission amplitudes have identical arguments (extreme media are not absorbing).

Now, as far as energy is concerned, we have

$$\frac{T}{T'} = \frac{\frac{\Re[\tilde{n}_s]}{\Re[\tilde{n}_0]}|t|^2}{\frac{\Re[\tilde{n}_0]}{\Re[\tilde{n}_s]}|t'|^2} = \left[\frac{\tilde{n}_s}{\tilde{n}_0}\right]^2 \left|\frac{t}{t'}\right|^2 = 1 , \tag{6.203}$$

which shows that the transmittance does not depend on the nature (progressive or retrograde) of the illumination.

## 6.8 Phase Closure Relation

We saw that the principle of conservation of energy imposes a closure relation between energetic-related variables describing the optical properties of a thin-film stack, namely $R + T + A = 1$. The same relation can be written for an illumination coming from the substrate, i.e., $R' + T' + A' = 1$. As demonstrated in the previous section, transmittance $T$ and $T'$ are equal. So, for transparent materials ($A = 0$), we have $R + T = 1 = R' + T$ and reflectance $R$ and $R'$ are also equal.

Still with transparent materials, the same fundamental principle actually imposes another closure relation to the arguments of the complex coefficients $r$, $r'$, $t$ and $t'$, which is less known.

To establish this phase closure relation, we can consider two different ways, the first being analytical and the second phenomenological.

### 6.8.1 Analytical Approach

Let $X$ and $X'$ be the variables (fields or admittances) that relate respectively to illumination from the incident medium (or superstrate) and from the substrate. To establish the relationship between these variables, it is useful to turn our attention to the *reciprocity* relations that emerge when the divergence of field vector products is expanded.

Recall that the complex harmonic fields $(\mathbb{E}, \vec{\mathcal{H}})$ and $(\vec{\mathcal{E}'}, \vec{\mathcal{H}'})$ satisfy Maxwell's equations in the same multilayer media; these can be written in the form

$$\mathbf{curl}\ \vec{\mathcal{E}} = i\omega\tilde{\mu}\vec{\mathcal{H}} \quad ; \quad \mathbf{curl}\ \vec{\mathcal{H}} = -i\omega\tilde{\epsilon}\vec{\mathcal{E}} \qquad (6.204)$$

$$\mathbf{curl}\ \vec{\mathcal{E}'} = i\omega\tilde{\mu}\vec{\mathcal{H}'} \quad ; \quad \mathbf{curl}\ \vec{\mathcal{H}'} = -i\omega\tilde{\epsilon}\vec{\mathcal{E}'}\,. \qquad (6.205)$$

Now consider the two vector quantities defined by

$$\vec{P}_1 = \vec{\mathcal{E}} \wedge \vec{\mathcal{H}'} - \vec{\mathcal{E}'} \wedge \vec{\mathcal{H}} \quad ; \quad \vec{P}_2 = \vec{\mathcal{E}}^* \wedge \vec{\mathcal{H}'} + \vec{\mathcal{E}'} \wedge \vec{\mathcal{H}}^*\,. \qquad (6.206)$$

Recall that

$$\mathrm{div}(\vec{A} \wedge \vec{B}) = \vec{B}.\mathbf{curl}\ \vec{A} - \vec{A}.\mathbf{curl}\ \vec{B}\,. \qquad (6.207)$$

Combining relations (6.204), (6.205), (6.206), and (6.207), we get

$$\mathrm{div}\vec{P}_1 = 0 \qquad (6.208)$$

$$\mathrm{div}\vec{P}_2 = i\omega(\tilde{\mu} - \tilde{\mu}^*)\vec{\mathcal{H}}^*.\vec{\mathcal{H}'} + i\omega(\tilde{\epsilon} - \tilde{\epsilon}^*)\vec{\mathcal{E}}^*.\vec{\mathcal{E}'}\,. \qquad (6.209)$$

Equation (6.208) is true in the general case (absorbing multilayer), while the divergence of the second vector is 0 only for a transparent multilayer. We shall assume from now on that this transparency condition is fulfilled, so we can then write

$$\mathrm{div}\vec{P}_1 = 0 \quad ; \quad \mathrm{div}\vec{P}_2 = 0\,. \qquad (6.210)$$

Hence, the fluxes of $\vec{P}_1$ and $\vec{P}_2$ are conservative. For a multilayer with a planar structure we obtain, by integrating over a volume $\Omega$ bounded by a closed surface $\Sigma$ whose sides are parallel with the interfaces of the multilayer,

$$\iiint_\Omega \mathrm{div}\vec{P}_i\,dxdydz = \iint_\Sigma \vec{z}.\vec{P}_i\,dxdy = 0 \quad i = 1, 2\,. \qquad (6.211)$$

Ultimately, this means that the quantities $\vec{z}.\vec{P}_i$ are constant in the stack of thin layers, i.e.,

$$\vec{z}.\vec{P}_1 = \vec{z}.[\vec{\mathcal{E}} \wedge \vec{\mathcal{H}'} - \vec{\mathcal{E}'} \wedge \vec{\mathcal{H}}] = C_1 \qquad (6.212)$$

$$\vec{z}.\vec{P}_2 = \vec{z}.[\vec{\mathcal{E}}^* \wedge \vec{\mathcal{H}'} + \vec{\mathcal{E}'} \wedge \vec{\mathcal{H}}^*] = C_2\,. \qquad (6.213)$$

In this chapter (Section 6.3.3), we saw that in the projections of vector products onto the $z$-axis, only the field tangential components were involved. On the other hand, in the interests of simplification, we shall henceforth assume that the incident light is perfectly collimated (with no divergence) and therefore contain just one spatial frequency. This avoids our having to work with a spatial wave packet and means that we can make direct use of the notion of admittance as defined in Section 6.3.5.

Hence it is possible at this stage to write

$$\vec{H}_{tg} = Y\,\vec{z} \wedge \vec{E}_{tg} \quad ; \quad \vec{H}'_{tg} = Y'\,\vec{z} \wedge \vec{E}'_{tg}\,. \tag{6.214}$$

In these expressions, admittance $Y$ is given for illumination on the superstrate side; it is therefore related to an outgoing (progressive) wave condition in the substrate, and is calculated by recurrence starting from an initial value corresponding to the effective index of this substrate, i.e., $Y(z) = Y(z, \tilde{n}_s)$. In a symmetrical manner, the admittance $Y'$ is given for illumination on the substrate side; it is therefore related to an outgoing (retrograde) wave condition in the superstrate, and is calculated by recurrence starting from an initial value corresponding to the effective index of the superstrate for this direction of propagation, i.e., $Y'(z) = Y'(z, -\tilde{n}_0)$.

If we expand relations (6.212) and (6.213) using (6.214), we get

$$[Y'(z) - Y(z)]\,\vec{E}_{tg}(\vec{\nu}, z).\vec{E}'_{tg}(\vec{\nu}, z) = C_1 \tag{6.215}$$

$$[Y'(z) + Y^*(z)]\,\vec{E}^*_{tg}(\vec{\nu}, z).\vec{E}'_{tg}(\vec{\nu}, z) = C_2\,. \tag{6.216}$$

Using the first of these two equations recalls the reciprocity theorem that we demonstrated in Section 6.7 and that (as we saw) is satisfied even when absorption is present. The second equation leads to a new result, which is exactly the phase closure relation we were looking for, satisfied uniquely in the transparent case.

On the interface of rank 0 ($z = 0$), we have

$$\begin{aligned} Y(0) &= Y_0 \;\; ; \;\; Y'(0) = Y'_0 = -\tilde{n}_0 \\ \vec{E}_{tg}(\vec{\nu}, 0) &= (1 + r)\vec{A}_{tg} \;\; ; \;\; \vec{E}'_{tg}(\vec{\nu}, 0) = t'\vec{A}_{tg}\,, \end{aligned} \tag{6.217}$$

where $\vec{A}_{tg}$ is the amplitude of the incident wave for the two illuminations, while $(r, t)$ and $(r', t')$ respectively are the associated reflection and transmission coefficients (illumination of the superstrate and substrate sides respectively). Substituting these equations into (6.216), we get

$$[Y'(0) + Y^*(0)]\,\vec{E}^*_{tg}(\vec{\nu}, 0).\vec{E}'_{tg}(\vec{\nu}, 0) = [-\tilde{n}_0 + Y_0^*]\,(1 + r^*)t'|\vec{A}_{tg}|^2\,. \tag{6.218}$$

Similarly, on the interface of rank $p$ ($z = z_p$), we have

$$\begin{aligned} Y(z_p) &= Y_p = \tilde{n}_s \;\; ; \;\; Y'(z_p) = Y'_p \\ \vec{E}_{tg}(\vec{\nu}, z_p) &= t\vec{A}_{tg} \;\; ; \;\; \vec{E}'_{tg}(\vec{\nu}, z_p) = (1 + r')\vec{A}_{tg}\,, \end{aligned} \tag{6.219}$$

and hence

$$[Y'(z_p) + Y^*(z_p)] \, \vec{\mathbb{E}}_{\text{tg}}^*(\vec{\nu}, z_p).\vec{\mathbb{E}}_{\text{tg}}'(\vec{\nu}, z_p) = [Y_p' + \tilde{n}_p^*]t^*(1 + r')|\vec{\mathbb{A}}_{\text{tg}}|^2. \quad (6.220)$$

Using (6.218) and (6.220), equation (6.216) enables us to write

$$[-\tilde{n}_0 + Y_0^*]\,(1 + r^*)t'|\vec{\mathbb{A}}_{\text{tg}}|^2 = [Y_p' + \tilde{n}_p^*]t^*(1 + r')|\vec{\mathbb{A}}_{\text{tg}}|^2. \quad (6.221)$$

On the other hand, the reflection coefficients $r$ and $r'$ can be expressed as

$$r = \frac{\tilde{n}_0 - Y_0}{\tilde{n}_0 + Y_0} \quad ; \quad r' = \frac{-\tilde{n}_s - Y_p'}{-\tilde{n}_s + Y_p'}, \quad (6.222)$$

which leads immediately to

$$Y_0 = \tilde{n}_0 \frac{1 - r}{1 + r} \quad ; \quad Y_p' = -\tilde{n}_s \frac{1 - r'}{1 + r'}. \quad (6.223)$$

Substituting these expressions into (6.221), and bearing in mind that there is no absorption in the end media, we get

$$-2\tilde{n}_0 r^* t' = 2\tilde{n}_s r' t^*, \quad (6.224)$$

and

$$\frac{r'}{r^*} = -\frac{\tilde{n}_0}{\tilde{n}_s} \frac{t'}{t^*}. \quad (6.225)$$

Taking the modulus squared of this last equation, we obtain

$$\frac{R}{R'} = \left[\frac{\tilde{n}_0}{\tilde{n}_s}\right]^2 \left|\frac{t'}{t}\right|^2. \quad (6.226)$$

Using relation (6.203) established in Section 6.7, this leads to

$$\frac{R}{R'} = 1. \quad (6.227)$$

If we now turn our attention to the arguments of the complex numbers involved in equation (6.225) and defined by

$$t = |t|\,e^{i\tau} \quad ; \quad t' = |t'|\,e^{i\tau'} \quad ; \quad r = |r|\,e^{i\rho} \quad ; \quad r' = |r'|\,e^{i\rho'}, \quad (6.228)$$

we obtain

$$\rho' + \rho = \pi + \tau' + \tau. \quad (6.229)$$

Relation (6.203) requires that $\tau = \tau'$, leading to

$$\rho + \rho' - 2\tau = \pi. \quad (6.230)$$

This is the **phase closure relation**, which connects the phases of reflection and transmission when the illumination is reversed on the coating.

### 6.8.2 Phenomenological Approach

Let us now consider a thought experiment involving interference, as illustrated Figure 6.23. A monochromatic unidirectional beam is impinging on

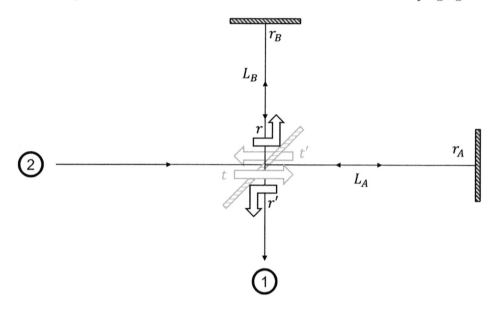

Figure 6.23 Thought experiment involving a multilayer stack and a Michelson interferometer.

a coating assumed to be deposited on air. The incidence angle of the illumination beam is $\theta$ (45 degrees on the drawing). Part of the beam is reflected (coefficient of reflection $r$), while the remaining part is transmitted by the multilayer stack (coefficient of transmission $t$).

The beam reflected by the coating is back-reflected by a perfect plane mirror (coefficient of reflection $r_B = -1$) and then crosses the coating with the same angle of incidence (coefficient of transmission again equal to $t$), while the beam transmitted by the coating is back-reflected by the same kind of ideal plane mirror (coefficient of reflection $r_A = r_B$) before being reflected by the coating (coefficient of reflection $r'$; indeed the wave is now retrograde).

The amplitude resulting from the interference between these two elementary waves in the output port 1 of the interferometer is then defined by

$$A_1 = A_0 t\, r_A r' e^{2ikL_A} + A_0 r\, r_B t e^{2ikL_B}\,, \tag{6.231}$$

where $L_A$ (respectively $L_B$) is the distance between the coating and the mirror A (respectively B). The quantities $r$, $r'$, and $t$ are complex, so we can always write

$$r = \sqrt{R}e^{i\rho} \quad ; \quad r' = \sqrt{R'}e^{i\rho'}, \tag{6.232}$$

and since the extreme media are identical

$$t = \sqrt{T}e^{i\tau} \quad ; \quad t' = \sqrt{T'}e^{i\tau'}. \tag{6.233}$$

In this way, we can compute the irradiance recorded by a photodetector located in port 1, i.e.,

$$E_1 \propto |A_1|^2 = |A_0|^2 \left\{ TR' + TR + 2T\sqrt{RR'}\cos(k\Delta + \rho' - \rho) \right\}, \tag{6.234}$$

where $\Delta$ is the optical path difference of the interferometer $[\Delta = 2(L_A - L_B)]$.

Furthermore, in this thought experiment we can also imagine computing the field amplitude $A_2$, and then the irradiance $E_2$ recorded by a photodetector located in port 2 of the same interferometer, i.e.,

$$A_2 = A_0 t\, r_A t' e^{2ikL_A} + A_0 r_B r^2 e^{2ikL_B}$$

$$E_2 \propto |A_0|^2 \left\{ TT' + R^2 + 2R\sqrt{TT'}\cos(k\Delta + \tau + \tau' - 2\rho) \right\}. \tag{6.235}$$

For a transparent coating, we have $R = R'$ and $T = T'$; moreover, the energy balance relative to the interferometer can be written as: $E_1 + E_2 = E_0$. This leads to

$$2RT + 2RT\cos(k\Delta + \rho' - \rho) + T^2 + R^2 + 2RT\cos(k\Delta + \tau + \tau' - 2\rho) = 1. \tag{6.236}$$

Using $R + T = 1$ we can write

$$2RT\left\{ \cos(k\Delta + \rho' - \rho) + \cos(k\Delta + \tau + \tau' - 2\rho) \right\} = 0, \tag{6.237}$$

which is equivalent to

$$\cos\{k\Delta + \frac{1}{2}(\rho' - 3\rho + \tau + \tau')\} \times \cos\{\frac{1}{2}[\rho + \rho' - (\tau + \tau')]\} = 0. \tag{6.238}$$

Since $\Delta$ may take arbitrary values, we get

$$\frac{1}{2}[\rho + \rho' - (\tau + \tau')] = \frac{\pi}{2}\,[\pi], \tag{6.239}$$

or

$$\rho + \rho' - (\tau + \tau') = \pi\,[2\pi]. \tag{6.240}$$

As shown in Section 6.7 (and also by left graph of Figure 6.24), the thin-film formalism requires that $\tau = \tau'$; so we have

$$\rho + \rho' - 2\tau = \pi\,[2\pi]. \tag{6.241}$$

Hence, we again find the relation demonstrated analytically in the previous section. As established through this same analytical approach, this relation remains valid when the superstrate and the substrate of the coating are different. However, it fails in the presence of absorption. We can see in

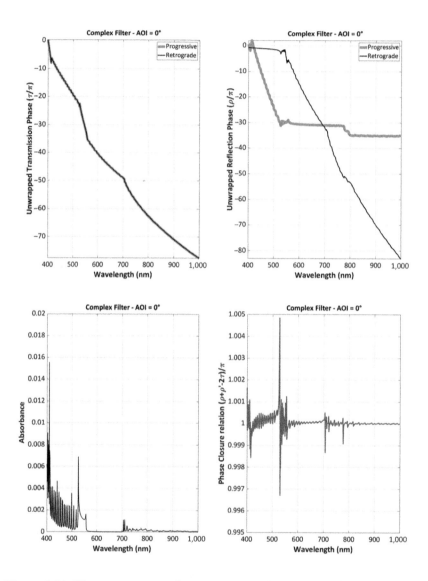

Figure 6.24  Phase properties of a complex interference filter (left top graph, unwrapped transmission phase for progressive and retrograde illumination; right top graph, unwrapped reflection phase for progressive and retrograde illumination; left bottom graph, filter absorbance for a progressive illumination; right bottom graph: deviation from the closure relation).

Figure 6.24 what happens for a complex interference filter when alternating dielectric layers are slightly absorbing: $\tau$ and $\tau'$ are precisely the same (see left graph), while the deviation with respect to the phase closure relation is shown on the right graph and can reach $0.005\pi$ for an absorbance of about 1%.

Finally, when the multilayer stack has a symmetrical structure with respect to a median plane, $\rho$ and $\rho'$ have to be equal, and in this case, the phase closure relation becomes

$$\rho = \tau + \frac{\pi}{2}[\pi],$$

(6.242)

which means that reflection and transmission are in quadrature.

## 6.9 Time Response of a Multilayer

Wavelength properties of optical coatings are most often given to characterize the multilayer design. However as discussed in Chapter 2, time and frequency behaviors carry similar information. Therefore in some situations we can use the time response of multilayers as an alternative characteristics. Such an alternative has frequently been used when working with short pulse lasers, but also in microcavities (see Chapter 8) to introduce the lifetime of molecules. Note also that in some frequency ranges far from the visible, such as the THz domain, it can be much easier to acquire a short pulsed source, rather than a continuous wave source in a broad spectrum; in this case the time response of the multilayer can be used to characterize the refractive index of materials over a wide spectrum.

Let us now consider an ultrashort pulse ideally represented by a Dirac $\delta(t)$ distribution impinging on a multilayer stack at normal incidence. To compute the time dependence of the pulse transmitted (or reflected) by this multilayer stack, we have to follow the same approach as that described in Section 3.5, namely

- To carry out a spectral decomposition of this incident pulse, i.e.,

$$\mathcal{E}_0(f) = \int\limits_{-\infty}^{+\infty} E_0(t)\, e^{2i\pi ft}\, dt = \int\limits_{-\infty}^{+\infty} \delta(t)\, e^{2i\pi ft}\, dt = 1 \,;$$

(6.243)

- Then to calculate the amplitude of the field transmitted at $z = z_p$ by the multilayer stack at each spectral component of the pulse:

$$\mathcal{E}_T(f) = t(f)\mathcal{E}_0(f) = t(f) \,;$$

(6.244)

- Before proceeding to the temporal reconstruction of the transmitted field, i.e.,

$$E_T(t) = 2\Re \left[ \int\limits_0^{+\infty} \mathcal{E}_T(f) e^{-2i\pi ft} \, dt \right] = 2\Re \left[ \int\limits_0^{+\infty} t(f) e^{-2i\pi ft} \, dt \right]. \qquad (6.245)$$

According to Section 2.4, $t(f)$ has Hermitian symmetry, so $t(-f) = t^*(f)$, which allows us to write

$$E_T(t) = \int\limits_{-\infty}^{+\infty} t(f) e^{-2i\pi ft} \, dt. \qquad (6.246)$$

So the time response of a multilayer at the output surface is directly proportional to the frequency Fourier transform of its spectral transmission (or reflection). Such a response is analogous to a **transfer function**. Note that the same result can also be directly recovered if we use relation (6.15) given at the beginning of this chapter.

To illustrate that in a realistic way, let us consider a Fabry–Perot filter consisting of two mirrors with $2q+1$ alternating H and L layers surrounding a 2L cavity spacer. This bandpass filter is centered at $\lambda_0 = 600$ nm, and we assume that the refractive indices of the layers are real (no absorption) and not dispersive ($n_H = 2.4$ and $n_L = 1.5$).

We can see in Figure 6.25 the time dependence of the modulus of the incident field, as well as those of the fields transmitted and reflected by the Fabry–Perot filter (left graphs). The right graph is a log scale view of the time dependence of the electric current provided by a photodiode used to detect the reflected field (quadratic detection).

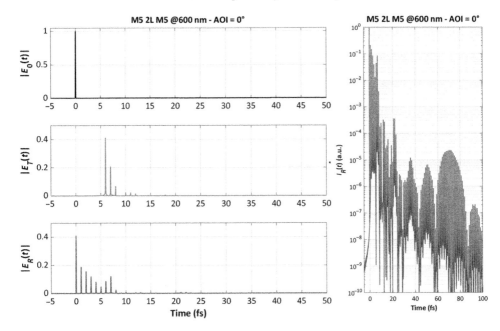

Figure 6.25 Time response of a Fabry–Perot filter [left top graph, modulus of the incident electric field; left middle graph, modulus of the transmitted electric field; left bottom graph, modulus of the reflected electric field; right graph: time dependence of the electric current provided by a detector (reflected field)].

# 7

# Multilayers in Total Reflection

## 7.1 Introduction

As mentioned in Chapter 6, the formalism of complex admittance applies just as much to the low spatial frequencies of the electromagnetic field ($\sigma <$ $k$) that are characteristic of plane propagative waves as to the high spatial frequencies of this same field ($\sigma > k$) that are characteristic of evanescent waves (see Chapter 4).

However, as most of the applications of multilayer coatings concern filtering in the far field (in reflection or transmission), for which only propagative waves are taken into account, we are often interested only in those expressions of admittance that correspond to low spatial frequencies; hence we frequently leave aside the variables associated with high spatial frequencies because applications aimed at controlling the near field, in the immediate neighborhood of this same multilayer component, are considerably less common.

However, there is one particular case, that of **total reflection**, that not only simply illustrates the notion of an evanescent wave, but also introduces the idea of resonance and that of the giant field enhancement of a harmonic component of the electromagnetic field; this is accessible due to a suitable optimization of the structure of a multilayer stack used in this particular regime. The approach we shall develop to describe how this works will also allow us to link the notion of resonance in total reflection in a multilayer stack with that of the guided mode propagating in this type of planar structure; it will also demonstrate the role played in both cases by the poles of the reflection coefficient of the multilayer component being considered.

## 7.2 Total Reflection at a Plane Optical Interface

As shown diagrammatically in Figure 7.1, consider a plane monochromatic progressive wave whose wave vector makes an angle $\theta_0$ with the normal to a plane optical interface separating two semi-infinite transparent media; the refractive indices are respectively $n_0$ in the case of the incident medium (or superstrate) and $n_s$ in the case of the substrate. Regardless of the angle of incidence $\theta_0$, we can associate with this plane wave a spatial frequency $\sigma$ defined by

$$\sigma = k_0 \sin \theta_0 = \frac{2\pi n_0}{\lambda_0}.$$ (7.1)

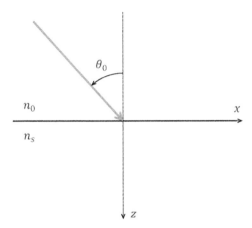

Figure 7.1 Plane wave encountering an optical interface at an angle of incidence $\theta_0$.

In each of these two media (superstrate or substrate), it is therefore possible to calculate the value taken by the normal component $\alpha_j$ of the corresponding wave vector, using the relation

$$\alpha_j^2 = k_j^2 - \sigma^2 \quad \text{where} \quad k_j = \frac{2\pi n_j}{\lambda} \quad j = 0, s.$$ (7.2)

In the case of the superstrate, this equation has two solutions corresponding respectively to the incident wave ($\alpha_0 = k_0 \cos \theta_0$) and to the wave reflected by the plane optical interface ($-\alpha_0 = k_0 \cos \theta_0$). The situation is different in the case of the substrate, and the result depends on the sign of the quantity $k_s - \sigma$:

• If $\sigma \leqslant k_s$, the wave in the substrate is **propagative**, and it is possible to define an angle of refraction $\theta_s$ ($\sigma = k_s \sin \theta_s$), and then deduce the value of the normal component of the wave vector associated with the progressive wave transmitted by the plane optical interface ($\alpha_s = k_s \cos \theta_s$).

- If $\sigma > k_s$, the wave in the substrate is **evanescent**, and it is no longer possible to define an angle of refraction (the equation $\sin \theta_s = \sigma / k_s > 1$ has no real solution).

However, we can continue to calculate the value of the normal component $\alpha_s$ of the wave vector in the substrate using equation (7.2), i.e.,

$$\alpha_s^2 = k_s^2 - \sigma^2 < 0 \quad \Rightarrow \quad \alpha_s \text{ pure imaginary} . \tag{7.3}$$

This guarantees that there is no transport of energy in the medium. Indeed we showed in Chapter 4 that the flux transported is given by the real part of the flux of the Poynting vector in medium $j$, and that this is proportional to the real part of $\alpha_j$. In the substrate, this real part is 0, and therefore implies that the transmission $T$ from the optical interface is 0 and that its reflection coefficient $R$ is equal to 1 (the media involved are lossless). For this reason, this particular situation is known generically as **total reflection**.

For completeness, note that the solution to equation (7.3) is of the type

$$\alpha_s = +i \sqrt{\sigma^2 - k_s^2} .$$

The solution of opposite sign is rejected because it leads to a divergence in the amplitude of the evanescent wave along the $z$-axis, a solution that has no physical meaning.

In order to fulfill this condition of total reflection, we must therefore have $\sigma > k_s$. Since the sine of an angle is always less than or equal to 1, we can write

$$k_0 \geqslant \sigma = k_0 \sin \theta_0 > k_s \quad \Rightarrow \quad n_0 > n_s . \tag{7.4}$$

Hence the refractive index of the superstrate must be greater than that of the substrate; this can be achieved by using a superstrate of high refractive index or by adopting a component geometry in which air takes the role of substrate, as shown in Figure 7.2.

From now on we shall assume that the refractive index $n_0$ of the superstrate is greater than the index $n_s$ of the substrate, the two media being supposed lossless, as before. We can therefore define a limiting angle $\theta_l$, where at values greater than this there will be total reflection of the plane wave at the plane optical interface, i.e.,

$$\sigma_l = k_0 \sin \theta_l = k_s \quad \Rightarrow \quad \sin \theta_l = \frac{n_s}{n_0} < 1 . \tag{7.5}$$

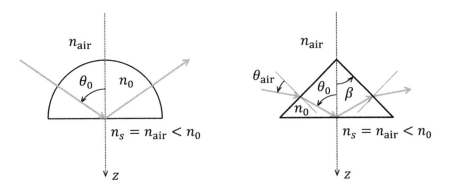

Figure 7.2 Component geometries using air as substrate.

In order to describe this situation succinctly, it is customary to introduce the **normalized spatial frequency** $\bar{\nu}$, defined by

$$\bar{\nu} = \lambda\nu = \lambda\frac{\sigma}{2\pi} . \qquad (7.6)$$

In the situation shown diagramatically in Figure 7.1, $\sigma$ can take all values between 0 (for $\theta_0 = 0$) and $k_0$ (for $\theta_0 = \pi/2$), including the particular value $k_s$ (for $\theta_0 = \theta_l$). Consequently, the associated normalized spatial frequency $\bar{\nu}$ varies between 0 and $n_0$, including $n_s$, but never exceeds $n_0$ in this particular illumination regime.

Moreover, we showed in Chapter 6 that the reflection coefficient $R$ at the interface between superstrate and substrate was given by

$$R = |r|^2 \quad \text{where} \quad r = \frac{\tilde{n}_0 - \tilde{n}_s}{\tilde{n}_0 + \tilde{n}_s} . \qquad (7.7)$$

The effective refractive index of each of the two media depends on the polarization state of the incident wave, and is defined by

$$\tilde{n}_j = \frac{1}{\eta_v \mu_{r,j}} \begin{cases} n_j \alpha_j / k_j & \text{TE polarization} \\ n_j k_j / \alpha_j & \text{TM polarization} \end{cases}, \qquad (7.8)$$

where $\eta_v$ is the impedance of vacuum and $\mu_{r,j}$ is the relative permeability of layer $j$, assumed nonmagnetic ($\mu_{r,j} = 1$).

Using equations (7.7) and (7.8), we can calculate the value of the reflection coefficient at any incidence, and hence plot its behavior as a function of the normalized spatial frequency $\bar{\nu}$ (see Figure 7.3).

It can be verified analytically that this reflection coefficient is indeed equal to 1 within the frequential window between $n_s$ and $n_0$. In this case, the wave is evanescent within the substrate, which means that the quantity $\alpha_s$ is pure

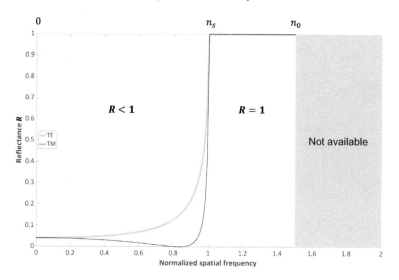

Figure 7.3 Behavior of the reflection coefficient $R$ as a function of normalized spatial frequency $\bar{\nu}$ ($n_0 = 1.5$ and $n_s = 1.0$), in TE (gray curve) and TM polarization (black curve). Total reflection occurs in the frequential window between $n_s$ and $n_0$.

imaginary, the same being true for the effective refractive index $\tilde{n}_s$ of the substrate in either TE or TM polarization. The effective refractive index $\tilde{n}_0$ of the superstrate is real, regardless of the polarization state of the incident wave (recall that the two media are lossless). The reflection coefficient $r$ then has the form

$$r = \frac{a - ib}{a + ib} \quad \text{where} \quad a, b \in \mathbb{R}. \tag{7.9}$$

This immediately leads to the desired result for $R$, i.e.,

$$R = \left| \frac{a - ib}{a + ib} \right|^2 = \frac{a^2 + b^2}{a^2 + b^2} = 1. \tag{7.10}$$

If, however, in the frequential window $[n_s, n_0]$, no energy is transmitted into the substrate in the far field ($R = 1 \Rightarrow T = 0$), this does not mean that the amplitude of the field is 0. In fact, it decays exponentially along the $z$-axis as

$$\mathbb{A}_s^+(z) = \mathbb{A}_s^+ e^{-z\sqrt{\sigma^2 - k_s^2}}, \tag{7.11}$$

where $\mathbb{A}_s^+$ is the value of the field at the interface. Consequently, some electromagnetic energy remains confined within the substrate near the interface and may be decoupled as, for example, the end of an optical fiber whose core refractive index is greater than that of the substrate is approached.

This confinement of energy therefore extends over a characteristic depth $z_s$, defined by

$$z_s = \frac{1}{2\sqrt{\sigma^2 - k_s^2}} = \frac{\lambda}{4\pi}\frac{1}{\sqrt{\bar{\nu}^2 - n_s^2}}, \tag{7.12}$$

and which depends on the value taken by the normalized spatial frequency within the window $[n_s, n_0]$. In particular, this confinement depth tends to infinity in the neighborhood of the limiting angle. In other practical situations, it will be around a few wavelengths.

In all our previous discussions, we made the assumption that the superstrate and substrate were lossless. Note that the condition of transparency of the superstrate (real refractive index $n_0$) is imposed by the fact that the wave transmitted by a source at infinity in the direction $z < 0$ must be able to reach the plane optical interface at $z = 0$. Conversely, it is entirely possible that the substrate might be absorbent: reflection is thus not 100% ($R < 1$), since the real part of $\alpha_s$ is not 0. The existence of this nonzero real part generates losses through absorption in the substrate, and hence in this case, the evanescent wave (said to be dissociated) contributes to the energy balance (via this absorption), even if it transports no energy in the far field.

Figure 7.2 shows operational geometries using air as the substrate. In the case of the hemispherical lens (on the left) the incident wave vector is undiverted in crossing the air/superstrate interface, thus avoiding any alteration of the incident spatial frequency; in this case, the accessible frequential window is the same as that shown in Figure 7.3, i.e., all values between 0 and $n_0$.

However, the use of a prism, as shown on the right of Figure 7.2, is often preferred because it is simpler to implement. In this case, the angle of incidence $\theta_{air}$ at the interface between the external medium and the prism, and the angle $\theta_0$ at the superstrate/substrate interface, are related as follows:

$$n_{air} \sin\theta_{air} = n_0 \sin\left(\theta_0 + \beta - \frac{\pi}{2}\right), \tag{7.13}$$

where $\beta$ is the half-angle at the apex of the prism, assumed isosceles. In this case, the range of accessible spatial frequencies on the superstrate/substrate interface lies between $n_0\sin\{\pi/2 - \beta - \arcsin(n_{air}/n_0)\}$ and $n_0\sin\{\pi/2 - \beta + \arcsin(n_{air}/n_0)\}$ and hence depends on the refractive index $n_0$ of the prism and the apex half-angle $\beta$. If the prism is rectangular ($2\beta = \pi/2$), the range of spatial frequencies $[0, n_0]$ can be covered provided a refractive index can be chosen that satisfies the condition

$$\arcsin(n_{\text{air}}/n_0) = \beta = \pi/4 \quad \Rightarrow \quad n_0 = \frac{n_{\text{air}}}{\sin(\pi/4)} = \sqrt{2}n_{\text{air}} \approx 1.41\,, \quad (7.14)$$

a value very close to that of silica.

Total reflection can also be obtained by injecting light into a thick substrate through the side of the substrate, provided the side face is sufficiently well polished (see Figure 7.4). In this case, the relation between the internal

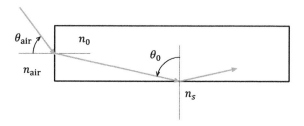

Figure 7.4 Operational geometry using illumination through the polished side face of a substrate.

and external angles can be written

$$n_0 \sin \theta_0 = \sqrt{n_0^2 - n_{\text{air}}^2 \sin^2 \theta_{\text{air}}}\,, \quad (7.15)$$

which shows that the accessible frequential window at the superstrate/substrate interface lies between $\sqrt{n_0^2 - 1}$ and $n_0$.

## 7.3 Frustrated Total Reflection

As mentioned in Section 7.2, in the total reflection regime it is possible to locally decouple the energy confined under the superstrate/substrate interface when the end of an optical fiber is approached. However, it is also possible to frustrate this total reflection by allowing the hypotenuse of the prism in Figure 7.2 to approach a second prism that is upside down with respect to the first (see Figure 7.5). The resulting component then functions as a beam splitter whose reflection (or transmission) coefficient can be set by adjusting the thickness of the air gap $d$ separating the two hypotenuse faces.

To keep things simple, if we restrict our interest to the propagation of light waves within just one component, this component can be assumed to be a single layer of air (thickness $d$, refractive index $n_{\text{air}} = 1$) between two prism media of index $n_0$ (see Figure 7.5).

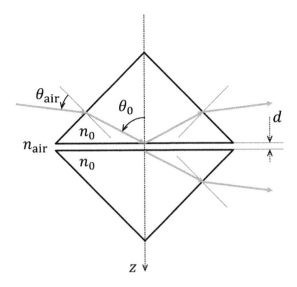

Figure 7.5 Beam splitter in frustrated total reflection mode.

The illumination frequency on the air layer is chosen enough high to prevent the existence of a refraction angle in this layer:

$$k_0 \sin \theta_0 = \sigma > k_{\text{air}} = 2\pi/\lambda \quad \Rightarrow \quad n_0 \sin \theta_0 > 1 . \tag{7.16}$$

Such a relation would correspond, in a case where the layer is semi-infinite, to a condition of total reflection at the top of the layer. However, this is not the case and we will see that we speak of a regime of frustrated total internal reflection.

We can then calculate the values taken by the quantity $\alpha_j$ in each of the three media (superstrate, air layer, and substrate), i.e.,

$$\alpha_0 = \sqrt{k_0^2 - \sigma^2} = k_0 \cos \theta_0$$
$$\alpha_1 = i\sqrt{\sigma^2 - k_{\text{air}}^2} = ik_v \sqrt{n_0^2 \sin^2 \theta_0 - 1} = ik_v \gamma \tag{7.17}$$
$$\alpha_2 = \alpha_s = \sqrt{k_s^2 - \sigma^2} = \alpha_0 ,$$

where $k_v = k_{\text{air}} = 2\pi/\lambda$, and then those of the associated effective refractive indices, regardless of the polarization state of the incident wave (TE or TM):

$$\tilde{n}_1 = ib \quad \text{where} \quad b = \begin{cases} \gamma/\eta_v & \text{TE} \\ -1/\gamma\eta_v & \text{TM} \end{cases} , \tag{7.18}$$

and

$$\tilde{n}_s = \tilde{n}_0 = \frac{1}{\eta_v} \begin{cases} n_0 \cos\theta_0 & \text{TE} \\ n_0/\cos\theta_0 & \text{TM} \end{cases}. \tag{7.19}$$

To calculate the complex admittances of the various interfaces of which this particular stacking arrangement is composed, we make use of the general rules set out in Chapter 6, namely:

- Initialization on the air/ lower prism interface:

$$Y_1 = \tilde{n}_s = \tilde{n}_0. \tag{7.20}$$

- Use of the recurrence relation between complex admittances to determine that associated with the upper prism/air interface:

$$Y_0 = \frac{Y_1 \cos\delta_1 - i\tilde{n}_1 \sin\delta_1}{\cos\delta_1 - i(Y_1/\tilde{n}_1)\sin\delta_1} \quad \text{where} \quad \delta_1 = \alpha_1 d = ik_v\gamma d. \tag{7.21}$$

Since the phase term $\delta_1$ is pure imaginary, the trigonometric functions appearing in the recurrence relation become hyperbolic functions, since

$$\cos\delta_1 = \cos(ik_v\gamma d) = \cosh(k_v\gamma d)$$
$$\sin\delta_1 = \sin(ik_v\gamma d) = i\sinh(k_v\gamma d). \tag{7.22}$$

The complex admittance $Y_0$ then takes the following particular form:

$$Y_0 = \frac{\tilde{n}_0 \cosh(k_v\gamma d) + \tilde{n}_1 \sinh(k_v\gamma d)}{\cosh(k_v\gamma d) + (\tilde{n}_0/\tilde{n}_1)\sinh(k_v\gamma d)}. \tag{7.23}$$

If we know $Y_0$ we can then calculate the amplitude reflection coefficient $r$, i.e.,

$$r = \frac{\tilde{n}_0 - Y_0}{\tilde{n}_0 + Y_0} = \frac{(\tilde{n}_0/\tilde{n}_1 - \tilde{n}_1/\tilde{n}_0)\sinh(k_v\gamma d)}{2\cosh(k_v\gamma d) + (\tilde{n}_0/\tilde{n}_1 + \tilde{n}_1/\tilde{n}_0)\sinh(k_v\gamma d)}, \tag{7.24}$$

and deduce from this its energy equivalent $R$:

$$R = \frac{(b/\tilde{n}_0 + \tilde{n}_0/b)^2 \sinh^2(k_v\gamma d)}{4\cosh^2(k_v\gamma d) + (b/\tilde{n}_0 - \tilde{n}_0/b)^2 \sinh^2(k_v\gamma d)}. \tag{7.25}$$

Using the identity $\cosh^2 x - \sinh^2 x = 1$, this can be put in the following form:

$$R = \frac{(b/\tilde{n}_0 + \tilde{n}_0/b)^2 \sinh^2(k_v\gamma d)}{4 + (b/\tilde{n}_0 + \tilde{n}_0/b)^2 \sinh^2(k_v\gamma d)}. \tag{7.26}$$

This expression easily shows that the coefficient $R$ is 0 when the thickness $d$ of the air gap is 0 (the component reduces to a uniform block of glass) and that it tends to 1 when the thickness of this same gap becomes large compared with the wavelength (the air gap playing the role of a semi-infinite substrate satisfying the condition of total reflection).

Figure 7.6 shows how the coefficient $R$ of this beam-splitter, in TE and TM polarization, varies as a function of the gap $d$, when the prisms used are right-angled isosceles triangles, transparent at the wavelength used (600 nm) and made of standard glass such as N-BK7 ($n_0 = 1.5163$). Furthermore, we chose a 0 angle of incidence in air, so that the angle of incidence in the glass is 45 degrees and the normalized spatial frequency $\bar{\nu}$ is 1.07, slightly greater than the refractive index of air; this guarantees total reflection in the absence of the lower prism.

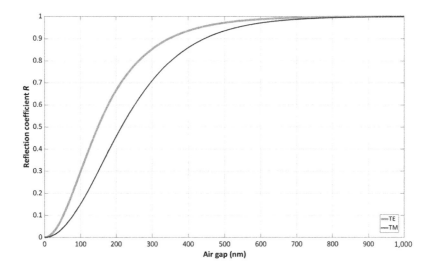

Figure 7.6 Variation of the reflection coefficient of a beam-splitter operating in frustrated total reflection mode as a function of the thickness of the air gap (gray curve, TE polarization; black curve, TM polarization).

As expected, it is observed that the reflection coefficient varies continuously between 0 and 1 when the gap $d$ varies between 0 and 1,000 nm and that it is greater than 90% when the gap thickness is equal to a wavelength. Since the media are lossless, the transmission coefficient $T$ of this beam-splitter is just $1 - R$.

In conclusion, our investigation of this frustrated total reflection regime has shown that the *hyperbolic* character of a wave in a layer does not prevent it from having a *trigonometric* (propagative) character in the media that follow it.

## 7.4 Total Reflection at a Multilayer

If we now replace the single optical interface shown in Figure 7.1 by a multilayer stack (see Figure 7.7), and keep in mind that total reflection means that there is no flux transport in the substrate, the situation thus created is not fundamentally different from that studied in Section 7.2. Indeed the total reflection condition ($\sigma > k_s$) relates only to the refractive index of the substrate, while the incident pulsation $\sigma$ is unchanged (the illuminating conditions are identical).

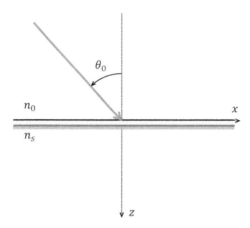

Figure 7.7 Plane wave encountering a multilayer stack at an incidence $\theta_0$.

Under the assumption that the condition $n_0 \sin \theta_0 > n_s$ for total reflection is satisfied, the question as to at which interface this total reflection effectively takes place does not apply (indeed, it has no meaning). Note also that the field in layer $j$ can be *trigonometric* ($\alpha_j$ real) or *hyperbolic* ($\alpha_j$ pure imaginary).

The condition for total reflection requires that the effective refractive index in the substrate $\tilde{n}_s$ is pure imaginary. Since the complex admittance of the optical interface $p$ is equal to this effective index, the quantity $Y_p$ is therefore also pure imaginary ($p$ is the number of layers). The complex admittance $Y_{p-1}$ can be calculated using the recurrence relation given in Section 7.3, i.e., here

$$Y_{p-1} = \frac{Y_p \cos \delta_p - i\tilde{n}_p \sin \delta_p}{\cos \delta_p - i(Y_p/\tilde{n}_p) \sin \delta_p} = \frac{\tilde{n}_s \cos \delta_p - i\tilde{n}_p \sin \delta_p}{\cos \delta_p - i(\tilde{n}_s/\tilde{n}_p) \sin \delta_p}. \tag{7.27}$$

Two cases can be distinguished:

- Either the field in layer $p$ is trigonometric in nature ($\sigma < k_p$); the effective index $\tilde{n}_p$ and the phase term $\delta_p$ are therefore real, such that the numerator of relation (7.27) is pure imaginary, while the denominator is real: $Y_{p-1}$ is thus pure imaginary,
- or the field in layer $p$ is hyperbolic in nature ($\sigma > k_p$); the effective index $\tilde{n}_p$ and the phase term $\delta_p$ are then pure imaginary, such that $\cos \delta_p$ is real and $\sin \delta_p$ is pure imaginary. The numerator in equation (7.27) is again pure imaginary and its denominator is again real: $Y_{p-1}$ is again pure imaginary.

Therefore the initialization of the recurrence relation for a pure imaginary quantity requires that the complex admittances of all interfaces are also pure imaginary, regardless of the refractive indices of the various layers making up the stack, provided they continue to be assumed lossless. This is also the case of admittance $Y_0$ ($Y_0 = ib$, $b \in \mathbb{R}$), such that the amplitude reflection coefficient $r$ can again be put in the form

$$r = \frac{\tilde{n}_0 - Y_0}{\tilde{n}_0 + Y_0} = \frac{a - ib}{a + ib} \quad \text{where } a, b \in \mathbb{R}. \tag{7.28}$$

This leads immediately to $R = 1$, as set out in (7.10). This result confirms that the condition $\sigma > k_s$ is sufficient to obtain total reflection in the multilayer case.

However, if the superstrate and substrate are transparent, but one or more of the layers in the stack are absorbing, then contrary to the case of a single optical interface, there will be a dissipation of energy in the multilayer, and we must then write $R = 1 - A$.

To illustrate this observation consider, for example, the component defined by the formula

$$\text{N-BK7 / H 5L 2H 6L 2H 6L / Air},$$

where H and L are quarter-wave layers of $Ta_2O_5$ ($n_H = 2.1489$, $d_H = 69.80$ nm) and $SiO_2$ ($n_L = 1.4922$, $d_L = 100.52$ nm) respectively, at a wavelength of 600 nm. The spectral dependence of the imaginary parts of the refractive indices of these two materials is identical to that used in Section 6.4.4. The angle of incidence in the glass is taken to be 45 degrees, and the polarization is TE. The normalized spatial frequency $\bar{\nu}$ associated with the incident wave is 1.07, showing that the total reflection regime applies ($\bar{\nu} > n_s$) comfortably.

Figure 7.8 shows the spectral dependence of the reflection and transmission coefficients of this component (left-hand graph), plus the spectral dependence of its absorption coefficient (middle graph, logarithmic units), and

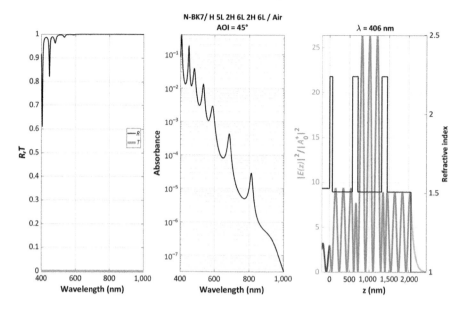

Figure 7.8 Optical properties of the component N-BK7 / H 5L 2H 6L 2H 6L / Air used in the total reflection regime (AOI = 45 degrees).

finally the $z$ dependence of the modulus squared of the electric field normalized with respect to the incident field (right-hand graph) at a wavelength of 406 nm.

Note that

- The transmission coefficient $T$ is 0 over the entire range of wavelengths.
- The absorption coefficient $A$, calculated using the second approach set out in Section 7.114, shows peaks at certain particular wavelengths (the strongest peak is centered on 406 nm).
- The modulus squared of the normalized field within the thickness of the component shows significant amplification factors (greater than 25 at this particular wavelength of 406 nm); this explains the high value of absorption with which it is associated.
- The modulus squared of the same normalized field shows an exponential decrease in the substrate (in light gray on the right-hand graph in Figure 7.8), consistent with the evanescent character of the wave in this medium.
- The reflection coefficient $R$ is indeed equal to $1 - A$, as expected.

These amplifications of the electromagnetic field are known as *resonances* and we shall now show how to determine their properties using a systematic approach.

## 7.5 Resonances in a Multilayer

We saw in the previous section that, at certain wavelengths, the existence of elevated field intensities within the thickness of the component has the effect of increasing the absorption at those same wavelengths. A simple means of identifying such resonances consists of assigning the constitutive materials in the multilayer a nonzero extinction coefficient (e.g., $10^{-4}$), and then exa-mining the variations in the reflection coefficient of the component with the spatial frequency $\nu = (n_0/\lambda)\sin\theta_0$. If this reflection coefficient $R(\nu)$ (which we know is equal to $1 - A(\nu)$ within the frequential window corresponding to total reflection) shows a marked decrease at some particular value $\nu_r$, then that means that absorption is high and that the corresponding field is amplified [recall that absorption is related to the integral of the modulus squared of the field within the thickness of the multilayer; see Section 6.3.11, equation (6.141)].

Figure 7.9 shows the variation of this coefficient $R$ as a function of the normalized spatial frequency $\bar{\nu}$ for the six-layer component considered in the previous section. The illuminating wavelength is taken to be 500 nm, and the angle of incidence is allowed to vary. The total reflection regime corresponds to the frequential window between $n_s = 1$ and the index of the N-BK7 at that wavelength, i.e., $n_0 = 1.5214$.

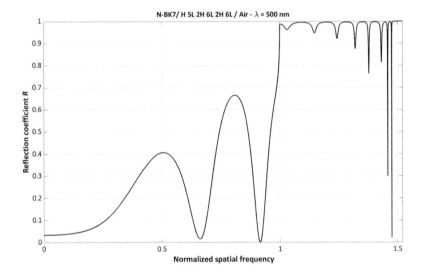

Figure 7.9 Variation of reflection coefficient $R$ of component N-BK7 / H 5L 2H 6L 2H 6L / Air as a function of normalized spatial frequency $\bar{\nu}$ at a wavelength of 500 nm (TE polarization). Total reflection takes place in the frequential window between $n_s = 1$ and $n_0 = 1.5214$.

In TE polarization eight distinct resonances can be seen, classified by decreasing value of the normalized spatial frequency; by analogy with the convention used in guided-wave optics, we shall refer to resonance of order 0 for the greatest angle of incidence (which would correspond to the direction of propagation closest to the axis of the waveguide) and then more generally to resonance of order $n$ depending on the number of its order in the list. Table 7.1 shows the main characteristics of these eight resonances (order $n$, normalized spatial frequency $\bar{\nu}$, total mid-height width $\Delta\bar{\nu}$, angle of incidence $\theta_0$, and absorbance $A$). Note here that the stronger the absorbance,

Table 7.1 Main characteristics of the eight resonances appearing in component N-BK7 / H 5L 2H 6L 2H 6L / Air ($\lambda = 500$ nm, TE polarization)

| Order | $\bar{\nu}$ | $\Delta\bar{\nu}$ | $\theta_0$ | $A$ |
|---|---|---|---|---|
| 0 | 1.4746 | 0.0004 | 75.8 | 97.8% |
| 1 | 1.4576 | 0.0011 | 73.3 | 70.2% |
| 2 | 1.4301 | 0.0044 | 70.0 | 18.7% |
| 3 | 1.3766 | 0.0040 | 64.8 | 23.8% |
| 4 | 1.3179 | 0.0077 | 60.0 | 12.6% |
| 5 | 1.2418 | 0.0126 | 55.1 | 8.0% |
| 6 | 1.1466 | 0.0186 | 48.9 | 5.5% |
| 7 | 1.0306 | 0.0261 | 42.6 | 3.8% |

the narrower the resonance. Hence the total mid-height width $\Delta\bar{\nu}$ of the resonance of zero order is $4 \times 10^{-4}$, which corresponds to 0.06 degree. At the same time, the absorbance goes from 97.8% at the peak of this same resonance to 0.3% in the zone intermediate between the two resonances of orders 0 and 1. The field enhancement within the component is therefore very strongly dependent on the value of the normalized spatial frequency, as shown by the two graphs in Figure 7.10. The graph on the left corresponds to the normalized spatial frequency associated with the zero-order resonance ($\bar{\nu} = 1.4746$, $A = 97.8\%$), while that on the right is plotted for a normalized spatial frequency between the resonances of order 0 and order 1 ($\bar{\nu} = 1.4682$, $A = 0.3\%$).

Note that at the zero-order resonance, the maximum value of the modulus squared of the normalized field approaches 120, while away from resonance it never exceeds 0.35. This ratio of greater than 300 explains the observed deviations in absorbance. We shall see later on (Section 7.8) how to optimize or control these resonances, but before that we need to discuss the existence of guided waves in multilayers. Then we will relate these resonances to the complex poles and zeros of the reflection coefficient.

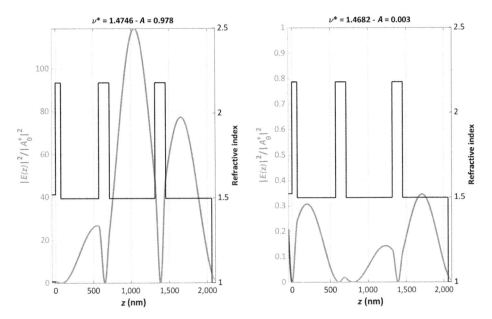

Figure 7.10 Variation with depth $z$ of the modulus squared of the normalized tangential field within the component N-BK7 / H 5L 2H 6L 2H 6L / Air ($\lambda = 500$ nm, TE polarization). Left-hand graph for resonance of zero order ($\bar{\nu} = 1.4746$, $A = 97.8\%$); right-hand graph away from resonance ($\bar{\nu} = 1.4682$, $A = 0.3\%$).

## 7.6 Planar Optics

### 7.6.1 Guided Modes

Up to this point we have always considered configurations in which the source was **located at infinity in the superstrate**. As mentioned on several occasions, an immediate consequence of this assumption is the absence of a retrograde wave within the substrate, and means that the admittance of the interface between the last layer and the substrate can be identified with its effective refractive index, i.e.,

$$Y_p = \tilde{n}_s . \tag{7.29}$$

To calculate the admittance $Y_0$ associated with the upper interface of the first layer, it remains only to apply the recurrence relation $p$ times:

$$Y_{j-1} = \frac{Y_j \cos \delta_j - i \tilde{n}_j \sin \delta_j}{\cos \delta_j - i(Y_j/\tilde{n}_j) \sin \delta_j} \quad j = 1, ..., p . \tag{7.30}$$

In the case of planar (or guided) optics – our next topic of interest – we look for solutions to the same type of problem (multilayer deposited on the

surface of a plane substrate), but **in the absence of any incident field
in the superstrate.** From a practical point of view, this situation corresponds to the diagram shown in Figure 7.11, where the wave penetrates the
multilayer due to a coupling process at the left of the figure at some very
large distance from the plane $x = 0$ (coupling by grating, prism, or focusing on the side face of the multilayer if it is sufficiently thick). Under these

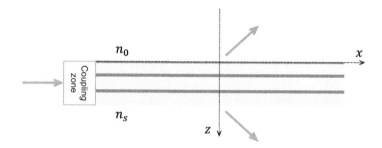

Figure 7.11 Operational configuration in a planar optics situation.

conditions, if we reformulate Maxwell's equations (boundary conditions included) in the region $x > 0$, these are identical to those considered up to
now, except that a progressive wave no longer exists in the superstrate. The
main characteristic of planar optics is therefore that the only wave present
in the superstrate is **retrograde,** so the admittance of the upper interface
of the multilayer can be identified with the effective index of the superstrate
for this particular direction of propagation, i.e.,

$$Y_0 = -\tilde{n}_0 \,. \tag{7.31}$$

It may be observed that the particular frequencies $\nu_g$ for which this condition
is satisfied are none other than the zeros of the function $1/r(\nu)$ [or the poles
of the function $r(\nu)$], since

$$\frac{1}{r(\nu_g)} = \frac{\tilde{n}_0(\nu_g) + Y_0(\nu_g)}{\tilde{n}_0(\nu_g) - Y_0(\nu_g)} = \frac{\tilde{n}_0(\nu_g) - \tilde{n}_0(\nu_g)}{\tilde{n}_0(\nu_g) + \tilde{n}_0(\nu_g)} = 0 \,. \tag{7.32}$$

Equation (7.32) may exhibit (when existing) a discrete series of solutions $\nu_g$,
which are the modal (or guided wave) frequencies of the stack. In the case
of a transparent multilayer, note that these spatial frequencies must be real
in order to guarantee that there is no loss. Moreover, the associated modes
must propagate in the multilayer without loss by radiation in the superstrate
and substrate, which requires that the waves must be evanescent in nature
in these extreme media and that the corresponding quantities $\alpha_0$ and $\alpha_s$
must therefore be pure imaginary. Hence we deduce immediately that

$$\sigma_g > \max(k_0, k_s) \quad \text{or} \quad \bar{\nu}_g > \max(n_0, n_s)\,. \tag{7.33}$$

Note that this condition could not be fulfilled in the case of a source at infinity in the superstrate, since we showed in Section 7.4 that the normalized spatial frequency was factored up by $n_0$. Hence we need a particular coupling process corresponding to one of the two shown schematically in Figure 7.11.

All these elements show that it might be inappropriate to talk here of a *reflection coefficient* for the function $r(\nu)$, since there is no longer incident wave in the superstrate. Instead we consider the extension of this function in a particular domain of frequencies beyond the bound physically accessible to a far-field illumination. The modal frequencies therefore correspond to the poles of the extension of $r(\nu)$.

### 7.6.2 Modal Window

We showed in the previous section that, in order to avoid radiation losses in the end media, the spatial frequencies $\nu_g$ we sought had to be factored down by equation (7.33). We shall now show that they are also factored up by the following relation:

$$\sigma_g < \max(k_j) \quad j = 1, ..., p \quad \text{or} \quad \bar{\nu}_g < \max(n_j) \quad j = 1, ..., p\,. \tag{7.34}$$

Combining these two conditions, i.e.,

$$\max(n_0, n_s) < \bar{\nu}_g < \max(n_j) \quad j = 1, ..., p\,, \tag{7.35}$$

allows a **modal window** to be defined, i.e., the range of frequencies over which zeros of the function $1/r$ can be sought. Clearly this does not guarantee that any solutions exist, but merely their location in the case where such solutions do exist.

To demonstrate equation (7.34), we start from the Helmholtz equation satisfied by the elementary components of the field in each medium $j$ (see Chapter 4), namely

$$\frac{\partial^2 \vec{\mathbb{E}}_j(\nu, z)}{\partial z^2} + \alpha_j^2 \vec{\mathbb{E}}_j(\nu, z) = \vec{0}\,, \tag{7.36}$$

where

$$\vec{\mathbb{E}}_j(\nu, z) = \vec{\mathbb{A}}_j^+ e^{i\alpha_j z} + \vec{\mathbb{A}}_j^- e^{-i\alpha_j z}\,. \tag{7.37}$$

Multiplying (7.36) by the complex conjugate of the field, we get

$$\vec{\mathbb{E}}_j^*(\nu, z) \cdot \frac{\partial^2 \vec{\mathbb{E}}_j(\nu, z)}{\partial z^2} + \alpha_j^2 |\vec{\mathbb{E}}_j(\nu, z)|^2 = 0\,. \tag{7.38}$$

Integrating this expression over the thickness of the layer $j$, we get

$$\left[\vec{\mathbb{E}}_j^*(\nu, z) \cdot \frac{\partial \vec{\mathbb{E}}_j(\nu, z)}{\partial z}\right]_{z_{j-1}}^{z_j} - \int_{z_{j-1}}^{z_j} \left|\frac{\partial \vec{\mathbb{E}}_j(\nu, z)}{\partial z}\right|^2 dz + \alpha_j^2 \int_{z_{j-1}}^{z_j} |\vec{\mathbb{E}}_j(\nu, z)|^2 dz = 0. \quad (7.39)$$

After summation over all the constitutive media of the component (superstrate, multilayer, and substrate), we get

$$\sum_{j=0}^{p+1} \left[\vec{\mathbb{E}}_j^*(\nu, z) \cdot \frac{\partial \vec{\mathbb{E}}_j(\nu, z)}{\partial z}\right]_{z_{j-1}}^{z_j} + \sum_{j=0}^{p+1} \alpha_j^2 \int_{z_{j-1}}^{z_j} |\vec{\mathbb{E}}_j(\nu, z)|^2 dz$$

$$= \sum_{j=0}^{p+1} \int_{z_{j-1}}^{z_j} \left|\frac{\partial \vec{\mathbb{E}}_j(\nu, z)}{\partial z}\right|^2 dz. \quad (7.40)$$

First we consider the case corresponding to TE polarization. Since the electric field and its derivative are continuous in nonmagnetic media, we can write the left-hand part of (7.40) in the following equivalent form:

$$\vec{\mathbb{E}}_s^*(\nu, z) \cdot \frac{\partial \vec{\mathbb{E}}_s(\nu, z)}{\partial z}\bigg|_{+\infty} - \vec{\mathbb{E}}_0^*(\nu, z) \cdot \frac{\partial \vec{\mathbb{E}}_0(\nu, z)}{\partial z}\bigg|_{-\infty} + \sum_{j=0}^{p+1} \alpha_j^2 \int_{z_{j-1}}^{z_j} |\vec{\mathbb{E}}_j(\nu, z)|^2 dz. \quad (7.41)$$

In guided optics, the waves are evanescent in the two extremal media, retrograde in the superstrate and progressive in the substrate. Consequently, in the substrate we have

$$\vec{\mathbb{E}}_s(\nu, z) = \vec{\mathbb{A}}_s^+ e^{i\alpha_s z} = \vec{\mathbb{A}}_s^+ e^{-\sqrt{\sigma_g^2 - k_s^2}\, z}$$

$$\vec{\mathbb{E}}_s^*(\nu, z) \cdot \frac{\partial \vec{\mathbb{E}}_s(\nu, z)}{\partial z}\bigg|_{+\infty} = -\sqrt{\sigma_g^2 - k_s^2}\, |\vec{\mathbb{A}}_s^+|^2 e^{-2\sqrt{\sigma_g^2 - k_s^2}\, z}\bigg|_{+\infty} = 0, \quad (7.42)$$

and similarly for the superstrate:

$$\vec{\mathbb{E}}_0(\nu, z) = \vec{\mathbb{A}}_0^- e^{-i\alpha_0 z} = \vec{\mathbb{A}}_0^- e^{\sqrt{\sigma_g^2 - k_s^2}\, z}$$

$$\vec{\mathbb{E}}_0^*(\nu, z) \cdot \frac{\partial \vec{\mathbb{E}}_0(\nu, z)}{\partial z}\bigg|_{-\infty} = \sqrt{\sigma_g^2 - k_0^2}\, |\vec{\mathbb{A}}_0^-|^2 e^{2\sqrt{\sigma_g^2 - k_s^2}\, z}\bigg|_{-\infty} = 0. \quad (7.43)$$

Substituting these two results into equation (7.41), then into (7.40), we finally get

$$\sum_{j=0}^{p+1} (k_j^2 - \sigma_g^2) \int_{z_{j-1}}^{z_j} |\vec{\mathbb{E}}_j(\nu, z)|^2 dz = \sum_{j=0}^{p+1} \int_{z_{j-1}}^{z_j} \left|\frac{\partial \vec{\mathbb{E}}_j(\nu, z)}{\partial z}\right|^2 dz > 0, \quad (7.44)$$

and, bringing out the terms associated with the superstrate and substrate, we get

$$\sum_{j=1}^{p}(k_j^2 - \sigma_g^2) \int_{z_{j-1}}^{z_j} |\vec{\mathbb{E}}_j(\nu, z)|^2 dz > (\sigma_g^2 - k_0^2) \int_{-\infty}^{0} |\vec{\mathbb{E}}_0(\nu, z)|^2 dz$$

$$+ (\sigma_g^2 - k_s^2) \int_{z_p}^{+\infty} |\vec{\mathbb{E}}_s(\nu, z)|^2 dz \,. \tag{7.45}$$

Given that the propagation constants of the guided modes satisfy relation (7.33), we can transform the inequality (7.45) into

$$\sum_{j=1}^{p}(k_j^2 - \sigma_g^2) \int_{z_{j-1}}^{z_j} |\vec{\mathbb{E}}_j(\nu, z)|^2 dz > 0 \,, \tag{7.46}$$

which requires that

$$\exists j \quad \sigma_g < k_j \quad \Rightarrow \quad \sigma_g < \max(k_j) \quad j = 1, ..., p \,. \tag{7.47}$$

Hence, in order for guided propagation to take place in a multilayer, there must be at least one medium $j$ in which it is possible to define an angle of refraction $\theta_j$ at the spatial frequency of the mode, i.e.,

$$\sigma_g = k_j \sin \theta_j \,. \tag{7.48}$$

Moreover, (7.35) requires that one of the constitutive media of the multilayer must have a refractive index greater than those of the substrate and superstrate; this condition can be interpreted by saying that there must be a layer of high refractive index in the stack in order for a guided wave to propagate.

Thus far, we have established this result only for TE polarization, but an entirely similar procedure can be carried out for TM polarization by replacing the elementary components of the electric field $\vec{\mathbb{E}}_j(\nu, z)$ with those of the magnetic field, using the two continuous variables $\vec{\mathbb{H}}(\nu, z)$ and $(1/\tilde{\epsilon}) \, \partial \vec{\mathbb{H}}/\partial z$.

### *7.6.3 Attenuation in Guided Propagation*

Contrary to the case of free space, it is helpful to point out that the spatial frequency can be complex for guided waves. This difference is because of the formalism used.

In free space we are working with a spatial wave packet that results from a Fourier transform in terms of the transverse space variable, a transform justified by the decay of the wave in that particular direction. The spatial frequency is therefore real, since this variable is the conjugate of the transverse space variable in this Fourier transform.

Conversely, in the case of a guided wave the same Fourier transform is no longer used, since the wave propagates in the transverse direction along which there is no decay in amplitude. It is therefore preferable to use another decomposition, in the forms of guided and radiative modes respectively, to show how it is possible to progress from an integral expression for the coupling zone to a series of modes at real or complex frequencies, provided the distance from that zone is sufficiently great.

If we now consider guided modes whose frequency is complex, this frequency will then be defined by an expression of the form $\sigma_g = \sigma'_g + i\sigma''_g$, and the poles must therefore be sought in the complex plane. The $x$ dependence of the modulus squared of the electric field associated with one of these modes will have the form $\exp(-2\sigma''_g x)$, thus defining an attenuation coefficient $\text{Att}_g$; this is usually expressed in dB/cm, i.e.,

$$e^{-2\sigma''_g x} = 10^{-10\text{Att}_g x} \quad \Rightarrow \quad \text{Att}_g \,(\text{dB/cm}) = \frac{2\sigma''_g}{10\ln(10)} . \qquad (7.49)$$

## 7.7 Relationships between Resonances and Guided Modes

### 7.7.1 Introduction

We showed in Sections 7.4 and 7.6.2 that, for a transparent multilayer component, the frequential windows related to total reflection (with resonance frequencies $\sigma_r$) and planar optics (with modal frequencies $\sigma_g$) were written respectively as

$$k_s < \sigma_r \leqslant k_0$$
$$\max(k_0, k_s) < \sigma_g < \max(k_j) \quad j = 1, ..., p. \qquad (7.50)$$

As things stand, it is clear that these two conditions are incompatible; this has already been highlighted in Section 7.6.2 to justify the modification to the mode of illumination necessitated by passing to a guided waves situation. In order for the same frequency $\sigma$ to be involved in both operational regimes, the configuration used in total reflection must be modified by introducing a new, high refractive index $(n_a)$ superstrate at a distance $d_0$ from the upper interface of the multilayer, as shown on the right of Figure 7.12. In this new structure, where the superstrate of index $n_a$ has replaced the initial superstrate of index $n_0$, the frequency associated with the incident wave can

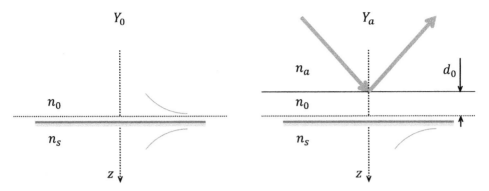

Figure 7.12 Operational configurations corresponding to planar optics (left) and total reflection (right), using in this case an additional material of high refractive index $n_a$.

be written $\sigma = k_a \sin \theta_a$ and is able to satisfy the condition for guided waves provided $n_a > \max(n_0, n_s)$. This means that, within the residual thickness $d_0$ of the initial superstrate of index $n_0$, the wave is now hyperbolic in nature. In other words, $\tilde{n}_0$ and $\tilde{n}_s$ are pure imaginary, while $\tilde{n}_a$ is real.

In practice, in the case of an initial air superstrate of index $n_0$, this amounts to approaching a prism of high refractive index $n_a$ from the upper interface of the multilayer, maintaining an air gap of thickness $d_0$ between this interface and the base of the prism.

However, to observe an absorption peak resulting from a resonance, the constituent materials of a multilayer must have a certain absorption, as mentioned in Section 7.5. This involves replacing the real refractive indices of these materials (assumed transparent) with complex values defined by

$$n_j \quad \longrightarrow \quad n_j + i\kappa \quad j = 1, ..., p, \tag{7.51}$$

where $\kappa$ is some low arbitrary value (e.g., $10^{-4}$ as chosen in Section 7.5), identical for all layers.

Since the principle of our thought experiment (use of a coupling prism of high refractive index separated from the multilayer by an air gap of thickness $d_0$ and introduction of an imaginary part $\kappa$ into the refractive index of the multilayers constitutive materials) has now been justified, we are in a position to examine whether or not there is a connection between the frequencies of the guided modes characteristic of the multilayer being considered (left-hand diagram in Figure 7.12) and those of the observed resonances when a prism of high index is approached from this structure (right-hand diagram in Figure 7.12). We must also analyze the role of the two free parameters, $d_0$ and $\kappa$, that are necessary for implementing this thought experiment.

### 7.7.2 Analytical Approach

*Transparent Case* $(\kappa = 0)$

We showed in Section 7.6.1 that the frequencies of the guided modes in the multilayer structure could be identified with the poles of the function $r_0(\bar{\nu})$, such that, in the case of a simple pole of characteristic frequency $\bar{\nu}_{g,0}$, we can write

$$r_0(\bar{\nu}) = \frac{f_0(\bar{\nu})}{\bar{\nu} - \bar{\nu}_{g,0}}, \tag{7.52}$$

where $f_0(\bar{\nu})$ is a slowly varying function of $\bar{\nu}$ in the neighborhood of $\bar{\nu}_{g,0}$. Recall here that $r_0$ is calculated for the geometry of the left drawing of Figure 7.12.

Furthermore, the function $r_0(\bar{\nu})$ is also classically defined by

$$r_0(\bar{\nu}) = \frac{\tilde{n}_0(\bar{\nu}) - Y_0(\bar{\nu})}{\tilde{n}_0(\bar{\nu}) + Y_0(\bar{\nu})} \quad \text{where} \quad \tilde{n}_0(\bar{\nu}) = \frac{i}{\eta_v}\sqrt{\bar{\nu}^2 - 1} = i\Im[\tilde{n}_0(\bar{\nu})]. \tag{7.53}$$

Combining (7.52) and (7.53), we can deduce an expression for the admittance $Y_0(\bar{\nu})$, i.e.,

$$Y_0(\bar{\nu}) = -\tilde{n}_0(\bar{\nu})\frac{f_0(\bar{\nu}) + (\bar{\nu} - \bar{\nu}_{g,0})}{f_0(\bar{\nu}) - (\bar{\nu} - \bar{\nu}_{g,0})}, \tag{7.54}$$

where we note that, in the absence of absorption, the function $f_0(\bar{\nu})$ is real ($r_0$ is real, since $\tilde{n}_0$ and $Y_0$ are pure imaginary).

Now that we have an expression for the admittance $Y_0$ in the neighborhood of the characteristic frequency of a guided mode in the multilayer (see Figure 7.12, left drawing), we are in a position to deduce the admittance of the lower face of the high-index superstrate. Indeed the two geometries shown in Figure 7.12 are similar with identical substacks (below the air gap layer), while we know that only these substacks are involved in the $Y_0$ admittance calculation, which is therefore the same for the left and right figures. Hence the two admittances are related by the expression

$$Y_a = \tilde{n}_0\frac{Y_0 \cos\delta_0 - i\tilde{n}_0\sin\delta_0}{\tilde{n}_0\cos\delta_0 - iY_0\sin\delta_0}, \tag{7.55}$$

where

$$\begin{aligned}
\delta_0 &= ik_v d_0 \Im[\eta_v \tilde{n}_0(\bar{\nu})] = ix \quad x \in \mathbb{R} \\
\cos\delta_0 &= \cosh x \\
\sin\delta_0 &= i\sinh x\,.
\end{aligned} \tag{7.56}$$

We immediately get

$$Y_a = i\Im[\tilde{n}_0]\frac{Y_0 \cosh x + i\Im[\tilde{n}_0]\sinh x}{Y_0 \sinh x + i\Im[\tilde{n}_0]\cosh x}, \qquad (7.57)$$

showing that this admittance $Y_a$ is also pure imaginary. Consequently the reflection coefficient $r_a$ of the Prism / Air Gap / Multilayer structure, defined by

$$r_a = \frac{\tilde{n}_a - Y_a}{\tilde{n}_a + Y_a}, \qquad (7.58)$$

has a modulus of 1 ($|r_a| = 1$), since $n_a$ is real. In the absence of losses in the multilayer, reflection is total.

We shall now examine whether the function $r_a$ has a pole in the neighborhood of $\bar{\nu}_{g,0}$, i.e., whether there exists a spatial frequency $\bar{\nu}_{g,a}$, close to $\bar{\nu}_{g,0}$, that is a solution of the equation $\tilde{n}_a(\bar{\nu}) + Y_a(\bar{\nu}) = 0$. Notice that this pole cannot be real because the reflection coefficient $R_a = |r_a|^2$ is involved in an energy balance, contrary to $R_0 = |r_0|^2$. Combining (7.54) and (7.57), we show that this solution has the form

$$\bar{\nu}_{g,a}(x) = \bar{\nu}_{g,0} + f_0(\bar{\nu}_{g,0})\, e^{i\varphi_a(\bar{\nu}_{g,0})}\frac{1 - \tanh x}{1 + \tanh x}, \qquad (7.59)$$

where

$$e^{i\varphi_a(\bar{\nu}_{g,0})} = \frac{\sqrt{n_a^2 - \bar{\nu}_{g,0}^2} - i\sqrt{\bar{\nu}_{g,0}^2 - 1}}{\sqrt{n_a^2 - \bar{\nu}_{g,0}^2} + i\sqrt{\bar{\nu}_{g,0}^2 - 1}} \qquad \text{and} \qquad x = k_v d_0\sqrt{\bar{\nu}_{g,0}^2 - 1}. \quad (7.60)$$

Equation (7.59) shows that as the thickness $d_0$ of the air gap grows, the frequency of the pole $\bar{\nu}_{g,a}$ tends asymptotically to the real value $\bar{\nu}_{g,0}$, corresponding to the frequency of the guided mode of the multilayer structure alone. Intuitively, this result is related to the fact that, when the air gap increases, the configuration of the right drawing of Figure 7.12 approaches that of the left drawing of the same figure. Indeed we recall again that the field calculation is performed by recurrence with a starting value in the substrate, so that both fields in Figure 7.12 (left and right drawings) are identical up to the Superstrate/Air Gap interface. Moreover, in this air gap the field is hyperbolic with a strong decrease toward the superstrate. To summarize, it is as if the superstrate was absent. Note that convergence is rapid: being driven by the quantity $x$, the limit is reached in practice for air gap thicknesses on the order of a wavelength.

Similarly, it can be shown that the complex resonant frequency $\nu_{r,a}$ of this same Prism / Air Gap / Multilayer structure, i.e., the frequency that causes the function $r_a(\bar{\nu})$ to fall to 0, and which is the solution to the equation $\tilde{n}_a(\bar{\nu}) - Y_a(\bar{\nu}) = 0$, is equal to the complex conjugate of the corresponding pole $\bar{\nu}_{g,a}$ i.e.,

$$\bar{\nu}_{r,a}(x) = \bar{\nu}_{g,0} + f_0(\bar{\nu}_{g,0})\, e^{-i\varphi_a(\bar{\nu}_{g,0})} \frac{1 - \tanh x}{1 + \tanh x}. \tag{7.61}$$

This property is general in scope and originates in the fact that the reflection coefficient $r_a$ has a modulus of 1. It can therefore be represented by an expression of the form

$$r_a = e^{i\psi} \frac{\bar{\nu} - \bar{\nu}_{r,a}}{\bar{\nu} - \bar{\nu}_{g,a}}, \tag{7.62}$$

where $\bar{\nu}_{r,a}$ and $\bar{\nu}_{g,a}$ are complex conjugates of one another.

For an arbitrary air gap thickness, these poles and zeros of $r_a$ are complex and therefore cannot be reached by an incident illumination frequency, though their real part can be approached to create field enhancement in the absence of absorption. When this gap thickness grows, the frequencies of the poles ($\nu_{g,a}$) and zeros ($\nu_{r,a}$) can be identified asymptotically with the same real value, namely the frequency ($\nu_{g,0}$) of the guided mode. This last result makes the connection between the guided modes and the resonances under the total reflection regime.

### *Absorbing Case ($\kappa > 0$)*

We now assume that the imaginary part of the refractive index of the ensemble of layers constituting the stack are equal to one and the same value $\kappa$, much smaller than 1 (e.g., between $10^{-6}$ and $10^{-4}$); this is true for the imaginary indices currently measured for dielectrics in thin layers. To a first approximation, it is therefore possible to take account of this low extinction coefficient by expanding $\eta_v Y_0$ about $\kappa = 0$, i.e.,

$$\eta_v Y_0(\bar{\nu}, \kappa) \approx \eta_v Y_0(\bar{\nu}, 0) + \kappa \eta_v \left.\frac{\partial Y_0}{\partial \kappa}\right|_{\kappa=0} = \eta_v Y_0(\bar{\nu}, 0) + \kappa B_0. \tag{7.63}$$

As we shall examine the variations in $\eta_v Y_0$ in the neighborhood of frequency $\bar{\nu}_{g,0}$, we shall approximate this quantity as being linearly dependent so we can again expand, but this time about $\bar{\nu}_{g,0}$, i.e.,

$$\eta_v Y_0(\bar{\nu}, 0) \approx \eta_v Y_0(\bar{\nu}_{g,0}, 0) + (\bar{\nu} - \bar{\nu}_{g,0})\eta_v \left.\frac{\partial Y_0}{\partial \bar{\nu}}\right|_{\bar{\nu}_{g,0}}$$

$$= -i\sqrt{\bar{\nu}_{g,0}^2 - 1} + i(\bar{\nu} - \bar{\nu}_{g,0})A_g, \tag{7.64}$$

where $A_g$ is real. Combining (7.63) and (7.64), we obtain, to a first approximation,

$$\eta_v Y_0(\bar{\nu}, \kappa) \approx -i\sqrt{\bar{\nu}_{g,0}^2 - 1} + i(\bar{\nu} - \bar{\nu}_{g,0})A_g + \kappa B_0 \,. \qquad (7.65)$$

Rigorously speaking, the coefficient $B_0$ is complex, but as its imaginary part is added to the principal term $-i\sqrt{\bar{\nu}_{g,0}^2 - 1}$ after multiplying by $\kappa$, this contribution can be neglected, and we shall henceforth assume that $B_0$ is real.

Expression (7.57) defining the admittance $Y_a$ is still valid, so we can inquire as to whether there exists a real value $\bar{\nu}_{r,a}(\kappa, d_0)$ of the normalized spatial frequency that corresponds to a 0 of the function $r_a(\bar{\nu})$ in the presence of absorption in the layers, i.e.,

$$r_a(\nu_{r,a}) = 0 \quad \Rightarrow \quad \tilde{n}_a = Y_a \,. \qquad (7.66)$$

By combining (7.57), (7.65), and (7.66), we get

$$\eta_v \tilde{n}_a \left\{ \left[-i\sqrt{\bar{\nu}_{g,0}^2 - 1} + iA_g(\bar{\nu} - \bar{\nu}_{g,0}) + B_0\kappa\right] \sinh x + i\sqrt{\bar{\nu}^2 - 1}\cosh x \right\}$$
$$= i\sqrt{\bar{\nu}^2 - 1} \left\{ \left[-i\sqrt{\bar{\nu}_{g,0}^2 - 1} + iA_g(\bar{\nu} - \bar{\nu}_{g,0}) + B_0\kappa\right] \cosh x + i\sqrt{\bar{\nu}^2 - 1}\sinh x \right\},$$

and, separating the real and imaginary parts,

$$\left[\sqrt{\bar{\nu}^2 - 1} + \frac{\sqrt{n_a^2 - \bar{\nu}^2}}{\sqrt{\bar{\nu}^2 - 1}} B_0\kappa\right] \sinh x = \left[\sqrt{\bar{\nu}_{g,0}^2 - 1} - A_g(\bar{\nu} - \bar{\nu}_{g,0})\right] \cosh x \,, \quad (7.67)$$

$$\left[\sqrt{\bar{\nu}_{g,0}^2 - 1} - A_g(\bar{\nu} - \bar{\nu}_{g,0})\right] \sinh x = \left[\sqrt{\bar{\nu}^2 - 1} - \frac{\sqrt{\bar{\nu}^2 - 1}}{\sqrt{n_a^2 - \bar{\nu}^2}} B_0\kappa\right] \cosh x \,. \quad (7.68)$$

Combining these two expressions, we obtain

$$\tanh x = \frac{\sqrt{\bar{\nu}_{g,0}^2 - 1} - A_g(\bar{\nu} - \bar{\nu}_{g,0})}{\sqrt{\bar{\nu}^2 - 1} + \frac{\sqrt{n_a^2 - \bar{\nu}^2}}{\sqrt{\bar{\nu}^2-1}} B_0\kappa} = \frac{\sqrt{\bar{\nu}^2 - 1} - \frac{\sqrt{\bar{\nu}^2-1}}{\sqrt{n_a^2-\bar{\nu}^2}} B_0\kappa}{\sqrt{\bar{\nu}_{g,0}^2 - 1} - A_g(\bar{\nu} - \bar{\nu}_{g,0})} \,. \qquad (7.69)$$

To a first approximation in $\kappa$ and $\bar{\nu} - \bar{\nu}_{g,0}$, we then calculate the optimum thickness $\bar{d}_0$ of the air gap:

$$\tanh[k_v \bar{d}_0 \sqrt{\bar{\nu}_{g,0}^2 - 1}] = 1 - \frac{B_0\kappa}{2\sqrt{\bar{\nu}_{g,0}^2 - 1}} \left\{ \frac{\sqrt{n_a^2 - \bar{\nu}_{g,0}^2}}{\sqrt{\bar{\nu}_{g,0}^2 - 1}} + \frac{\sqrt{\bar{\nu}_{g,0}^2 - 1}}{\sqrt{n_a^2 - \bar{\nu}_{g,0}^2}} \right\}, \qquad (7.70)$$

and the value $\bar{\nu}_{r,a}$ of the corresponding resonant frequency:

$$\bar{\nu}_{r,a} = \bar{\nu}_{g,0} - \frac{B_0 \kappa}{2 \left[ A_g + \frac{\bar{\nu}_{g,0}}{\sqrt{\bar{\nu}_{g,0}^2 - 1}} \right]} \left\{ \frac{\sqrt{n_a^2 - \bar{\nu}_{g,0}^2}}{\sqrt{\bar{\nu}_{g,0}^2 - 1}} - \frac{\sqrt{\bar{\nu}_{g,0}^2 - 1}}{\sqrt{n_a^2 - \bar{\nu}_{g,0}^2}} \right\}. \qquad (7.71)$$

Note that equation (7.70) has a solution only if $B_0$ is **positive**. These results show that, in this case, if the function $r_0(\bar{\nu})$ associated with the transparent multilayer guide has a pole in $\bar{\nu}_{g,0}$, then a convenient choice for the thickness $d_0$ of the air gap between a high-index superstrate and this guide, now assumed to be weakly absorbing, allows an absorption of 100% to be obtained at a frequency $\bar{\nu}_{r,a}$ close to $\bar{\nu}_{g,0}$. Conversely, if $B_0$ is negative, it will no longer be possible to obtain total absorption by altering the thickness of the air gap. As mentioned in Section 7.5, this absorption (whether it is total or not) reveals the existence of a resonance, i.e., an increase in the modulus squared of the field within the thickness of the multilayer at that particular frequency.

For completeness, note that these results were obtained assuming a linear approximation, and that their use will require that this constructive assumption is properly satisfied (tanh $x$ close to 1, small difference between $\bar{\nu}_{r,a}$ and $\bar{\nu}_{g,0}$).

### 7.7.3 Numerical Example

To confirm the conclusions reached by the analytical approach developed in the previous section, we shall now consider a particular case of a multilayer planar waveguide, and shall analyze numerically its modal properties and potentially resonant behavior in the total reflection regime. For clarity, we choose a simple Air / 8H / Air monolayer structure, where H denotes a $Ta_2O_5$ quarter-wave layer at a wavelength of 633 nm ($n_H = 2.1417$, $d_H = 561.12$ nm).

#### Guided Modes

For the transparent case ($\kappa = 0$) the normalized spatial frequencies $\bar{\nu}_{g,0}$ corresponding to the guided modes of this structure are given by the zeros of the function $1/R_0(\bar{\nu})$; its variations are shown in Figure 7.13 in logarithmic units (TE polarization) as $\bar{\nu}$ traverses the modal window $[1, n_H]$.

Note the presence of four guided modes, numbered according to the convention defined in Section 7.5; the corresponding characteristic frequencies are set out in Table 7.2.

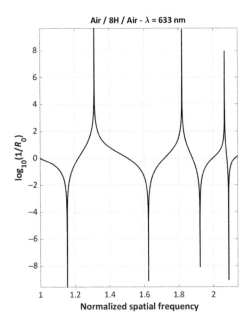

Figure 7.13 Variation of the function $1/R_0$ with the normalized spatial frequency $\bar{\nu}$ for the structure Air / 8H / Air (illuminating wavelength equal to the design wavelength, i.e., 633 nm).

Table 7.2 Normalized spatial frequencies associated with the four modes of the Air / 8H / Air structure (TE polarization, $\lambda = 633$ nm, $\kappa = 0$)

| Order | 3 | 2 | 1 | 0 |
|---|---|---|---|---|
| $\bar{\nu}_{g,0}$ | 1.1550 | 1.6252 | 1.9237 | 2.0887 |

Note also the presence of three zeros of the function $R_0(\bar{\nu})$ that correspond to the particular spatial frequencies where the H layer is *absent* at the illuminating wavelength.

### *Resonances*

The locations of resonances in the structure high-index Superstrate / Air gap / 8H / Air are determined from the plot of the function $R_a(\bar{\nu})$ when $\bar{\nu} > n_s = 1$ and $\kappa > 0$. Figure 7.14 compares the variations in the functions $R_a$ and $1/R_0$ over the range of normalized spatial frequencies between 1 and 2.5 for an arbitrarily chosen quarter-wave air gap ($d_0 = 158.25$ nm) and an extinction coefficient $\kappa$ of $10^{-4}$.

Note that the number of resonances (four) is identical to that of the guided modes. Note also that the normalized spatial frequencies associated with the

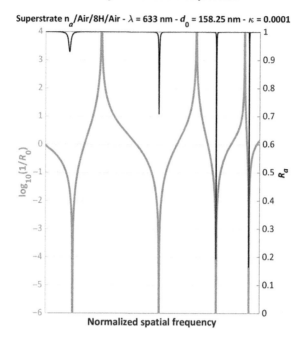

Figure 7.14 Comparison of variations in functions $R_a$ (in black) and $1/R_0$ (in gray) with the value of the normalized spatial frequency $\bar{\nu}$ for the structure Air / 8H / Air.

resonances of the prism/air gap structure are, as expected, very close to those that characterize the guided modes of the structure Air / 8H / Air (they would be identical for an infinite air gap). For greater clarity, Table 7.3 gives the values of normalized spatial frequencies corresponding both to these resonances and to the previously determined guided modes; we have also indicated the corresponding value of absorbance for each resonance.

Table 7.3 Comparison between the normalized spatial frequencies associated respectively with the resonances and the guided modes of the structure Air / 8H / Air, showing also for each resonance the corresponding value of absorbance (TE polarization, $\lambda = 633$ nm, $n_a = 4$, $d_0 = 158.25$ nm, $\kappa = 10^{-4}$)

| Order | 3 | 2 | 1 | 0 |
|---|---|---|---|---|
| $\bar{\nu}_{r,a}$ | 1.1359 | 1.6235 | 1.9235 | 2.0887 |
| $\bar{\nu}_{g,0}$ | 1.1550 | 1.6252 | 1.9237 | 2.0887 |
| $\bar{\nu}_{r,a} - \bar{\nu}_{g,0}$ | −0.0191 | −0.0017 | −0.0002 | 0.0000 |
| $A(\bar{\nu}_{r,a})$ | 7.1% | 29.0% | 80.4% | 84.0% |

We shall now analyze numerically whether there is, for each mode, a particular value of air gap that yields an absorbance of 100%, and shall compare this optimum thickness $\bar{d}_0$, as well as the value of $\bar{\nu}_{r,a}$ of the associated resonant frequency, with those defined by our analytical approach.

To this end we studied the variations in the characteristic frequency of each resonance and the variations in the associated absorbance over a wide range of air gap thicknesses (between 0 and 1,000 nm). The results of this numerical investigation are shown in Figure 7.15 for resonances of order 0 (left) and order 3 (right).

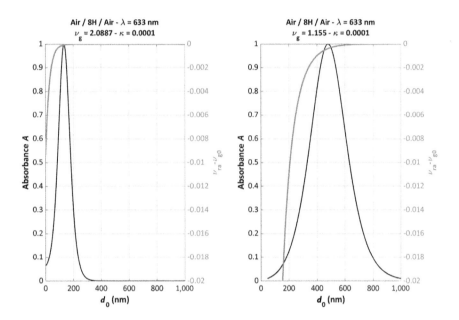

Figure 7.15 Variations in the normalized spatial frequency $\bar{\nu}_{r,a}$ of resonances [gray] and associated absorbance $A(\bar{\nu}_{r,a})$ [black] as a function of thickness $d_0$ of the air gap in the case of the structure Superstrate of index 4 / Air gap / 8H / Air (left, mode of order 0; right, mode of order 3).

For each of the resonances in the structure being considered, this investigation confirms that there is indeed a particular value for the thickness of air gap for which the absorbance is 100% (all coefficients $B_0$ are positive). Moreover, the results obtained can be compared with those obtained using the analytical approach described in Section 7.7.2 [see equations (7.70) and (7.71)], as summarized in Table 7.4.

The agreement between the analytical and numerical values is quite excellent, both for the optimal air gaps and the associated resonant frequencies.

Table 7.4 Comparison between optimum air gap thicknesses and the associated resonant frequencies determined analytically ($\bar{d}_0^{\text{ana}}$, $\bar{\nu}_{r,a}^{\text{ana}}$) and numerically ($\bar{d}_0^{\text{num}}$, $\bar{\nu}_{r,a}^{\text{num}}$) for the four modes of the structure Air / 8H / Air ($\lambda = 633$ nm, $\kappa = 10^{-4}$, $n_a = 4$)

| Mode | $\bar{\nu}_{g,0}$ | $A_g$ | $B_0$ | $\bar{d}_0^{\text{ana}}$ | $\bar{d}_0^{\text{num}}$ | $\bar{\nu}_{r,a}^{\text{ana}} - \bar{\nu}_{g,0}$ | $\bar{\nu}_{r,a}^{\text{num}} - \bar{\nu}_{g,0}$ |
|---|---|---|---|---|---|---|---|
| 0 | 2.0887 | 221.6 | 226.0 | 134.6 | 134.7 | $-67 \times 10^{-6}$ | $-68 \times 10^{-6}$ |
| 1 | 1.9237 | 51.68 | 56.23 | 187.1 | 197.3 | $-89 \times 10^{-6}$ | $-88 \times 10^{-6}$ |
| 2 | 1.6252 | 20.10 | 24.81 | 254.3 | 254.5 | $-145 \times 10^{-6}$ | $-144 \times 10^{-6}$ |
| 3 | 1.1550 | 9.503 | 13.92 | 479.1 | 480.6 | $-391 \times 10^{-6}$ | $-390 \times 10^{-6}$ |

However, all these verifications were made for a particular value of the extinction coefficient $\kappa$ ($10^{-4}$). If this value is reduced, e.g., by a factor of 10 or 100, we may wonder whether this reduction might still allow an absorbance of 100% to be attained. Our analytical approach guarantees this, since our linearity assumption can only get better and better and, e.g., in the case of the order 3 mode for the Air / 8H / Air structure, its application leads to optimum air gap thicknesses of 479.6 nm, 680.1 nm, and 880.8 nm for values of $\kappa$ equal respectively to $10^{-4}$, $10^{-5}$, and $10^{-6}$.

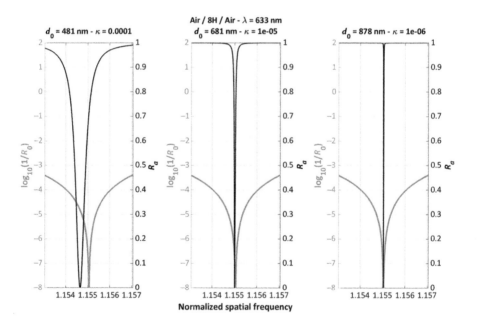

Figure 7.16 Variations of the reflection coefficient $R_a$ as a function of the normalized spatial frequency $\bar{\nu}$ for optimum air gap thicknesses corresponding to three different values of $\kappa$ for the third-order mode of the Air / 8H / Air structure.

Note also that the optimum thickness of the air gap increases as the value of $\kappa$ gets smaller. This result, confirmed by a numerical check, might appear surprising since the effect of increasing this air gap is to reduce the effective amplitude of the incident wave in the immediate proximity of the 8H monolayer. The explanation of this apparent contradiction is provided by examining the curves of the reflection coefficient $R_a$ in the neighborhood of the normalized spatial frequency of the associated mode, i.e., $\bar{\nu}_{g,0} = 1.1550$ (see Figure 7.16).

Observe that the mid-height width of the resonance varies in proportion to this same coefficient $\kappa$, while the associated normalized frequency $\bar{\nu}_{r,a}$ approaches the characteristic frequency $\bar{\nu}_{g,0}$ of the corresponding mode, as per relation (7.71). This means that physically, i.e., when the illuminating beam comprises a wave packet, the apparent absorption will be a decreasing function of air gap thickness, since the angular aperture (of the beam) associated with this unit absorbance will be smaller and smaller until, in the case of a transparent material, it becomes a Dirac distribution.

### Field Profiles

We now propose to plot the variation in $z$ of the modulus squared of the field in the structure high-index superstrate / Air gap / 8H / Air, when the normalized spatial frequency is a resonant frequency and when the air gap thickness is equal to the corresponding optimum value. The imaginary part of the refractive index is taken to be $10^{-4}$.

Figure 7.17 shows the tangential field profiles obtained for the four resonances of the structure Index-4 superstrate / Air Gap / 8H / Air (curves in black).

Furthermore, we have also plotted (in gray, but the curves are superimposed to the black ones) the same field profiles determined using the formulae from the theory of symmetric planar waveguides, i.e., for all TE modes of even order (0 and 2),

$$\frac{|\vec{E}_{tg}(z)|^2}{|\vec{E}_{tg}|^2_{max}} = \begin{cases} \cos^2\left[\frac{2\pi}{\lambda}(z - z_H)\sqrt{n_H^2 - \bar{\nu}^2_{g,0}}\right] & |z - z_H| \leqslant \dfrac{d_H}{2} \\[2em] \cos^2\left[\frac{2\pi}{\lambda}d_H\sqrt{n_H^2 - \bar{\nu}^2_{g,0}}\right] \dfrac{e^{-\frac{4\pi}{\lambda}|z-z_H|\sqrt{\bar{\nu}^2_{g,0}-1}}}{e^{-\frac{2\pi}{\lambda}d_H\sqrt{\bar{\nu}^2_{g,0}-1}}} & |z - z_H| > \dfrac{d_H}{2} \end{cases},$$

and for TE modes of odd order (1 and 3),

$$\frac{|\vec{E}_{tg}(z)|^2}{|\vec{E}_{tg}|^2_{max}} = \begin{cases} \sin^2\left[\frac{2\pi}{\lambda}(z - z_H)\sqrt{n_H^2 - \bar{\nu}^2_{g,0}}\right] & |z - z_H| \leqslant \dfrac{d_H}{2} \\[2em] \sin^2\left[\frac{2\pi}{\lambda}d_H\sqrt{n_H^2 - \bar{\nu}^2_{g,0}}\right] \dfrac{e^{-\frac{4\pi}{\lambda}|z-z_H|\sqrt{\bar{\nu}^2_{g,0}-1}}}{e^{-\frac{2\pi}{\lambda}d_H\sqrt{\bar{\nu}^2_{g,0}-1}}} & |z - z_H| > \dfrac{d_H}{2} \end{cases}.$$

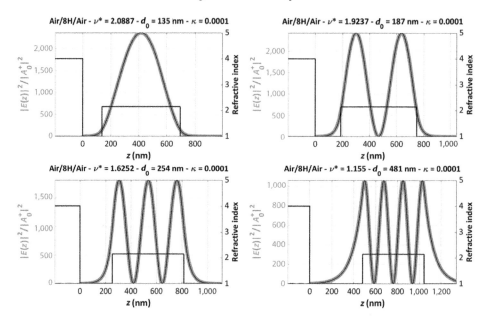

Figure 7.17 Field profiles associated with the four resonances of the structure Index-4 superstrate / Air gap / 8H / Air when the thickness of the
air gap is equal to $\bar{d}_0$, leading to total absorption (black); in gray are the
field profiles corresponding to the guided modes in the same structure.

Note in Figure 7.17 that the field profiles calculated numerically using the
formalism of complex admittances coincide perfectly with those determined
analytically using the theory of symmetric planar waveguides.

Note also that the enhancement factors vary between 2430 (mode of order
1, $\bar{\nu}_{r,a} = 1.9237$) and 1060 (mode of order 3, $\bar{\nu}_{r,a} = 1.1550$) and that the
confinement of the mode (related to its integral in medium H) falls with the
number of its order.

## *Conclusion*

The conclusions we can draw at the end of this investigation into the links
between resonances and guided modes are as follows:

- In the total internal reflection (TIR) regime, the reflection coefficient $r_a$
  may support poles ($\nu_{g,a}$) and zeros ($\nu_{r,a}$) whose frequencies are conjugated.
  These poles frequencies cannot be real because reflection is bounded.
- For each resonant frequency $\bar{\nu}_{r,a}$ and each extinction coefficient $\kappa$ in the
  layers, there can exist an air gap thickness for which the ab-
  sorbance $A(\bar{\nu}_{r,a})$ is equal to 1.
- When the air gap between the prism and the multilayer increases, the

complex frequencies of these poles and zeros asymptotically tend to the real frequencies of the guided modes of the stack (considered without the prism). This makes the connection between guided optics (modal frequencies) and the TIR regime (resonance frequencies).

- The resonance frequencies allow strong field enhancement (several decades) to occur in the multilayer, which create strong absorption in case of a dissipative stack.

We can also observe that, in the absence of absorption, detecting resonances experimentally is not immediately obvious unless we know the phase variations of the reflection coefficient $r_a$; indeed the phase shows strong variations in the resonance neighborhood (due to a sign change of the reflection coefficient), and they can be determined using ellipsometric or interferometric techniques, for example.

More generally, these results are a reminder of the level of precision (size of air gap and associate flatness, divergence of illumination, etc.) required when carrying out a coupling experiment using a prism in a multilayer guide. Such an experiment involves using a prism cut at a right angle to allow the light from the superstrate to penetrate into the waveguide (see Figure 7.18). Intuitively, this coupling can be explained as follows: the left-hand figure

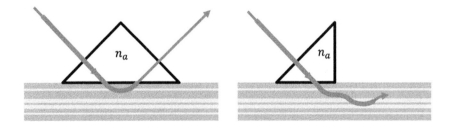

Figure 7.18 Principle of coupling with a prism.

shows the distance between the barycenters of the incident and reflected beams (known as the *Goos–Hanchen shift*), and it can be shown that this distance increases at the resonant frequency for an optimal gap. In the right-hand figure, the incident beam falls in the neighborhood of the prism's edge. Under these conditions, the barycenter of the reflected beam is in the zone where the prism is no longer present: the supposedly reflected light hence remains trapped in the guide because it sees a superstrate of low refractive index. Although an important issue, calculation of the coupling efficiency is not tackled here.

Finally, by way of introduction to the next section, we have up to this point been content to show the existence of resonances and associated enhancements for a single layer. We shall now examine whether it is possible to find a constructive method for choosing the frequency, polarization, and amplitude of these enhancements in multilayer systems.

## 7.8 Synthesis of Giant Field Enhancements

This section is therefore dedicated to the synthesis of multilayer stacks of dielectric materials that allow giant electromagnetic field enhancements to be obtained. For preestablished illumination conditions (incidence, wavelength, and polarization), it is convenient to be able to control the level of electromagnetic field enhancement in this type of stack, since it then becomes possible to develop low-threshold microsources or high sensitivity optical sensors. The method we shall describe is based on the use of **zero admittance layers**.

### 7.8.1 Preliminary Remarks

Electromagnetic field enhancements (or high-intensity or resonance phenomena) are common in interference structures, whether stratified, guided, or even spherical. In the case of a Fabry–Perot multilayer (see Chapter 6, Section 6.4.4), we know that heightened intensity appears in the median layer (the *spacer*) that acts as a cavity. It is worth observing (see Figure 6.16) that, generally speaking, the field is a maximum in those places where the admittance is low.

It is easy to explain this phenomenon using the absorbance relation established in Chapter 6 (Section 7.114); for a transparent medium, the quantity below describing the flux transported across a plane of side $z_j$,

$$\Phi(z_j) = \frac{1}{2}\Re[Y_j]|\vec{\mathbb{E}}_{j,\mathrm{tg}}(z_j)|^2 \,,$$

is constant in the stack. Therefore maxima of the field are obtained at $z$ locations where the real part of admittance is low. The optimal situation would be that of a zero admittance, but this is not possible in the free space regime (in opposition to the TIR regime); otherwise the flux would be 0 throughout the whole stack, including the substrate. Still using the Fabry–Perot example in free space, these considerations explain why low values of admittance in the spacer are obtained asymptotically as a function of the number of layers.

Furthermore, with this type of Fabry–Perot structure, the field enhancement stays confined within the core of the stack; this is unsuitable for sensor type applications where we seek to probe external media (generally the substrate). This observation is essential since, if we wish to probe an external medium at high sensitivity, the field in this medium must be intense (several orders of magnitude greater than the incident field). However, this output field is proportional to the coefficient of transmission $t$, which is bounded; indeed in a propagative configuration (see Chapter 6), we have

$$T = \frac{\Phi_s^+}{\Phi_0^+} = \frac{\Re[\tilde{n}_s]|\vec{\mathbb{A}}_s^+|^2}{\Re[\tilde{n}_0]|\vec{\mathbb{A}}_0^+|^2} = \frac{\tilde{n}_s}{\tilde{n}_0}|t|^2 \leqslant 1 \quad \Rightarrow \quad |t|^2 \leqslant \frac{\tilde{n}_0}{\tilde{n}_s}. \tag{7.72}$$

This leads to the idea that the total reflection regime could be used to get round this restriction. In this regime, we have seen that the effective index of the substrate is pure imaginary, i.e.,

$$\Re[\tilde{n}_s] = 0 \quad \rightarrow \quad T = \frac{\Re[\tilde{n}_s]|\vec{\mathbb{A}}_s^+|^2}{\Re[\tilde{n}_0]|\vec{\mathbb{A}}_0^+|^2} = 0. \tag{7.73}$$

Hence expression (7.73), established for total reflection, replaces the previous expression (7.72) and places no restrictions on the value of the modulus squared in the lower interface of the multilayer ($|\vec{\mathbb{A}}_{s,tg}|^2$). Indeed the zero transmission in energy is valid whatever the amplitude $t$ of the stack in the TIR regime.

In addition, we will show that this same regime will also make it possible to obtain zero values of complex admittance within an optimized stack.

### *7.8.2 Field Distribution in a Quarter-Wave Stack*

Here we consider a stack for which all the layers are both transparent and quarter-wave for a specific incidence (we say that they are *matched*), which requires the following condition to be met:

$$\delta_j = \alpha_j d_j = \frac{\pi}{2} \quad \forall j = 1, ..., p. \tag{7.74}$$

The quantities $\alpha_j$ are therefore real, such that the field variation within the thickness of all layers is trigonometric ($\sigma < k_j$); this means that (7.74) can be put in the following equivalent form:

$$n_j d_j \cos \theta_j = \frac{\lambda_0}{4} \quad \forall j = 1, ..., p, \tag{7.75}$$

where $\lambda_0$ is the design wavelength.

The characteristic matrix $M_j$ associated with each of these layers is written (see Chapter 6) as

$$M_j = \begin{bmatrix} \cos \delta_j & -\frac{i}{\tilde{n}_j} \sin \delta_j \\ -i\tilde{n}_j \sin \delta_j & \cos \delta_j \end{bmatrix} = -i \begin{bmatrix} 0 & \frac{1}{\tilde{n}_j} \\ \tilde{n}_j & 0 \end{bmatrix}, \qquad (7.76)$$

where

$$\begin{bmatrix} \vec{z} \wedge \vec{\mathbb{E}}_{j-1,\text{tg}}(z_{j-1}) \\ \vec{\mathbb{H}}_{j-1,\text{tg}}(z_{j-1}) \end{bmatrix} = M_j \begin{bmatrix} \vec{z} \wedge \vec{\mathbb{E}}_{j,\text{tg}}(z_j) \\ \vec{\mathbb{H}}_{j,\text{tg}}(z_j) \end{bmatrix}. \qquad (7.77)$$

Given these basic recalls, we now turn our interest to four different quarter-wave stacks $(Q_i)$ defined by the following formulas:

$$Q_1 = (\text{LH})^N \text{S} \quad Q_2 = (\text{HL})^N \text{S} \quad Q_3 = \text{H}(\text{LH})^N \text{S} \quad Q_4 = \text{L}(\text{HL})^N \text{S},$$

where H and L are the quarter-wave layers of high and low refractive index respectively for the incidence and illuminating wavelength, and S is the substrate, assumed semi-infinite. As is customary, the index of the incident medium is $n_0$.

Using equation (7.76), it is easy to show that the matrix associated with a pair of layers (LH or HL) can be written as

$$M_{\text{LH}} = M_{\text{L}} M_{\text{H}} = -\begin{bmatrix} \beta & 0 \\ 0 & 1/\beta \end{bmatrix} \qquad M_{\text{HL}} = M_{\text{H}} M_{\text{L}} = -\begin{bmatrix} 1/\beta & 0 \\ 0 & \beta \end{bmatrix}, \qquad (7.78)$$

where $\beta = \tilde{n}_H/\tilde{n}_L$ is the ratio of effective indices in the two materials being used.

We can immediately deduce an expression for the characteristic matrices associated with the four configurations envisaged, namely

$$M_1 = (-1)^N \begin{bmatrix} \beta^N & 0 \\ 0 & 1/\beta^N \end{bmatrix} \quad M_3 = (-1)^{N+1} \begin{bmatrix} 0 & i/(\tilde{n}_H \beta^N) \\ i\tilde{n}_H \beta^N & 0 \end{bmatrix}$$

$$M_2 = (-1)^N \begin{bmatrix} 1/\beta^N & 0 \\ 0 & \beta^N \end{bmatrix} \quad M_4 = (-1)^{N+1} \begin{bmatrix} 0 & i\beta^N/\tilde{n}_L \\ i\tilde{n}_L/\beta^N & 0 \end{bmatrix}. \qquad (7.79)$$

Using (7.77) and (7.79), for each configuration $Q_i$ we can express the ratio of the moduli squared of the fields associated with the two extremal interfaces:

$$\frac{|\vec{\mathbb{E}}_{s,\text{tg}}|^2}{|\vec{\mathbb{E}}_{0,\text{tg}}|^2} = \begin{cases} 1/\beta^{2N} & \text{for } Q_1 \\ \beta^{2N} & \text{for } Q_2 \\ \beta^{2N}|\tilde{n}_H/Y_s|^2 & \text{for } Q_3 \\ (1/\beta^{2N})|\tilde{n}_L/Y_s|^2 & \text{for } Q_4 \end{cases} \qquad (7.80)$$

However, the field enhancement is defined as the ratio of the moduli squared of the transmitted and incident fields. For the first, clearly

$$\vec{\mathbb{E}}_{s,\text{tg}} = \vec{\mathbb{A}}^+_{s,\text{tg}} = t\,\vec{\mathbb{A}}^+_{0,\text{tg}}, \tag{7.81}$$

while for the second, the relation can be written

$$\vec{\mathbb{E}}_{0,\text{tg}} = (1+r)\vec{\mathbb{A}}^+_{0,\text{tg}} \quad \text{where} \quad r = \frac{\tilde{n}_0 - Y_0}{\tilde{n}_0 + Y_0}. \tag{7.82}$$

We also deduce immediately that

$$\frac{|\vec{\mathbb{A}}^+_{s,\text{tg}}|^2}{|\vec{\mathbb{A}}^+_{0,\text{tg}}|^2} = |1+r|^2\frac{|\vec{\mathbb{E}}_{s,\text{tg}}|^2}{|\vec{\mathbb{E}}_{0,\text{tg}}|^2} = \left|\frac{2\tilde{n}_0}{\tilde{n}_0 + Y_0}\right|^2\frac{|\vec{\mathbb{E}}_{s,\text{tg}}|^2}{|\vec{\mathbb{E}}_{0,\text{tg}}|^2}. \tag{7.83}$$

Hence we need to determine an expression for the admittance $Y_0$ for the different configurations envisaged. The simplest way to do this is to make use of the definition of admittance at the extremal interfaces, i.e.,

$$\vec{\mathbb{H}}_{0,\text{tg}} = Y_0\,[\vec{z} \wedge \vec{\mathbb{E}}_{0,\text{tg}}] \qquad \vec{\mathbb{H}}_{s,\text{tg}} = Y_s\,[\vec{z} \wedge \vec{\mathbb{E}}_{s,\text{tg}}]. \tag{7.84}$$

To establish a relationship between this admittance $Y_0$ and the admittance $Y_s$ of the substrate, we use equations (7.79). Adopting the same order as that given in (7.80), we get

$$Y_0 = \begin{cases} Y_s/\beta^{2N} \\ Y_s\beta^{2N} \\ \tilde{n}_H^2\beta^{2N}/Y_s \\ \tilde{n}_L^2/\beta^{2N}Y_s \end{cases}. \tag{7.85}$$

For the four configurations envisaged, relations (7.80), (7.83), and (7.85) now allow us to express the ratio between the moduli squared of the transmitted and incident fields:

$$\frac{|\vec{\mathbb{A}}^+_{s,\text{tg}}|^2}{|\vec{\mathbb{A}}^+_{0,\text{tg}}|^2} = |t|^2 = \begin{cases} \beta^{2N}\left|\dfrac{2\tilde{n}_0}{Y_s + \tilde{n}_0\beta^{2N}}\right|^2 & \text{for } Q_1 \\[3mm] \beta^{2N}\left|\dfrac{2\tilde{n}_0}{\tilde{n}_0 + Y_s\beta^{2N}}\right|^2 & \text{for } Q_2 \\[3mm] \tilde{n}_H^2\beta^{2N}\left|\dfrac{2\tilde{n}_0}{\tilde{n}_0 Y_s + \tilde{n}_H^2\beta^{2N}}\right|^2 & \text{for } Q_3 \\[3mm] \tilde{n}_L^2\beta^{2N}\left|\dfrac{2\tilde{n}_0}{\tilde{n}_L^2 + \tilde{n}_0 Y_s\beta^{2N}}\right|^2 & \text{for } Q_4 \end{cases}, \tag{7.86}$$

where the admittance $Y_s$ is identified with $\tilde{n}_s$, the effective index of the substrate. This last equation (7.86) shows that, regardless of the structure

$Q_i$ being considered, the quantity $|t|^2$ tends to 0 as the number of layers increases:

$$\lim_{N \to \infty} |t|^2 = 0. \qquad (7.87)$$

This confirms that electromagnetic field enhancement cannot be obtained in the substrate by following such an approach. The profile of the modulus squared of the field was plotted in Chapter 6 for quarter-wave stacks and confirms this conclusion. Moreover, this conclusion is valid regardless of the value of the effective index ratio ($\beta = \tilde{n}_H/\tilde{n}_L < 1$ or $\beta > 1$).

### 7.8.3 Zero Admittance Layer

At this stage the only assumptions made so far are that the layers can be matched ($\sigma < k_L < k_H$) and that the substrate was a semi-infinite transparent material ($Y_s = \tilde{n}_s$). We introduced no other conditions characteristic of the propagative ($\sigma < k_s$) or total reflection ($\sigma > k_s$) regimes. Equations (7.86) therefore apply either way to both regimes.

However, the conclusions we drew (decrease in $|t|^2$ with the number of layers) would change if the **admittance $Y_s$ of the substrate were zero**. As things stand, that would not be the case (recall that this admittance is identified with the effective index of the substrate), but suppose we were to insert, between this semi-infinite substrate and the stack corresponding to one of the $Q_i$ arrangements, a *preparation* layer. Under these conditions, the relationship between $Y_s$ and $\tilde{n}_s$ would become

$$Y_s = \frac{\tilde{n}_s \cos \delta_z - i\tilde{n}_z \sin \delta_z}{\cos \delta_z - i(\tilde{n}_s/\tilde{n}_z) \sin \delta_z} \qquad \text{where} \quad \delta_z = \alpha_z d_z, \qquad (7.88)$$

where all the quantities identified by the subscript $z$ relate to this preparation layer.

To obtain zero admittance the following condition must be satisfied:

$$Y_s = 0 \quad \Rightarrow \quad \tan \delta_z = -i\tilde{n}_s/\tilde{n}_z \quad \text{where} \quad \sigma < k_z. \qquad (7.89)$$

Recall that the effective indices $\tilde{n}_j$ of the multilayer are all real, since we are working under the condition that the fields are trigonometric ($\sigma < k_L < k_H$). In the propagative regime ($\sigma < k_s$), it is the same for the substrate, so condition (7.89) cannot be satisfied. Our conclusion would be identical in the event that this preparation layer might be replaced by a more complex stack as indicated in Section 7.8.1.

Conversely, for total reflection ($\sigma > k_s$), $\tilde{n}_s$ is pure imaginary, so condition (7.89) now admits solutions given by

$$\tan \delta_z = \Im[\tilde{n}_s]/\tilde{n}_z . \tag{7.90}$$

This relation makes it possible to determine the characteristics (thickness and index) of this layer for which (in the total reflection regime) deposition onto the substrate allows the admittance $Y_s$ to become zero. For this reason, this layer will from now on have the generic label ZAL (*Zero Admittance Layer*).

Putting $Y_s = 0$ in (7.86), we obtain the ratio of the modulus squared of the field at the interface between the stack and the ZAL layer, to the modulus squared of the incident field, i.e.,

$$\frac{|\vec{\mathbb{E}}_{z,\mathrm{tg}}|^2}{|\vec{\mathbb{A}}_{0,\mathrm{tg}}^+|^2} = \begin{cases} \dfrac{4}{\beta^{2N}} & \text{for } Q_1 \\[2mm] 4\beta^{2N} & \text{for } Q_2 \\[2mm] \dfrac{4}{\beta^{2N}} \left(\dfrac{\tilde{n}_0}{\tilde{n}_H}\right)^2 & \text{for } Q_3 \\[2mm] 4\beta^{2N} \left(\dfrac{\tilde{n}_0}{\tilde{n}_L}\right)^2 & \text{for } Q_4 \end{cases} . \tag{7.91}$$

It remains for us to account for the ZAL to express the relationship between the field with subscript $z$ and the field with subscript $s$, i.e.,

$$\begin{aligned} \vec{z} \wedge \vec{\mathbb{E}}_{z,\mathrm{tg}} &= \cos \delta_z [\vec{z} \wedge \vec{\mathbb{E}}_{s,\mathrm{tg}}] - i/\tilde{n}_z \sin \delta_z \vec{\mathbb{H}}_{s,\mathrm{tg}} \\ &= \{\cos \delta_z - i(\tilde{n}_s/\tilde{n}_z) \sin \delta_z\} [\vec{z} \wedge \vec{\mathbb{A}}_{s,\mathrm{tg}}^+] , \end{aligned} \tag{7.92}$$

and, including the condition for the definition of the ZAL,

$$\vec{\mathbb{E}}_{z,\mathrm{tg}} = \cos \delta_z \left\{ 1 + \left(\frac{\Im[\tilde{n}_s]}{\tilde{n}_z}\right)^2 \right\} \vec{\mathbb{A}}_{s,\mathrm{tg}}^+ , \tag{7.93}$$

we immediately deduce that

$$\begin{aligned} \frac{|\vec{\mathbb{A}}_{s,\mathrm{tg}}^+|^2}{|\vec{\mathbb{A}}_{0,\mathrm{tg}}^+|^2} &= \frac{|\vec{\mathbb{E}}_{z,\mathrm{tg}}|^2}{|\vec{\mathbb{A}}_{0,\mathrm{tg}}^+|^2} \frac{1}{\cos^2 \delta_z \{1 + (\Im[\tilde{n}_s]/\tilde{n}_z)^2\}^2} \\ &= \frac{1}{1 + (\Im[\tilde{n}_s]/\tilde{n}_z)^2} \frac{|\vec{\mathbb{E}}_{z,\mathrm{tg}}|^2}{|\vec{\mathbb{A}}_{0,\mathrm{tg}}^+|^2} . \end{aligned} \tag{7.94}$$

Combining this result with (7.91), we finally get

$$
\frac{|\vec{A}_{s,tg}^+|^2}{|\vec{A}_{0,tg}^+|^2} = |t|^2 = \frac{1}{1 + (\Im[\tilde{n}_s]/\tilde{n}_z)^2}
\begin{cases}
\dfrac{4}{\beta^{2N}} & \text{for } (LH)^N ZS \\[2ex]
4\beta^{2N} & \text{for } (HL)^N ZS \\[2ex]
\dfrac{4}{\beta^{2N}} \left(\dfrac{\tilde{n}_0}{\tilde{n}_H}\right)^2 & \text{for } H(LH)^N ZS \\[2ex]
4\beta^{2N} \left(\dfrac{\tilde{n}_0}{\tilde{n}_L}\right)^2 & \text{for } L(HL)^N ZS
\end{cases}
\cdot \quad (7.95)
$$

The asymptotic behavior of the transmission is therefore significantly altered, since for each structure it is possible from now on to increase the transmission with the number of layers. Notice that the field enhancement is not bounded; actually there are other effects that will limit this enhancement, such as damage-related or nonlinearity aspects, conditions of illumination (angular divergence and spectral line width) and manufacture (precision of the deposited layer).

For structures $Q_1$ and $Q_3$, we need $\beta < 1$, while for structures $Q_2$ and $Q_4$ we need $\beta > 1$. Recall that the value of $\beta$ depends on the incidence and polarization, since it is given by

$$
\beta = \frac{\tilde{n}_H}{\tilde{n}_L} =
\begin{cases}
\dfrac{n_H \alpha_H/k_H}{n_L \alpha_L/k_L} = \dfrac{\sqrt{n_H^2 - \bar{\nu}^2}}{\sqrt{n_L^2 - \bar{\nu}^2}} & \text{TE} \\[3ex]
\dfrac{n_H k_H/\alpha_H}{n_L k_L/\alpha_L} = \dfrac{n_H^2 \sqrt{n_L^2 - \bar{\nu}^2}}{n_L^2 \sqrt{n_H^2 - \bar{\nu}^2}} & \text{TM}
\end{cases}
, \quad (7.96)
$$

where $\bar{\nu} = n_0 \sin\theta_0 < n_L$. The variations in the ratio $\beta$ as a function of the normalized spatial frequency $\bar{\nu}$ are shown in Figure 7.19 for the two polarization states TE and TM.

Note that this ratio is always greater than 1 for TE polarization, so only structures $Q_2$ and $Q_4$ can be used for this polarization. In TM polarization, this ratio is equal to 1 for a specific value of the normalized spatial frequency $\bar{\nu}_c$ defined by

$$
\bar{\nu}_c = \frac{n_H n_L}{\sqrt{n_H^2 + n_L^2}}. \quad (7.97)
$$

This particular frequency corresponds to the Brewster angle at the HL or LH interfaces. We have

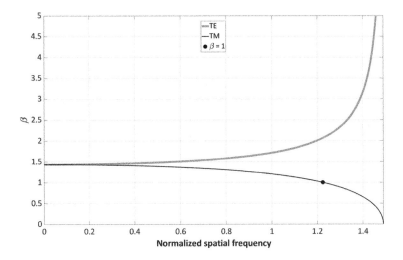

Figure 7.19 Variations of the effective index ratios $\beta = \tilde{n}_H/\tilde{n}_L$ as a function of the normalized spatial frequency $\bar{\nu}$ for polarization states TE (gray) and TM (black).

$$\sin\theta_L = \frac{n_H}{\sqrt{n_H^2 + n_L^2}} \quad \Rightarrow \quad \cos\theta_L = \frac{n_L}{\sqrt{n_H^2 + n_L^2}} \quad \Rightarrow \quad \tan\theta_L = \frac{n_H}{n_L}. \quad (7.98)$$

Consequently, in TM polarization we have to use structures $Q_1$ or $Q_3$ for $\bar{\nu} > \bar{\nu}_c$ ($\beta < 1$), while structures $Q_2$ or $Q_4$ must again be selected for $\bar{\nu} < \bar{\nu}_c$ ($\beta > 1$).

### *7.8.4 Example of a Structure*

#### *Design*

From now on we shall deal with TE polarization at wavelength $\lambda_0 = 633$ nm, and we shall consider a structure of type $Q_4$ used in total reflection on an air substrate ($n_s = 1$); the superstrate is assumed to be N-BK7 ($n_0 = 1.5151$). The angle of incidence $\theta_0$ is taken to be 45 degrees. We shall assume that the H and L materials are tantalum pentoxide (Ta$_2$O$_5$, $n_H = 2.1417$) and silica (SiO$_2$, $n_L = 1.4909$) respectively, and that neither material demonstrates any absorption.

The characteristics of the zero admittance layer (ZAL) are defined ensuring that condition (7.90) is met, which is independent of the number of quarter-wave layers making up the structure. There are two possible choices:

- This ZAL is made using high refractive index material, with a thickness $d_{z,H}$ defined by

$$d_{z,H} = \frac{\lambda_0}{2\pi\sqrt{n_H^2 - n_0^2 \sin^2\theta_0}} \arctan\left[\frac{\sqrt{n_0^2 \sin^2\theta_0 - 1}}{\sqrt{n_H^2 - n_0^2 \sin^2\theta_0}}\right] = 11.1 \text{ nm}. \quad (7.99)$$

The multiplying factor appearing in the enhancement ratio [see relation (7.95)] then has the value

$$\frac{1}{1 + (\Im[\tilde{n}_s]/\tilde{n}_H)^2} = 0.9588. \quad (7.100)$$

- This ZAL is made using low refractive index material, with a thickness $d_{z,L}$ defined by

$$d_{z,L} = \frac{\lambda_0}{2\pi\sqrt{n_L^2 - n_0^2 \sin^2\theta_0}} \arctan\left[\frac{\sqrt{n_0^2 \sin^2\theta_0 - 1}}{\sqrt{n_L^2 - n_0^2 \sin^2\theta_0}}\right] = 34.5 \text{ nm}, \quad (7.101)$$

where the multiplying factor has a value of

$$\frac{1}{1 + (\Im[\tilde{n}_s]/\tilde{n}_L)^2} = 0.8792. \quad (7.102)$$

The ratio of the values given in (7.100) and (7.102) is around unity, so that the choice of the ZAL material will be anecdotic because we search for several decades' enhancement. In the latter case, note that this amounts simply to increasing the thickness of the last layer of low-index material in the structure $Q_4$ by $d_{z,L}$.

### *Field Enhancement Factor*

Using the fourth equation in (7.95) we can calculate the change in the field enhancement factor of this structure having a number $N$ of HL pairs for the two possible choices of ZAL material. We observe (see Figure 7.20, black open circles and gray curve) that this enhancement factor can in theory reach extremely high values (up to $1.6 \times 10^{10}$ for 19 HL layer pairs).

However, the sensitivity of this enhancement factor with respect to the thickness of the ZAL becomes quite extreme as the number of HL pairs rises, making any practical realization highly improbable, unless we aim at an enhancement lower than 4 decades. This is illustrated in the same Figure 7.20, where the change in enhancement factor is shown as a function of the number of HL pairs given an error of 0.1 nm in the ZAL (gray and black dots). Observe that the configuration using a low-index ZAL is more tolerant and that the optimum is obtained using six HL pairs; in this case

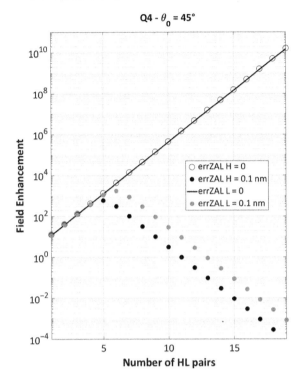

Figure 7.20 Change in the field enhancement factor for a structure of type $Q_4$ with number $N$ of HL pairs when the ZAL is made of a high-index (black) or low-index (gray) material [$\lambda = 633$ nm, $\theta_0 = 45$ degrees, Superstrate N-BK7, Substrate Air]. Calculations are made with, and without, a manufacturing error (see text).

the theoretical enhancement factor is in the order of 4000. At this step it is useful to stress on the fact that a too high number of layers would severely degrade the stack performance, depending on the ZAL thickness accuracy (here 1% for the 10 nm high-index layer).

From now on we shall adopt this particular stacking arrangement [L(HL)$^6$ ZAL$_L$] as the reference configuration for all our analyses.

### Field Profile

Figure 7.21 shows how the normalized modulus squared of the field varies with the $z$ coordinate for this particular configuration. Note that, as expected, maximum field enhancement is obtained on the upper interface of the zero admittance layer and that the modulus squared of the field retains high values over several hundreds of nanometers within the air substrate.

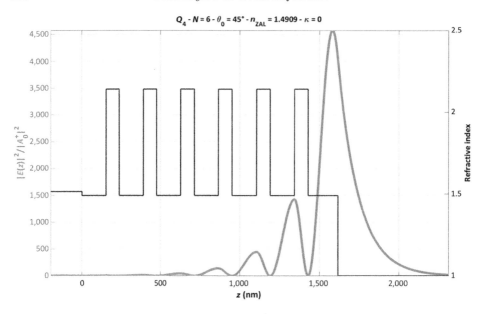

Figure 7.21 Profile of the field associated with a structure of type $Q_4$ ($N = 6$) incorporating a low-index zero admittance layer ($\lambda = 633$ nm, $\theta_0 = 45$ degrees, Superstrate N-BK7, Substrate Air).

## *Resonances/Guided Modes*

This electromagnetic field enhancement corresponds to a resonance of the structure N-BK7 / $L(HL)^6$ $ZAL_L$ / Air and hence must be related to the presence of a guided mode in the associated multilayer structure [ Air / $L(HL)^6$ $ZAL_L$ / Air ]. This is confirmed by examining the function $1/R_0(\bar{\nu})$ corresponding to this stack in planar (guided waves) optics; the variations of this function are shown in Figure 7.22 when the normalized frequency $\bar{\nu}$ spans the modal window $[1, n_H]$.

We observe that eight guided modes can be counted; we are interested in the mode of order 6 ($\bar{\nu}_{g,0} = 1.0718$). To confirm this identification in the presence of absorption in the layers ($\kappa = 10^{-4}$), it is sufficient to plot the changes in the reflection coefficient $R_a$ of the structure N-BK7 / $L(HL)^6$ $ZAL_L$ / Air (see Figure 7.23) when the angle of incidence in N-BK7 varies between $\arcsin[1/n_{\text{N-BK7}}] = 41.3$ and 90; this corresponds to a variation in the normalized spatial frequency of between 1 and the refractive index of N-BK7, i.e., 1.5151, for which reason only two resonances can be seen. In Figure 7.23 note the absorption peak at an incidence of 45 degrees, as expected; this peak is characterized by an absorbance of 91.2% when the extinction coefficient $\kappa$ is $10^{-4}$.

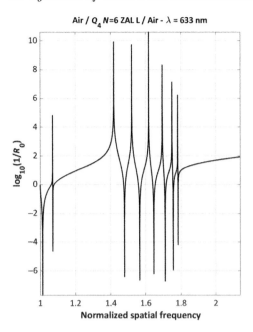

Figure 7.22 Variation of the function $1/R_0$ with the normalized spatial frequency $\bar{\nu}$ for the structure Air / $L(HL)^6$ ZAL$_L$ / Air (illuminating wavelength equal to the design wavelength, i.e., 633 nm, $\kappa = 0$).

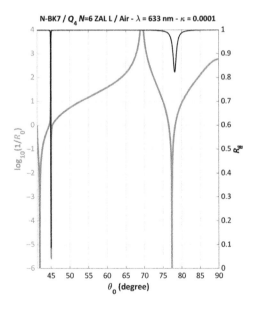

Figure 7.23 A comparison of the variations in the functions $R_a$ and $1/R_0$ with the angle of incidence $\theta_0$ in the total reflection regime, for the structure N-BK7 / $L(HL)^6$ ZAL$_L$ / Air.

We might wonder if the introduction of an appropriately sized air gap could bring this absorbance to an optimum value of 100%, as obtained in Section 7.7.3. The answer is yes, with the optimum thickness of the air gap being 202 nm (here $B_0$ is again positive).

It must be emphasized that there is no reason why, on the one hand, maximizing the enhancement factor at the interface between stack and zero admittance layer and on the other hand, obtaining an absorbance of 100%, can be realized simultaneously for a zero air gap thickness. The first condition is local (value of the modulus squared of the field at $z = z_p$), while the second is global (the integral of this same modulus squared over the thickness $\sum_{j=1}^{p} z_j$ of the multilayer).

### 7.8.5 Generalization to a Multilayer Substrate

The configuration described in Section 7.8.4 is very suitable for a sensor, since the field is maximized in the air substrate. For other applications, however, such as the development of a microsource, it might be beneficial to locate the field maximum in the core of the stack. This is possible provided the definition of a zero admittance layer can be extended to a sublayer considered as a *multilayer substrate*.

We can make use of the previous developments assuming that the interface $k$ is an optical interface delimiting a substack, considered here as the *multilayer substrate* (see Figure 7.24).

If we now introduce a layer of characteristics $[\tilde{n}_z, \delta_z]$ above this *substrate* (of admittance $Y_k$), the admittance of the upper interface of this layer will be defined by

$$Y_z = \frac{Y_k \cos \delta_z - i\tilde{n}_z \sin \delta_z}{\cos \delta_z - i(Y_k/\tilde{n}_z) \sin \delta_z}, \tag{7.103}$$

and will become 0 if

$$\tan \delta_z = -i\frac{Y_k}{\tilde{n}_z} \quad \text{where} \quad \sigma < k_z. \tag{7.104}$$

This equation admits a real solution, since $Y_k$ is pure imaginary, like all the admittances in the total reflection regime.

Satisfying this condition is sufficient to obtain an electromagnetic field enhancement as described by (7.95), since this is derived from (7.80) and (7.85), and whose only relationship with the lower stack is the value of admittance $Y_s$. Besides, recall that according to the formalism of thin films (see Chapter 6), the field distribution within the lower stack (the multilayer substrate) is not altered by the upper quarter-wave stack, or at least to a

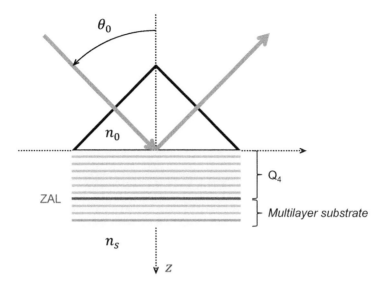

Figure 7.24 Generalization of the zero admittance (ZAL) technique to a *multilayer substrate.*

factor of proportionality. To summarize, everything happens as if the final component consisted of two quasi-independent substructures. Note finally that the field enhancement factor depends only on the number of layers in the upper stack.

By way of example, consider a structure consisting of the stacking arrangement defined in Section 7.8.4, i.e., N-BK7 / $\mathrm{L(HL)}^6$ and a multilayer substrate of type $\mathrm{(LH)}^M$ / Air matched quarter-wave mirror. This is a particular case for which two matched quarter-wave mirrors are separated by a ZAL in the total reflection regime. The admittance $Y_k$ can then be expressed as

$$Y_k = \tilde{n}_s \left( \frac{\tilde{n}_L}{\tilde{n}_H} \right)^{2M} = i \frac{\sqrt{n_0^2 \sin^2 \theta_0 - 1}}{\beta^{2M}} \qquad \beta > 1. \qquad (7.105)$$

From this the thickness of the zero admittance layer can be deduced, under the assumption that it is constructed using high-index material:

$$d_{z,H} = \frac{\lambda_0}{2\pi \sqrt{n_H^2 - n_0^2 \sin^2 \theta_0}} \arctan \left[ \frac{\sqrt{n_0^2 \sin^2 \theta_0 - 1}}{\beta^{2M} \sqrt{n_H^2 - n_0^2 \sin^2 \theta_0}} \right]. \qquad (7.106)$$

Note that this tends to 0 (modulo $\lambda_0/2$) as the number $M$ of LH pairs increases, meaning that the global structure tends asymptotically to a Fabry–Perot with L spacer. For $M = 3$, this thickness $d_{z,H}$ is only 0.34 nm. Figure

7.25 shows how the normalized modulus squared of the field varies with the coordinate $z$ for this particular configuration. We might wonder whether, given its small thickness (0.34 nm), removing this zero admittance layer might have an effect on the field enhancement coefficient. It does indeed, since this factor falls from 4500 to about 30 ! To lessen the profound impact of such a removal, the number $M$ of LH pairs used in constructing the multilayer substrate must imply a ZAL thickness smaller than the tolerance defined for depositing this layer, i.e., 0.1 nm. This is obtained for $M > 5$.

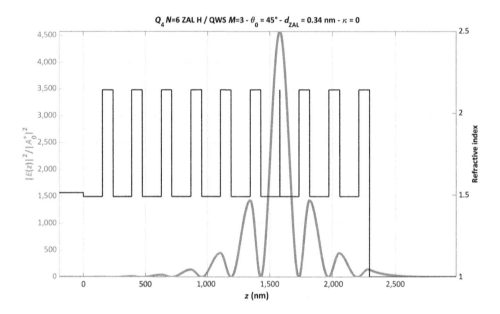

Figure 7.25 Profile of the field associated with a structure of type N-BK7 / L(HL)$^6$ ZAL$_H$ (LH)$^3$ / Air ($\lambda = 633$ nm, $\theta_0 = 45$, $\kappa = 0$).

Taking this illustration a little further, now suppose that the *multilayer substrate* is of type H(LH)$^M$. The admittance $Y_k$ can then be expressed as

$$Y_k = \frac{\tilde{n}_H^2}{\tilde{n}_s}\left(\frac{\tilde{n}_H}{\tilde{n}_L}\right)^{2M} = \frac{\tilde{n}_H^2}{\tilde{n}_s}\beta^{2M} = -i\frac{n_H^2 - n_0^2\sin^2\theta_0}{\sqrt{n_0^2\sin^2\theta_0 - 1}}\beta^{2M} \quad \beta > 1. \quad (7.107)$$

Again suppose that the zero admittance layer is made using high-index material; its thickness will then be defined by

$$d_{z,H} = \frac{\lambda_0}{2\sqrt{n_H^2 - n_0^2\sin^2\theta_0}}\left\{1 - \frac{1}{\pi}\arctan\left[\frac{\beta^{2M}\sqrt{n_H^2 - n_0^2\sin^2\theta_0}}{\sqrt{n_0^2\sin^2\theta_0 - 1}}\right]\right\}, \quad (7.108)$$

and tends asymptotically to $\lambda_0/4\tilde{n}_H$ as the number $M$ of LH pairs increases; this time the global structure tends to a Fabry–Perot with H *spacer*.

### 7.8.6 Bandwidth Effects

As already mentioned, the performance of these multilayer resonators is limited by effects both intrinsic (damage under intense illumination, non-linearities) and extrinsic (errors in fabrication, divergence, and spectral line width of the illuminating beam). Since the effect of errors in constructing the zero admittance layer have been analyzed in Section 7.8.4, we shall limit ourselves to the influence of the illumination bandwidth (divergence and line width). This analysis is essential if we wish to move on to experimentation.

### The Effect of Divergence

Consider a given illumination at a wavelength $\lambda_0$. Taking account of the extreme sensitivity to incidence of these components, we need to express the illuminating wave in the form of a packet of plane waves (see Chapter 4), i.e.,

$$\vec{\mathcal{E}}_i(\vec{r}, z) = \int_{\vec{\nu}} \vec{\mathbb{A}}_0^+ (\vec{\nu} - \vec{\nu}_i) \, e^{i[2\pi\vec{\nu}.\vec{r} + \alpha_0(\nu)z]} \, d^2\vec{\nu} \,. \tag{7.109}$$

In this expression, $\vec{\nu}_i$ is the incident spatial frequency, while beam divergence is accounted for by the profile of the function $\vec{\mathbb{A}}_0^+ (\vec{\nu})$. Note that the same function is used to describe the beam at all incidences. Although used very often, this procedure is not rigorously exact but it is, however, a very good approximation when the divergence of the beam being considered is low (we shall return to this point in greater detail in Section 8.2.7).

Under these conditions, the reflected field is obtained via a simple weighting of the elementary field component by the amplitude reflection coefficient $r(\nu)$ in the integral (7.109), i.e.,

$$\vec{\mathcal{E}}_r(\vec{r}, z) = \int_{\vec{\nu}} r(\nu) \vec{\mathbb{A}}_0^+ (\vec{\nu} - \vec{\nu}_i) \, e^{i[2\pi\vec{\nu}.\vec{r} - \alpha_0(\nu)z]} \, d^2\vec{\nu} \,. \tag{7.110}$$

Furthermore, Chapter 4 gave an expression for the flux transported by a wave packet through a plane of ordinate $z$, so for the incident and reflected beams,

$$\Phi_i = \frac{1}{2\omega\tilde{\mu}_0} \int_{\vec{\nu}} \alpha_0(\nu) \left| \vec{\mathbb{A}}_0^+ (\vec{\nu} - \vec{\nu}_i) \right|^2 d^2\vec{\nu} \,, \tag{7.111}$$

$$\Phi_r = \frac{1}{2\omega\tilde{\mu}_0} \int_{\vec{\nu}} \alpha_0(\nu) R(\nu) \left| \vec{\mathbb{A}}_0^+ (\vec{\nu} - \vec{\nu}_i) \right|^2 d^2\vec{\nu} \,. \tag{7.112}$$

Consequently, the reflection coefficient takes the form

$$
R = \frac{\Phi_r}{\Phi_i} = \frac{\displaystyle\int_{\vec{\nu}} \alpha_0(\nu) R(\nu) \left| \vec{\mathbb{A}}_0^+(\vec{\nu} - \vec{\nu}_i) \right|^2 d^2\vec{\nu}}{\displaystyle\int_{\vec{\nu}} \alpha_0(\nu) \left| \vec{\mathbb{A}}_0^+(\vec{\nu} - \vec{\nu}_i) \right|^2 d^2\vec{\nu}}.
\tag{7.113}
$$

This then leads to the following expression for the absorption coefficient in the total reflection regime $(T = 0)$:

$$
A = 1 - R = \frac{\displaystyle\int_{\vec{\nu}} \alpha_0(\nu)[1 - R(\nu)] \left| \vec{\mathbb{A}}_0^+(\vec{\nu} - \vec{\nu}_i) \right|^2 d^2\vec{\nu}}{\displaystyle\int_{\vec{\nu}} \alpha_0(\nu) \left| \vec{\mathbb{A}}_0^+(\vec{\nu} - \vec{\nu}_i) \right|^2 d^2\vec{\nu}}.
\tag{7.114}
$$

Note the product $[1 - R(\nu)]|\vec{\mathbb{A}}_0^+(\vec{\nu} - \vec{\nu}_i)|^2$ in (6.141). If we assume that the illumination, which we have supposed to be weakly divergent, excites just one resonance $\vec{\nu}_r$, this product can be written as $g(\nu - \nu_r)|\vec{\mathbb{A}}_0^+(\vec{\nu} - \vec{\nu}_i)|^2$, where $g$ describes the form of the resonance in absorption $[1 - R(\nu)]$, similar to those shown in Figure 7.23.

Hence in the neighborhood of the resonant frequency $\nu_r$ we get

$$
A = \frac{\displaystyle\int_{\vec{\nu}} \alpha_0(\nu) g(\nu - \nu_r) \left| \vec{\mathbb{A}}_0^+(\vec{\nu} - \vec{\nu}_i) \right|^2 d^2\vec{\nu}}{\displaystyle\int_{\vec{\nu}} \alpha_0(\nu) \left| \vec{\mathbb{A}}_0^+(\vec{\nu} - \vec{\nu}_i) \right|^2 d^2\vec{\nu}}.
\tag{7.115}
$$

It appears, therefore, that resonance can be observed only if the functions $g(\nu - \nu_r)$ and $|\vec{\mathbb{A}}_0^+(\vec{\nu} - \vec{\nu}_i)|^2$ overlap. Consider the optimum case in which the illumination is perfectly centered on the resonance $(\nu_i = \nu_r)$; under these conditions, the strength of this resonance will be controlled by the mid-height widths $\Delta\nu_i$ and $\Delta\nu_r$ of the two functions $|\vec{\mathbb{A}}_0^+|^2$ and $g$.

To illustrate this, let us use the reference structure defined in Section 7.8.4, namely N-BK7 / L(HL)$^6$ ZAL$_L$ / Air, with the H and L layers matched for a mean incidence of 45 degrees. Assume that the incident wave packet corresponds to a Gaussian beam of angular divergence $\Delta\theta$ and waist radius $w_0$. Under these conditions, it can be shown (see Section 8.2.7 for more detail) that the function $\vec{\mathbb{A}}_0^+(\vec{\nu} - \vec{\nu}_i)$ can be written

$$
\vec{\mathbb{A}}_0^+(\vec{\nu} - \vec{\nu}_i) = \frac{\pi^2 w_0^2}{\cos\theta_0^i} e^{-\pi^2 \frac{w_0^2}{\cos^2\theta_0^i}(\nu_x - \nu_i)^2} e^{-\pi^2 w_0^2 \nu_y^2} \vec{\mathcal{E}}_i,
\tag{7.116}
$$

where $w_0 = \lambda/(\pi \Delta\theta)$. In the interests of simplicity, adopting a 1D geometry ($xOz$ plane), the expression for the absorbance $A$ becomes

$$A = \frac{\int_{\bar{\nu}} \alpha_0(\nu) g(\nu - \nu_r) e^{-2\frac{\lambda^2}{(\Delta\theta\cos\theta_0^i)^2}(\nu - \nu_i)^2} d^2\bar{\nu}}{\int_{\bar{\nu}} \alpha_0(\nu) e^{-2\frac{\lambda^2}{(\Delta\theta\cos\theta_0^i)^2}(\nu - \nu_i)^2} d^2\bar{\nu}} \qquad \text{where} \quad \nu_i = \nu_r. \quad (7.117)$$

Figure 7.26 shows the change in this *effective* absorbance when the divergence $\Delta\theta$ of the Gaussian beam varies between 1 microradian and 10 milliradians. Observe that this absorbance remains extremely close to the nominal value calculated for a plane wave (i.e., 91.2%) if the divergence of the illuminating Gaussian beam is less than or equal to 0.1 milliradians, with the mid-height bandwidth being in the order of milliradians.

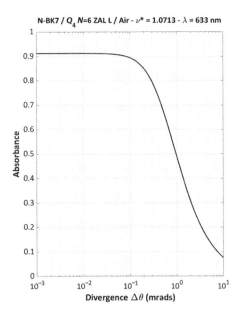

Figure 7.26 Variation of the absorption of a wave packet at resonance as a function of the divergence $\Delta\theta$ of this wave packet [structure N-BK7 / L(HL)$^6$ ZAL$_L$ / Air, TE polarization, $\lambda = 633$ nm, $\bar{\nu}_r = 1.0713$, $\kappa = 10^{-4}$].

A standard helium–neon laser of 1 mm diameter ($w_0 = 0.5$ mm, $\Delta\theta = 0.4$ mrad) is sufficient to make it possible to observe such a resonance.

## The Effect of Spectral Bandwidth

We can carry out the same development for a perfectly collimated beam ($\Delta\theta = 0$), but with a line width of $\Delta\lambda$. As we saw in Chapter 3, the incident wave can be written in the form of a frequential wave packet, i.e.,

$$\vec{E}_i(\vec{\rho}, t) = \int_f \vec{\mathcal{E}}_i(\vec{\rho}, f)\, e^{-2i\pi ft}\, df = \int_f \vec{A}_0^+ (f - f_0)\, e^{-2i\pi ft}\, df\,. \qquad (7.118)$$

Also, we saw in Chapter 5 for an integrating detector that the flux transported by this polychromatic beam was given by the sum of elementary monochromatic fluxes. Hence for the incident monochromatic beam:

$$\Phi_i = \int_f \frac{1}{2\omega\tilde{\mu}_0}\, \alpha_0(f)\, \left| \vec{A}_0^+ (f - f_0) \right|^2\, df \quad \text{where} \quad \omega = 2\pi f\,, \qquad (7.119)$$

and in TE polarization,

$$\Phi_i = \frac{1}{2} \int_f \tilde{n}_0(f)\, \left| \vec{A}_0^+ (f - f_0) \right|^2\, df\,. \qquad (7.120)$$

If we neglect dispersion in the index of the incident medium over the support of the function $|\vec{A}_0^+ (f - f_0)|^2$, the resonator's absorption can then be expressed as

$$A = \frac{\displaystyle\int_f h(f - f_r)\, \left| \vec{A}_0^+ (f - f_0) \right|^2\, df}{\displaystyle\int_f \left| \vec{A}_0^+ (f - f_0) \right|^2\, df}\,, \qquad (7.121)$$

where $h(f - f_r) = 1 - R(f - f_r)$ is the spectral profile of the resonance centered on frequency $f_r$. Making the change of variable $\lambda = c/f$ and assuming a Gaussian profile for the spectral distribution of the incident field, we get

$$A = \frac{\displaystyle\int_\lambda h_\lambda(\lambda - \lambda_r)\, e^{-2\left(\frac{\lambda - \lambda_0}{\Delta\lambda}\right)^2}\, d\lambda}{\displaystyle\int_\lambda e^{-2\left(\frac{\lambda - \lambda_0}{\Delta\lambda}\right)^2}\, d\lambda} \quad \text{where} \quad \lambda_0 = \lambda_r\,. \qquad (7.122)$$

Figure 7.27 shows the variation of effective absorbance with the spectral bandwidth of the incident beam. Note that this absorbance remains equal to the maximum value of 91.2% up to a spectral width of 20 picometers, with a mid-height bandwidth in the order of 0.3 nm. Again, a helium–neon laser will allow this resonance to be observed with no loss of contrast.

Note that these two bandwidths (angular and spectral) are actually connected by a very simple relation, which derives directly from the expression for the spatial frequency $\nu$. If we neglect dispersion in the refractive index of the incident medium, we get

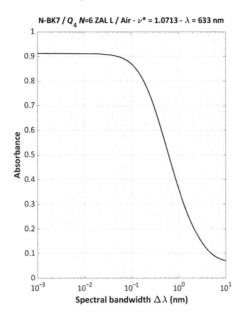

Figure 7.27 Variation in the absorption of a wave packet at resonance as a function of the spectral bandwidth $\Delta\lambda$ of this wave packet [structure N-BK7 / L(HL)$^6$ ZAL$_L$ / Air, TE polarization, $\lambda = 633$ nm, $\bar{\nu}_r = 1.0713$, $\kappa = 10^{-4}$].

$$\nu = \frac{n_0 \sin\theta_0}{\lambda} \Rightarrow d\nu = \frac{n_0 \cos\theta_0}{\lambda} d\theta_0 - \frac{n_0 \sin\theta_0}{\lambda^2} d\lambda = 0$$

$$\Rightarrow \Delta\theta = \tan\theta_0 \frac{\Delta\lambda}{\lambda}. \tag{7.123}$$

This explains why the two curves shown in Figures 7.26 and 7.27 are identical, to a scale factor defined by equation (7.123).

Finally, by way of conclusion, recall that a maximum absorption guarantees that the field enhancement in the multilayer guide preserves its nominal value.

## 7.9 Distortion of the Wave Front at Resonance

On several occasions it has already been pointed out that, in the absence of absorption, the coefficient of reflection $R$ is unity for the entire domain of total reflection, and that there is therefore no possibility of detecting any resonances or enhancements with photometric (noninterferometric) measurements performed with a monoblock (nonpixelated) receiver.

Under these conditions, and apart from the scattering phenomena (also amplified) to be discussed in Chapter 8, it is the shape of the reflected wave front that informs us of the presence of resonances, as we shall show in this section.

In a 1D geometry, in a harmonic regime and for illumination centered on the resonant frequency $\nu_r$, the amplitudes of the incident and reflected beams can be written as

$$\vec{\mathcal{E}}_i(x, z) = \int_\nu \vec{\mathbb{A}}_i(\nu - \nu_r)\, e^{i[2\pi\nu x + \alpha(\nu)z]}\, d\nu , \tag{7.124}$$

$$\vec{\mathcal{E}}_r(x, z) = \int_\nu r(\nu)\vec{\mathbb{A}}_i(\nu - \nu_r)\, e^{i[2\pi\nu x - \alpha(\nu)z]}\, d\nu , \tag{7.125}$$

where the frequency f has been omitted for greater clarity. It was shown in 7.7.2 that in the neighborhood of a resonance, the reflection coefficient could be put in the following general form:

$$r(\nu) = C(\nu)\frac{\nu - \nu_c^*}{\nu - \nu_c} \quad \text{where} \quad \nu_c = \nu_r + i\nu'' . \tag{7.126}$$

$C(\nu)$ denotes a slowly varying function of unit modulus, and the symbol * denotes the complex conjugate.

Given the sharp nature of resonance, we can expand $\alpha(\nu)$ about $\nu_r$, i.e.,

$$\alpha(\nu) = \sqrt{k^2 - 4\pi^2\nu^2} \approx \sqrt{k^2 - 4\pi^2\nu_r^2} - \frac{4\pi^2\nu_r(\nu - \nu_r)}{\sqrt{k^2 - 4\pi^2\nu_r^2}}$$

$$= \alpha(\nu_r) - \frac{4\pi^2\nu_r(\nu - \nu_r)}{\alpha(\nu_r)} , \tag{7.127}$$

but we can also evaluate $C(\nu)$ at $\nu_r$ in calculating the integral (7.125). Using these two approximations, and making the change of variable $\nu' = \nu - \nu_r$, we get

$$\vec{\mathcal{E}}_r(x, z) = C(\nu_r)\, e^{i[2\pi\nu_r x - \alpha(\nu_r)z]} \int_{\nu'} \vec{\mathbb{A}}_i(\nu')\frac{\nu' + i\nu''}{\nu' - i\nu''}\, e^{2i\pi\nu'[x + \frac{2\pi\nu_r}{\alpha_r}z]}\, d\nu' . \tag{7.128}$$

Let

$$X = x + \frac{2\pi\nu_r}{\alpha_r}z \quad \text{et} \quad F(\nu') = \frac{\nu' + i\nu''}{\nu' - i\nu''} = 1 + \frac{2i\nu''}{\nu' - i\nu''} , \tag{7.129}$$

so that (7.128) can be put in the form

$$|\vec{\mathcal{E}}_r(X)|^2 = \left| \int_{\nu'} \vec{\mathbb{A}}_i(\nu')F(\nu')\, e^{2i\pi\nu'X}\, d\nu' \right|^2 . \tag{7.130}$$

We now need to calculate the inverse Fourier transform of the two quantities $\vec{\mathbb{A}}_i$ and $F$, and then convolve them. Equation (7.124) gives us the inverse Fourier transform of $\vec{\mathbb{A}}_i$. Indeed

$$\vec{\mathcal{E}}_i(x,0) = \int_\nu \vec{\mathbb{A}}_i(\nu - \nu_r)\, e^{2i\pi\nu x}\, d\nu$$

$$\Rightarrow \quad \mathrm{FT}^{-1}[\vec{\mathbb{A}}_i](X) = \int_{\nu'} \vec{\mathbb{A}}_i(\nu')\, e^{2i\pi\nu' X}\, d\nu' = e^{-2i\pi\nu_r X}\, \vec{\mathcal{E}}_i(X,0)\,. \tag{7.131}$$

On the other hand,

$$\mathrm{FT}^{-1}[F](X) = \int_{\nu'} F(\nu')\, e^{2i\pi\nu' X}\, d\nu' = \int_{\nu'} \left\{ 1 + \frac{2i\nu''}{\nu' - i\nu''} \right\} e^{2i\pi\nu' X}\, d\nu'$$

$$= \delta(X) + 2i\nu'' \int_{\nu'} \frac{e^{2i\pi\nu' X}}{\nu' - i\nu''}\, d\nu'\,. \tag{7.132}$$

The last integral in (7.132) is calculated using the method of residues. We get

$$2i\nu'' \int_{\nu'} \frac{e^{2i\pi\nu' X}}{\nu' - i\nu''}\, d\nu' = -4\pi\nu'' H(X)\, e^{-2\pi\nu'' X}\,, \tag{7.133}$$

where $H(X)$ is the Heaviside unit step function.

From this an expression for the modulus squared of the reflected field can be deduced, i.e.,

$$|\vec{\mathcal{E}}_r(X)|^2 = \left| \left[ e^{-2i\pi\nu_r X} \vec{\mathcal{E}}_i(X,0) \right] \star \left[ \delta(X) - 4\pi\nu'' H(X)\, e^{-2\pi\nu'' X} \right] \right|^2$$

$$= \left| e^{-2i\pi\nu_r X} \vec{\mathcal{E}}_i(X,0) \right. \tag{7.134}$$

$$\left. -4\pi\nu'' \int_u H(u) e^{-2\pi\nu'' u}\, e^{-2i\pi\nu_r(X-u)}\, \vec{\mathcal{E}}_i(X-u,0)\, du \right|^2.$$

As we did in Section 7.8.6, we assume that the spatial profile of the incident field corresponds to that of a weakly divergent Gaussian beam of waist radius $w_0$. The spatial distribution of its amplitude is then defined (in a 1D configuration) by

$$\vec{\mathcal{E}}_i(x,0) = \vec{\mathcal{E}}_i\, e^{ik_0 x \sin\theta_0}\, e^{-\frac{(x\cos\theta_0)^2}{w_0^2}} \tag{7.135}$$

(see Section 8.2.7 for more details). Equation (7.134) describing an expression for the modulus squared of the reflected field becomes

$$|\vec{\mathcal{E}}_r(X)|^2 = |\vec{\mathcal{E}}_i|^2 \left| e^{-\frac{x^2}{w^2}} - 4\pi\nu'' \int_0^\infty e^{-\frac{(X-u)^2}{w^2}}\, e^{-2\pi\nu'' u}\, du \right|^2, \tag{7.136}$$

where $w = w_0/\cos\theta_0$. We make the following changes of variable:

$$X_w = \frac{X}{w} \quad ; \quad v = \frac{u}{w} \quad ; \quad s = wv'' \tag{7.137}$$

to get a general expression that depends on just one parameter ($s$), namely,

$$\frac{|\vec{\mathcal{E}}_r(X_w)|^2}{|\vec{\mathcal{E}}_i|^2} = \left| e^{-X_w^2} - 4\pi s \int_0^\infty e^{-(X_w - v)^2} e^{-2\pi s v}\, dv \right|^2 . \tag{7.138}$$

This parameter $s$ is the product of the effective waist radius ($w$) of the incident beam with the imaginary part of the complex frequency of the pole of the reflection function ($v''$). Note that this is a key parameter, since the beam is not distorted when this parameter is 0. It can also be shown that distortion of the reflected beam disappears for large values of $s$.

This parameter may also be considered in a different way. Since the reflection function $r(v)$ is of unit modulus, we can write, as per equation (7.126):

$$r(v) = e^{i\phi(v)} = e^{i\psi(v)}\frac{(v - v_r) + iv''}{(v - v_r) - iv''} . \tag{7.139}$$

The variation in phase $\phi(v)$ in the neighborhood of resonance $v_r$ can therefore be expressed as

$$\phi(v) \approx \psi(v_r) + 2\arctan\left[\frac{v''}{v - v_r}\right] = \psi(v_r) + \varphi(v) . \tag{7.140}$$

The phase $\varphi(v)$ is equal to $-\pi$ at $v = v_r$ and takes the values of $-\pi/2$ and $-3\pi/2$ respectively when $v - v_r$ is equal to $\mp v''$. Hence, we can assume that the quantity $2v''$ defines the total width $\Delta v_r$ of the resonance. The corresponding angular width $\Delta\theta_r$ can be deduced, i.e.,

$$\Delta\theta_r = \frac{\lambda}{n_0\cos\theta_0}\Delta v_r = \frac{2\lambda}{n_0\cos\theta_0}v'' . \tag{7.141}$$

On the other hand, the total angular divergence $\Delta\theta_w$ of the Gaussian beam is defined by

$$\Delta\theta_w = \frac{2\lambda}{\pi w} . \tag{7.142}$$

From this it can be deduced that the ratio of the angular width at resonance to the divergence of the Gaussian beam that is defined by

$$\frac{\Delta\theta_r}{\Delta\theta_w} = \frac{\pi}{n_0\cos\theta_0}wv'' = \frac{\pi}{n_0\cos\theta_0}s \tag{7.143}$$

is directly proportional to the parameter $s$ introduced earlier.

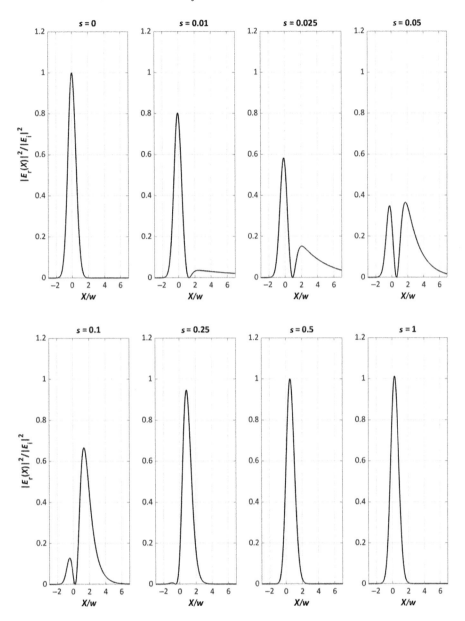

Figure 7.28 Change in the shape of the modulus squared of the reflected field as a function of the value of the parameter $s$ (see text).

Figure 7.28 shows the result of calculating the integral (7.138) for various values of the parameter $s$, namely

- $s = 0$: the result obtained corresponds to the incident beam profile (or to that of the reflected beam in the absence of any resonance).
- $s = 0.01$: we can see a black line (i.e., a local zero intensity) appearing on the right of the beam, as well as a trail covering values of $X$ greater than $2w$.
- $s = 0.025$, $s = 0.05$ and $s = 0.1$: the black line gradually crosses the beam from left to right, with the trail continuing to disrupt the symmetry of the spatial profile of the reflected beam.
- $s = 0.25$: the black line is now located on the left-hand edge of the beam; furthermore, the peak of the beam is shifted to the right by roughly one waist radius.
- $s = 0.5$ and $s = 1.0$: the beam has recovered its initial shape, but its peak is shifted to the right by an amount that is small if the parameter $s$ is large. This shift corresponds to the Goos–Hanchen effect in the presence of a resonance.

Note we have checked that the integral in $X_w$ of the modulus squared of the reflected field $|\vec{\mathcal{E}}_r(X_w)|^2$ is independent of parameter $s$ (conservation of energy).

In conclusion, recall that this calculation was made for the case of a simple pole. In the general case, the Fourier transform of the reflection function must be calculated numerically, taking account of the multilayer structure being considered.

## 7.10  Asymptotic Expression of the Wave Packet

In Section 7.9 we saw that, to analyze correctly the shape of the wave front at resonance, it was essential to invoke a description of the beam in the form of a wave packet, even in the far field.

This is not always the case, and it is useful here to recall the conditions that lead to an **asymptotic expression of the wave packet**, use of which allows the calculations to be simplified considerably.

By way of example, in Chapter 4 we showed an expression for the total incident flux in a transparent medium, namely

$$\Phi_i(z) = \frac{1}{2\omega\tilde{\mu}} \int_\nu \alpha(\nu) \, |\vec{\mathbb{A}}_i(\nu)|^2 \, d\nu , \qquad (7.144)$$

and then we used it to define the spectral density of that same flux, i.e.,

$$\frac{d\Phi_i(z)}{d\nu} = \frac{\alpha(\nu)}{2\omega\tilde{\mu}} \, |\vec{\mathbb{A}}_i(\nu)|^2 . \qquad (7.145)$$

However, one may wonder whether the flux density in (7.145) is a definition or a consequence of (7.144). Actually and strictly speaking, one cannot justify that (7.145) is a consequence of (7.144), in particular when large phase variations occur in the field. The case of resonance (where these phase variations can be huge) is a good illustration of this observation, since, if we were to use the spectral flux density to describe the spectral (or angular) distribution of the reflected beam, i.e.,

$$\frac{d\Phi_r(z)}{d\nu} = \frac{\alpha(\nu)}{2\omega\tilde{\mu}} R(\nu) \, |\vec{\mathbb{A}}_i(\nu)|^2 \, , \tag{7.146}$$

then, given that $R(\nu) = 1$ in the absence of absorption, this would lead to a spatial distribution of the reflected beam that was identical to that of the incident beam; however, we just showed in Section 7.9 that the distribution was profoundly altered by the complex poles of the reflection function.

In other words, equation (7.146) does not allow interference or diffraction of the reflected beam at resonance to be taken into account, even if its integral correctly gives the exact value of the reflected flux.

Also, to conclude this chapter, we shall show how using the well-known principle of *stationary phase* will allow us to specify the conditions leading to an asymptotic expression for the wave packet, thereby justifying the validity of the expressions given for the spectral flux density.

Consider the general expression for a spatial wave packet propagating in the direction $z < 0$:

$$\vec{\mathcal{E}}_r(x, z) = \int_{\nu} \vec{\mathbb{A}}_r(\nu) \, e^{i[2\pi\nu x - \alpha(\nu)z]} \, d\nu \, . \tag{7.147}$$

Let the argument of the amplitude $\vec{\mathbb{A}}_r(\nu)$ be $\phi(\nu)$ and put

$$\psi(\nu) = 2\pi\nu x - \alpha(\nu)z + \phi(\nu) \, . \tag{7.148}$$

We can then write equation (7.147) in the form

$$\vec{\mathcal{E}}_r(x, z) = \int_{\nu} |\vec{\mathbb{A}}_r(\nu)| \, e^{i\psi(\nu)} \, d\nu \, . \tag{7.149}$$

For large values of $x$ and/or $z$, i.e., in the far field, the complex exponential appearing in (7.149) varies very rapidly with the spatial frequency $\nu$ since its angular period is in the order of $\lambda/x$ or $\lambda/z$, if we first abstract the contribution of the term $\phi(\nu)$. These rapid phase oscillations significantly reduce the value of the resulting field $\vec{\mathcal{E}}_r$ and make us think that the value of the integral is driven by an integration domain in $\nu$ where these phase variations are slower. Provided they exist, these stationary frequencies are defined by

$$\frac{\partial \psi(\nu)}{\partial \nu} = 2\pi x - \frac{\partial \alpha(\nu)}{\partial \nu} z + \frac{\partial \phi(\nu)}{\partial \nu} = 2\pi x + \frac{4\pi^2 \nu}{\alpha(\nu)} z + \frac{\partial \phi(\nu)}{\partial \nu} = 0. \quad (7.150)$$

For some arbitrary derivative $\frac{\partial \phi(\nu)}{\partial \nu}$ there is no guarantee that solutions to condition (7.150) exist, or even that potential stationary frequencies can be identified. On the other hand, consider the case in which the phase variations of amplitude $\vec{A}_r(\nu)$ can be neglected, i.e., $\frac{\partial \phi(\nu)}{\partial \nu} \approx 0$. Under these conditions, we get a single stationary frequency $\nu_0$ defined by

$$2\pi \nu_0 = -\frac{x}{z} \alpha(\nu_0) \quad \Rightarrow \quad k \sin \theta_0 = -\frac{x}{z} k \cos \theta_0 \quad \Rightarrow \quad x = -z \tan \theta_0. \quad (7.151)$$

Note that this angle $\theta_0$ defines the position of the point $M(x, z)$ at which the value of the field is calculated, the polar coordinates of this point being defined by

$$x = \rho \cos \chi \quad ; \quad z = -\rho \sin \chi \quad \Rightarrow \quad \theta_0 = \frac{\pi}{2} - \chi(x, z). \quad (7.152)$$

We then proceed to a second-order expansion of the phase term $\psi(\nu)$ in the neighborhood of the stationary frequency $\nu_0$, i.e.,

$$\psi(\nu) \approx \psi(\nu_0) + (\nu - \nu_0) \left.\frac{\partial \psi}{\partial \nu}\right|_{\nu_0} + \frac{1}{2}(\nu - \nu_0)^2 \left.\frac{\partial^2 \psi}{\partial \nu^2}\right|_{\nu_0}$$

$$= \psi(\nu_0) + \frac{2\pi^2 k^2 z}{\alpha^3(\nu_0)}(\nu - \nu_0)^2, \quad (7.153)$$

where

$$\psi(\nu_0) = \phi(\nu_0) + 2\pi \nu_0 x - \alpha(\nu_0) z = \phi(\nu_0) - \frac{k^2 z}{\alpha(\nu_0)} = \phi(\nu_0) + k\rho, \quad (7.154)$$

The expression for the field becomes

$$\vec{\mathcal{E}}_r(x, z) = e^{ik\rho} \vec{A}_r(\nu_0) \int_\nu e^{i\frac{2\pi^2 k^2 z}{\alpha^3(\nu_0)}(\nu - \nu_0)^2} \, d\nu. \quad (7.155)$$

Rigorously, the integral in equation (7.155) should be calculated in the neighborhood of $\nu_0$, but it can be shown that it is possible to replace this restricted integration domain by the entire real axis, with the oscillations of this function increasing with distance from $\nu_0$. Consequently,

$$\int_\nu e^{i\frac{2\pi^2 k^2 z}{\alpha^3(\nu_0)}(\nu - \nu_0)^2} \, d\nu = \int_{-\infty}^{+\infty} e^{i\frac{2\pi^2 k^2 z}{\alpha^3(\nu_0)}\nu^2} \, d\nu = e^{i\frac{\pi}{4}} \sqrt{\frac{\alpha^3(\nu_0)}{2\pi k^2 z}}. \quad (7.156)$$

Finally, the field is written as

$$\vec{\mathcal{E}}_r(x, z) = e^{ik\rho}\vec{\mathbb{A}}_r(\nu_0)e^{i\frac{\pi}{4}}\sqrt{\frac{a^3(\nu_0)}{2\pi k^2 z}} = e^{-i\frac{\pi}{4}}\sqrt{\frac{k}{2\pi}}\,\vec{\mathbb{A}}_r(\nu_0)\frac{z}{\sqrt{\rho^3}}e^{ik\rho}. \quad (7.157)$$

Observe from this calculation that we were able to extract the elementary component $\vec{\mathbb{A}}_r(\nu_0)$ from the wave packet to approximate in the far field the value of the field in the direction given by $-z/x = \tan\chi$ (recall that the field propagates here in the direction $z < 0$). To get to such a result, we had to assume that it was possible to neglect variations in phase of amplitude $\vec{\mathbb{A}}_r(\nu)$.

Now that we have this asymptotic expression, clearly valid for $\vec{\mathcal{E}}$ and $\vec{\mathcal{H}}$, we are in a position to calculate the associated Poynting vector, i.e.,

$$\vec{\Pi} = \frac{1}{2}\left(\vec{\mathcal{E}}_r^* \wedge \vec{\mathcal{H}}_r\right) = \frac{1}{2}\frac{kz^2}{2\pi\rho^3}\left[\vec{\mathbb{A}}_r^*(\nu_0) \wedge \vec{\mathbb{B}}_r(\nu_0)\right]. \quad (7.158)$$

We now use the relation between elementary field components (see Chapter 4), namely

$$\vec{\mathbb{B}}_r(\nu_0) = \frac{1}{\omega\tilde{\mu}}\,k\,\vec{u}\wedge\vec{\mathbb{A}}_r(\nu_0) \quad \text{with} \quad \vec{u} = \frac{\vec{\rho}}{\rho}. \quad (7.159)$$

to obtain the final expression for the Poynting vector:

$$\vec{\Pi} = \frac{1}{2\omega\tilde{\mu}}\frac{1}{2\pi\rho}\left(\frac{kz}{\rho}\right)^2|\vec{\mathbb{A}}_r(\nu_0)|^2\,\vec{u}. \quad (7.160)$$

In 2D geometry, the flux transported through an elementary *surface* perpendicular to the Poynting vector can be written as

$$d\Phi = \vec{\Pi}.\vec{u}\,dl = |\vec{\Pi}|\,\rho\,d\theta = \frac{1}{2\omega\tilde{\mu}}\frac{1}{2\pi}\left(\frac{kz}{\rho}\right)^2|\vec{\mathbb{A}}_r(\nu_0)|^2\,d\theta, \quad (7.161)$$

which gives, in terms of spectral flux density,

$$\frac{d\Phi}{d\nu} = \frac{1}{2\omega\tilde{\mu}}\alpha(\nu_0)|\vec{\mathbb{A}}_r(\nu_0)|^2. \quad (7.162)$$

Hence we get the expression established in Chapter 4 and given in (7.146). This expression is therefore valid in the far field and where the amplitude of the field shows small frequential variations in phase.

# 8

# Light Scattering from Optical Multilayers

## 8.1 Introduction

In the preceding Chapters (6 and 7) we assumed that the interfaces within a multilayer stack were perfectly smooth.

This assumption was necessary to obtain a simple expression for the boundary conditions of the tangential field components, to establish a formalism for complex admittances, and to define a method for calculating the optical properties of an interference filter, either in terms of amplitude $(r, t)$ or energy (reflection $R$, transmission $T$, and absorption $A$); the principle of energy conservation required that these three quantities must be bound by the relation $R + T + A = 1$.

However, it is clear that this assumption of ideally plane surfaces fails to address reality, at least as far as the substrate is concerned. The surface of this substrate is created by polishing, and regardless of todays impressive quality for this process of making the final form of an optical component, the polished surfaces are somewhat irregular (we talk of *roughness* or *rugosity*): very small, to be sure, but nonzero (typically a few nanometers or fractions of a nanometer). Note that this concept of roughness differs from that of *flatness*, a phenomenon responsible for distorting the reflected wave front. In other words, roughness is essentially statistical in nature and involves high spatial frequencies, in contrast with flatness, which is deterministic by nature and characterized by low spatial frequencies.

When a plane wave encounters such a *rough* surface, part of the reflected and transmitted light is by nature *diffuse*, i.e., the wave vectors that characterize it differ from those deduced from the laws of reflection or refraction. If we let $D$ be the total power carried by this scattered light, then the equation for the conservation of energy must be put in the following modified form:

$$R + T + A + D = 1. \tag{8.1}$$

With this particular energy balance, it is often possible to ignore the absorbed quantity $A$; indeed the purity of the materials and the quality of the deposition processes used these days can guarantee values of imaginary refractive index as low as a few $10^{-6}$ for oxide-type dielectric materials. The scattering term $D$ is thus critical in terms of the ultimate filter performance, not only because a loss of energy reduces the reflection and transmission factors, but also because this parasitic scattered light will negatively affect the overall performance of an optical system. For example, this effect becomes particularly critical for the ring laser gyros or the interferometric detectors for gravitational waves (the components involved here are multilayer dielectric mirrors), or for the focal planes of Earth observation satellites (these involve bandpass filters placed in front of the detector matrices). It should be understood that all the interfaces of a multilayer stack demonstrate a certain roughness, and hence give rise to multiple sources of scattering within the filter; this can lead to an unwelcome emphasis of these scattering phenomena at certain wavelengths or in certain spatial directions.

For this reason we felt it necessary to write a chapter outlining a theoretical approach that reliably quantifies the value of this coefficient $D$, and that also predicts the angular and spectral dependencies that characterize this *scattered* light.

We shall first consider the canonical problem posed by a slightly rough plane interface separating two semi-infinite media of differing optical properties (the incident medium and the substrate), before treating the more complex problem corresponding to a stack of optical thin layers deposited on such an interface; we then finish with the most general configuration, namely that of a substrate of finite thickness whose two faces are multilayer coated.

We state at once that the formalism used is aimed at surfaces of optical quality, i.e., those for which the asperities are small compared with the wavelength of the illuminating light and the slopes are much smaller than 1. These conditions of validity correspond perfectly to the interfaces in optical interference filters (roughness on the order of a nanometer, slopes on the order of a few percent); such an observation has been widely validated in the scientific literature.

## 8.2 Scattering by a Single Rough Interface

### 8.2.1 Statement of the Problem

Consider a *slightly rough* interface described by the equation $z - h(x, y) = 0$ separating two semi-infinite media (incident medium identified by index 0 and substrate identified by index 1), as shown in Figure 8.1. Since the $z$ origin is arbitrary, we assume that the surface is centered on $z = 0$.

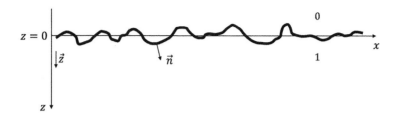

Figure 8.1  2D representation of a rough interface between two semi-infinite media.

To express the boundary conditions at this interface we will need to know the components of the vector $\vec{n}$ normal to this surface. To determine these components, it is sufficient to calculate the vector gradient of its cartesian equation $f(x, y, z) = 0$, and then to normalize the result obtained, i.e.,

$$
\vec{n} = \frac{1}{\sqrt{1 + \left(\dfrac{\partial h}{\partial x}\right)^2 + \left(\dfrac{\partial h}{\partial y}\right)^2}}
\begin{bmatrix} -\dfrac{\partial h}{\partial x} \\[2mm] -\dfrac{\partial h}{\partial y} \\[2mm] 1 \end{bmatrix}
\quad \Rightarrow \quad
\vec{n} = \frac{\vec{z} - \mathbf{grad}\, h}{\sqrt{1 + |\mathbf{grad}\, h|^2}}, \quad (8.2)
$$

where $\vec{z}$ denotes the unit vector associated with the $z$-axis. If the surface is perfectly plane ($h = h_0$), we find immediately that the normal to this surface is everywhere in the same direction as the $z$-axis.

Recall the form taken by Maxwell's equations in the harmonic regime in each of the two media ($j = 0, 1$), each assumed homogeneous, isotropic, and linear:

$$
\begin{cases}
\mathbf{curl}\, \vec{\mathcal{E}}_j(\vec{\rho}, f) = i\omega \tilde{\mu}_j(f) \vec{\mathcal{H}}_j(\vec{\rho}, f) \\
\mathbf{curl}\, \vec{\mathcal{H}}_j(\vec{\rho}, f) = -i\omega \tilde{\epsilon}_j(f) \vec{\mathcal{E}}_j(\vec{\rho}, f) \\
\mathrm{div}[\tilde{\epsilon}_j(f)\, \vec{\mathcal{E}}_j(\vec{\rho}, f)] = 0 \\
\mathrm{div}[\tilde{\mu}_j(f)\, \vec{\mathcal{H}}_j(\vec{\rho}, f)] = 0
\end{cases}
\qquad \text{where} \quad \vec{\rho} = (\vec{r}, z) = (x, y, z). \quad (8.3)
$$

As stated in Chapter 2, equations (8.3) are valid in the sense of functions, i.e., within each of the two media [medium 0 for $z < h(x, y)$, medium 1 for $z > h(x, y)$]. These must therefore be completed by the boundary conditions at the interface (assumed not charged), i.e., for the tangential field components at height $z = h(x, y) = h(\vec{r})$,

$$\begin{cases} \vec{n} \wedge \vec{\mathcal{E}}_1[\vec{r}, h(\vec{r}), f] - \vec{n} \wedge \vec{\mathcal{E}}_0[\vec{r}, h(\vec{r}), f] = \vec{0} \\ \vec{n} \wedge \vec{\mathcal{H}}_1[\vec{r}, h(\vec{r}), f] - \vec{n} \wedge \vec{\mathcal{H}}_0[\vec{r}, h(\vec{r}), f] = \vec{0} \end{cases} , \qquad (8.4)$$

and for their normal components,

$$\begin{cases} \tilde{\epsilon}_1(f)\{\vec{n} \cdot \vec{\mathcal{E}}_1[\vec{r}, h(\vec{r}), f]\} - \tilde{\epsilon}_0(f)\{\vec{n} \cdot \vec{\mathcal{E}}_0[\vec{r}, h(\vec{r}), f]\} = 0 \\ \tilde{\mu}_1(f)\{\vec{n} \cdot \vec{\mathcal{H}}_1[\vec{r}, h(\vec{r}), f]\} - \tilde{\mu}_0(f)\{\vec{n} \cdot \vec{\mathcal{H}}_0[\vec{r}, h(\vec{r}), f]\} = 0 \end{cases} . \qquad (8.5)$$

In the rest of this chapter we shall be interested only in the field components at frequency $f$ and shall therefore omit this frequency when writing the various equations. Note that we are here interested in *elastic* scattering mechanisms, i.e., with no change in frequency.

We can express the field $\vec{\mathcal{E}}_0(\vec{r}, z)$ governing medium 0 as the sum of an ideal field [written $\vec{\mathcal{E}}_{0,e}(\vec{r}, z)$], corresponding to the field that would govern medium 0 if the interface between media 0 and 1 was ideally smooth, and of a scattered field [written $\vec{\mathcal{E}}_{0,d}(\vec{r}, z)$] generated by the irregularities at this interface at the same height and in the same medium, i.e.,

$$\vec{\mathcal{E}}_0(\vec{r}, z) = \vec{\mathcal{E}}_{0,e}(\vec{r}, z) + \vec{\mathcal{E}}_{0,d}(\vec{r}, z) \quad \text{where} \quad z < h(\vec{r}). \qquad (8.6)$$

Note here that the ideal field $\vec{\mathcal{E}}_{0,e}$ is just the specular field studied in Chapter 6 and associated with ideally plane interfaces. Hence this field is perfectly known if the illumination conditions and the geometry of the multilayer are also known. We shall see in due course that it is responsible for exciting (fictitious) scattering currents. Finally, the scattered field $\vec{\mathcal{E}}_{0,d}$ can be seen as a perturbation of the ideal field brought about by the roughness; taking account of our assumptions about the quality of the scattering surfaces, this perturbation will therefore be assumed to be of first order.

### 8.2.2 First-order Approximation

Consider the quantity $\vec{n} \wedge \vec{\mathcal{E}}_0(\vec{r}, h)$ and expand it in the neighborhood of $h = 0$ (low interface roughness assumption):

$$\vec{n} \wedge \vec{\mathcal{E}}_0(\vec{r}, h) = \vec{n} \wedge \left\{ \vec{\mathcal{E}}_0(\vec{r}, 0) + h \left[ \frac{\partial \vec{\mathcal{E}}_0}{\partial z} \right]_{\vec{r}, 0} \right\}. \qquad (8.7)$$

Using equations (8.2) and (8.6), let us elaborate on the quantities $\vec{n}$ and $\vec{\mathcal{E}}_0$ which appear in the limited expansion (8.7). We get

$$\vec{n} \wedge \vec{\mathcal{E}}_0(\vec{r}, h) = \frac{\vec{z} - \mathbf{grad}\, h}{\sqrt{1 + |\mathbf{grad}\, h|^2}}$$

$$\wedge \left\{ \left[ \vec{\mathcal{E}}_{0,e}(\vec{r}, 0) + \vec{\mathcal{E}}_{0,d}(\vec{r}, 0) \right] + h \left[ \frac{\partial \vec{\mathcal{E}}_{0,e}}{\partial z} + \frac{\partial \vec{\mathcal{E}}_{0,d}}{\partial z} \right]_{\vec{r},0} \right\}. \quad (8.8)$$

It is important at this stage to recall the assumptions we made, namely:

- The roughness amplitude is small compared with the illuminating wavelength ($|h/\lambda| \ll 1$), and the slope is much less than 1 ($|\mathbf{grad}\, h| \ll 1$); as already pointed out, this corresponds to the interface characteristics of an optical interference filter. These two variables (slope and roughness-to-wavelength ratio) will therefore be assumed to be of first order. However, for the sake of rigor, we point out that the notions of roughness and slope must be associated with a spatial frequency window; by limiting this window to that of the far field (i.e. $\sigma < k$, see Chapter 4), we can show that these two assumptions are actually redundant.
- Taking account of the previous conditions concerning the roughness and slope of the scattering surfaces, the scattered field will naturally be assumed to be very small compared with the ideal field ($|\vec{\mathcal{E}}_{0,d}| \ll |\vec{\mathcal{E}}_{0,e}|$), and hence assumed to be a variable of first order; although this assumption is perfectly appropriate here, it is nonetheless worth pointing out that it could break down in certain extreme cases, e.g., when the number of layers is very large (a thousand or so) or in the case of giant enhancement of the ideal field (see Chapter 7).
- Finally, we note that the assumption of low roughness ($|h/\lambda| \ll 1$) allows us to neglect the variations of the scattered field in the troughs and furrows of the surface, i.e., when $|z| < |h|$. Rigorously, the scattered field should be written in the form of a spatial wave packet and then differentiated with respect to $z$ to obtain the term in $1/\lambda$. Then, by limiting the frequency window solely to propagative waves ($\sigma < k$) we can show that $h \left[ \frac{\partial \vec{\mathcal{E}}_{0,d}}{\partial z} \right]$ is of second order.

Following this last remark, we must therefore bear in mind that this formalism could be limited in cases in which the weight of evanescent waves (which transport no energy in transparent media; see Chapter 4) might dominate that of propagative waves. Since this case is extraordinarily rare, in particular for our far-field filter applications, we can finally rewrite equation (8.8) as an expansion to first order, i.e.,

$$\vec{n} \wedge \vec{\mathcal{E}}_0(\vec{r}, h) = \vec{z} \wedge [\vec{\mathcal{E}}_{0,e}(\vec{r}, 0) + \vec{\mathcal{E}}_{0,d}(\vec{r}, 0)]$$

$$+ h\vec{z} \wedge \left[ \frac{\partial \vec{\mathcal{E}}_{0,e}}{\partial z} \right]_{\vec{r},0} - \mathbf{grad}\, h \wedge \vec{\mathcal{E}}_{0,e}(\vec{r}, 0) \,. \qquad (8.9)$$

Similarly, in medium 1 we get

$$\vec{n} \wedge \vec{\mathcal{E}}_1(\vec{r}, h) = \vec{z} \wedge [\vec{\mathcal{E}}_{1,e}(\vec{r}, 0) + \vec{\mathcal{E}}_{1,d}(\vec{r}, 0)]$$

$$+ h\vec{z} \wedge \left[ \frac{\partial \vec{\mathcal{E}}_{1,e}}{\partial z} \right]_{\vec{r},0} - \mathbf{grad}\, h \wedge \vec{\mathcal{E}}_{1,e}(\vec{r}, 0) \,. \qquad (8.10)$$

We can now express the boundary condition (8.4) in this first-order approximation, i.e.,

$$\vec{n} \wedge \left[ \vec{\mathcal{E}}_1(\vec{r}, h) - \vec{\mathcal{E}}_0(\vec{r}, h) \right] = \vec{z} \wedge [\vec{\mathcal{E}}_{1,e}(\vec{r}, 0) - \vec{\mathcal{E}}_{0,e}(\vec{r}, 0)]$$

$$+ \vec{z} \wedge [\vec{\mathcal{E}}_{1,d}(\vec{r}, 0) - \vec{\mathcal{E}}_{0,d}(\vec{r}, 0)]$$

$$+ h\vec{z} \wedge \left[ \frac{\partial \vec{\mathcal{E}}_{1,e}}{\partial z} - \frac{\partial \vec{\mathcal{E}}_{0,e}}{\partial z} \right]_{\vec{r},0} \qquad (8.11)$$

$$- \mathbf{grad}\, h \wedge [\vec{\mathcal{E}}_{1,e}(\vec{r}, 0) - \vec{\mathcal{E}}_{0,e}(\vec{r}, 0)] \,.$$

Expression (8.4) shows that the first member of this equation is 0, and we therefore deduce that the boundary condition applicable to the tangential components of the scattered electric fields is

$$\vec{z} \wedge [\vec{\mathcal{E}}_{1,d} - \vec{\mathcal{E}}_{0,d}] = (\mathbf{grad}\, h - \vec{z}) \wedge [\vec{\mathcal{E}}_{1,e} - \vec{\mathcal{E}}_{0,e}] - h\vec{z} \wedge \left[ \frac{\partial \vec{\mathcal{E}}_{1,e}}{\partial z} - \frac{\partial \vec{\mathcal{E}}_{0,e}}{\partial z} \right], \quad (8.12)$$

where all quantities are written at $(\vec{r}, 0)$.

Furthermore, in each of these two media the ideal field satisfies the same Maxwell's equations (8.3) as the real field, but for a local normal $\vec{z}$ independent of position $(x, y)$. Hence we get for the tangential components of this ideal field (see Chapter 6),

$$\begin{cases} \vec{z} \wedge [\vec{\mathcal{E}}_{1,e}(\vec{r}, 0) - \vec{\mathcal{E}}_{0,e}(\vec{r}, 0)] = \vec{0} \\ \vec{z} \wedge [\vec{\mathcal{H}}_{1,e}(\vec{r}, 0) - \vec{\mathcal{H}}_{0,e}(\vec{r}, 0)] = \vec{0} \end{cases}, \qquad (8.13)$$

and for the normal components,

$$\begin{cases} \vec{z} \cdot [\tilde{\epsilon}_1 \vec{\mathcal{E}}_{1,e}(\vec{r}, 0) - \tilde{\epsilon}_0 \vec{\mathcal{E}}_{0,e}(\vec{r}, 0)] = \vec{0} \\ \vec{z} \cdot [\tilde{\mu}_1 \vec{\mathcal{H}}_{1,e}(\vec{r}, 0) - \tilde{\mu}_0 \vec{\mathcal{H}}_{0,e}(\vec{r}, 0)] = \vec{0} \end{cases}. \qquad (8.14)$$

Making use of the continuity of the tangential components of the ideal field in equation (8.12), we get finally

$$\vec{z} \wedge [\vec{\mathcal{E}}_{1,d} - \vec{\mathcal{E}}_{0,d}] = \mathbf{grad}\, h \wedge [\vec{\mathcal{E}}_{1,e} - \vec{\mathcal{E}}_{0,e}] - h\vec{z} \wedge \left[ \frac{\partial \vec{\mathcal{E}}_{1,e}}{\partial z} - \frac{\partial \vec{\mathcal{E}}_{0,e}}{\partial z} \right]. \quad (8.15)$$

Now considering the magnetic fields instead of the electric fields, the same procedure would have led to an equation similar to (8.15), namely

$$\vec{z} \wedge [\vec{\mathcal{H}}_{1,d} - \vec{\mathcal{H}}_{0,d}] = \mathbf{grad}\, h \wedge [\vec{\mathcal{H}}_{1,e} - \vec{\mathcal{H}}_{0,e}] - h\vec{z} \wedge \left[ \frac{\partial \vec{\mathcal{H}}_{1,e}}{\partial z} - \frac{\partial \vec{\mathcal{H}}_{0,e}}{\partial z} \right]. \quad (8.16)$$

If we take the classical case of nonmagnetic optical media ($\tilde{\mu}_1 = \tilde{\mu}_0 = \mu_V$), the normal and tangential components of the magnetic field are continuous, so the previous equation takes the following simplified form:

$$\vec{z} \wedge [\vec{\mathcal{H}}_{1,d} - \vec{\mathcal{H}}_{0,d}] = -h\vec{z} \wedge \left[ \frac{\partial \vec{\mathcal{H}}_{1,e}}{\partial z} - \frac{\partial \vec{\mathcal{H}}_{0,e}}{\partial z} \right]. \quad (8.17)$$

If we now consider the boundary conditions (8.5) relating to the normal components of the total field, we can carry out the same first-order expansion as that used for the tangential components.

For the electric field we get

$$\vec{z} \cdot [\tilde{\epsilon}_1 \vec{\mathcal{E}}_{1,d} - \tilde{\epsilon}_0 \vec{\mathcal{E}}_{0,d}] = \mathbf{grad}\, h \cdot [\tilde{\epsilon}_1 \vec{\mathcal{E}}_{1,e} - \tilde{\epsilon}_0 \vec{\mathcal{E}}_{0,e}] - h\vec{z} \cdot \left[ \tilde{\epsilon}_1 \frac{\partial \vec{\mathcal{E}}_{1,e}}{\partial z} - \tilde{\epsilon}_0 \frac{\partial \vec{\mathcal{E}}_{0,e}}{\partial z} \right], \quad (8.18)$$

and for the magnetic field,

$$\vec{z} \cdot [\vec{\mathcal{H}}_{1,d} - \vec{\mathcal{H}}_{0,d}] = -h\vec{z} \cdot \left[ \frac{\partial \vec{\mathcal{H}}_{1,e}}{\partial z} - \frac{\partial \vec{\mathcal{H}}_{0,e}}{\partial z} \right]. \quad (8.19)$$

We note that these discontinuities in the normal components are not generally necessary when calculating the scattered field.

### 8.2.3 Maxwell's Equations for the Scattered Field

In each of the two media and away from the surface furrows ($|z| > |h|$), the scattered electromagnetic field (the difference between the actual (total) field and the ideal field) satisfies the same Maxwell's equations as those applicable to both these types of field.

We have seen, furthermore, that the variations in this field are negligible in the furrows of the rough surface; we can therefore assume that it is defined everywhere in the regions $z > 0$ (medium 1) and $z < 0$ (medium 0). Under these conditions, we only have to know the discontinuities in this field for the problem to be completely defined. These have just been given at height $z = 0$, with respect to a surface of constant normal $z$. In summary, the scattered field satisfies the same equations as the ideal field in the sense of functions, with the same surface ($z = 0$) of discontinuities; however, these discontinuities are different.

It is possible, therefore, to summarize these results by appealing to the formalism of distributions (see Chapter 2), i.e.,

$$
\begin{cases}
\mathbf{curl}\,\vec{\mathcal{E}}_d(\vec{\rho}, f) = i\omega\tilde{\mu}(f)\vec{\mathcal{H}}_d(\vec{\rho}, f) + \vec{\mathcal{M}}\delta(z) \\
\mathbf{curl}\,\vec{\mathcal{H}}_d(\vec{\rho}, f) = -i\omega\tilde{\epsilon}(f)\vec{\mathcal{E}}_d(\vec{\rho}, f) + \vec{\mathcal{J}}\delta(z) \\
\mathrm{div}[\tilde{\epsilon}(f)\,\vec{\mathcal{E}}_d(\vec{\rho}, f)] = \mathcal{Q}\delta(z) \\
\mathrm{div}[\tilde{\mu}(f)\,\vec{\mathcal{H}}_d(\vec{\rho}, f)] = \mathcal{P}\delta(z)
\end{cases}
\tag{8.20}
$$

where $\delta(z)$ is the Dirac distribution centered on $z = 0$, i.e., at the interface between the two media, which is assumed to be perfectly plane; the currents $\vec{\mathcal{M}}$ and $\vec{\mathcal{J}}$ are defined in nonmagnetic media by

$$
\begin{cases}
\vec{\mathcal{M}} = \mathbf{grad}\,h \wedge [\vec{\mathcal{E}}_{1,e} - \vec{\mathcal{E}}_{0,e}] - h\vec{z} \wedge \left[\dfrac{\partial \vec{\mathcal{E}}_{1,e}}{\partial z} - \dfrac{\partial \vec{\mathcal{E}}_{0,e}}{\partial z}\right] \\[4mm]
\vec{\mathcal{J}} = -h\vec{z} \wedge \left[\dfrac{\partial \vec{\mathcal{H}}_{1,e}}{\partial z} - \dfrac{\partial \vec{\mathcal{H}}_{0,e}}{\partial z}\right]
\end{cases}
\tag{8.21}
$$

while charges $\mathcal{Q}$ and $\mathcal{P}$ are given by the following relations (still in nonmagnetic media):

$$
\begin{cases}
\mathcal{Q} = \mathbf{grad}\,h \cdot [\tilde{\epsilon}_1\vec{\mathcal{E}}_{1,e} - \tilde{\epsilon}_0\vec{\mathcal{E}}_{0,e}] - h\vec{z} \cdot \left[\tilde{\epsilon}_1\dfrac{\partial \vec{\mathcal{E}}_{1,e}}{\partial z} - \tilde{\epsilon}_0\dfrac{\partial \vec{\mathcal{E}}_{0,e}}{\partial z}\right] \\[4mm]
\mathcal{P} = -h\vec{z} \cdot \left[\dfrac{\partial \vec{\mathcal{H}}_{1,e}}{\partial z} - \dfrac{\partial \vec{\mathcal{H}}_{0,e}}{\partial z}\right]
\end{cases}
\tag{8.22}
$$

Equations (8.20), (8.21), and (8.22) show that the fields radiated by a slightly rough surface are identical to the fields radiated by an ideal plane surface, but supporting *fictitious* currents and charges, both electric ($\vec{\mathcal{J}}$, $\mathcal{Q}$) and magnetic ($\vec{\mathcal{M}}$, $\mathcal{P}$). These fictitious sources depend on the topography of the surface and vary with the ideal electromagnetic field; for this reason the ideal field is often known as the exciting field, and we have therefore chosen to identify it from the start in Section 8.2 with the subscript $e$.

### 8.2.4 Elementary Components of the Scattered Field

We have shown that the scattered electromagnetic field satisfied Maxwell's equations (8.20) relative to a charged plane optical interface. Each of the scattered fields is therefore composed of a wave packet satisfying (in the sense of functions) the homogeneous Helmholtz equation in each medium:

$$\Delta \vec{\mathcal{E}}_{j,d}(\vec{\rho}) + k_j^2 \vec{\mathcal{E}}_{j,d}(\vec{\rho}) = \vec{0} \quad \text{where} \quad k_j^2 = \omega^2 \mu_V \tilde{\epsilon}_j \quad \text{and} \quad j = 0, 1. \quad (8.23)$$

We showed in Chapter 4 that, quite generally, the solution of equation (8.23) has the form

$$\vec{\mathcal{E}}_{j,d}(\vec{\rho}) = \vec{\mathcal{E}}_{j,d}(\vec{r}, z) = \int_{\vec{\nu}_d} \vec{\mathbb{E}}_{j,d}(\vec{\nu}_d, z) e^{2i\pi \vec{\nu}_d \cdot \vec{r}} \, d^2 \vec{\nu}_d, \quad (8.24)$$

where

$$\begin{aligned} \vec{\mathbb{E}}_{j,d}(\vec{\nu}_d, z) &= \vec{\mathbb{A}}_{j,d}^+(\vec{\nu}_d) e^{i\alpha_j^d z} + \vec{\mathbb{A}}_{j,d}^-(\vec{\nu}_d) e^{-i\alpha_j^d z} \\ (\alpha_j^d)^2 &= k_j^2 - \sigma_d^2 = k_j^2 - 4\pi^2 \nu_d^2. \end{aligned} \quad (8.25)$$

This solution is therefore constructed by superposing progressive $(\vec{\mathbb{A}}_{j,d}^+)$ and retrograde $(\vec{\mathbb{A}}_{j,d}^-)$ elementary plane waves.

The particular features of our problem (i.e., a single interface) require that these scattered waves be retrograde in the incident medium and progressive in the substrate, both media being assumed semi-infinite. Hence, for the electric field,

$$\begin{cases} \vec{\mathbb{E}}_{0,d}(\vec{\nu}_d, z) = \vec{\mathbb{A}}_{0,d}^-(\vec{\nu}_d) e^{-i\alpha_0^d z} \\ \vec{\mathbb{E}}_{1,d}(\vec{\nu}_d, z) = \vec{\mathbb{A}}_{1,d}^+(\vec{\nu}_d) e^{i\alpha_1^d z} \end{cases}, \quad (8.26)$$

and for the magnetic field,

$$\begin{cases} \vec{\mathbb{H}}_{0,d}(\vec{\nu}_d, z) = \vec{\mathbb{B}}_{0,d}^-(\vec{\nu}_d) e^{-i\alpha_0^d z} \\ \vec{\mathbb{H}}_{1,d}(\vec{\nu}_d, z) = \vec{\mathbb{B}}_{1,d}^+(\vec{\nu}_d) e^{i\alpha_1^d z} \end{cases}. \quad (8.27)$$

To obtain an expression for the elementary components of the scattered fields, we need to take account not only of the relations (8.26) and (8.27) that define the propagative components of the wave packets associated with the scattered fields, but also the boundary conditions (8.15) and (8.17) relating to the tangential components of these same fields.

However, given that the equations (8.26) and (8.27) are expressed in the Fourier plane, we must also consider the (spatial) Fourier transformation of relations (8.15) and (8.17). Hence, for the elementary components of the electric fields,

$$\vec{z} \wedge [\vec{\mathbb{A}}_{1,d}^{+}(\vec{\nu}_d) - \vec{\mathbb{A}}_{0,d}^{-}(\vec{\nu}_d)] = \vec{\mathbb{M}}(\vec{\nu}_d), \tag{8.28}$$

and similarly for those of the magnetic fields,

$$\vec{z} \wedge [\vec{\mathbb{B}}_{1,d}^{+}(\vec{\nu}_d) - \vec{\mathbb{B}}_{0,d}^{-}(\vec{\nu}_d)] = \vec{\mathbb{J}}(\vec{\nu}_d), \tag{8.29}$$

where $\vec{\mathbb{M}}$ and $\vec{\mathbb{J}}$ correspond to the two-dimensional Fourier transforms of the quantities $\vec{\mathcal{M}}$ and $\vec{\mathcal{J}}$, i.e.,

$$\vec{\mathbb{M}}(\vec{\nu}_d) = \int_{\vec{r}} \vec{\mathcal{M}}(\vec{r}) e^{-2i\pi\vec{\nu}_d \cdot \vec{r}} d^2\vec{r} \quad \text{et} \quad \vec{\mathbb{J}}(\vec{\nu}_d) = \int_{\vec{r}} \vec{\mathcal{J}}(\vec{r}) e^{-2i\pi\vec{\nu}_d \cdot \vec{r}} d^2\vec{r}. \tag{8.30}$$

Let us first examine equation (8.29). We can replace the fields $\vec{\mathbb{B}}_{j,d}^{\pm}$ by their tangential components $\vec{\mathbb{B}}_{j,d}^{T\pm}$ (the superscript $T$ will henceforth be used to indicate the tangential component of a field); then we can express these tangential components $\vec{\mathbb{B}}_{j,d}^{T\pm}$ as functions of the components of fields $\vec{\mathbb{A}}_{j,d}^{\pm}$ and the effective refractive indices $\tilde{n}_j$ of the two media, using the defining relationship introduced in Chapter 6 (the dependence of the amplitudes of the elementary components on $\vec{\nu}_d$ will now be omitted for the sake of conciseness):

$$\vec{\mathbb{B}}_{j,d}^{T\pm} = \pm\tilde{n}_j^d(\vec{z} \wedge \vec{\mathbb{A}}_{j,d}^{T\pm}), \tag{8.31}$$

where

$$\tilde{n}_j^d = \frac{1}{\eta_v \mu_r} \begin{cases} n_j \alpha_j^d / k_j & \text{TE polarization} \\ n_j k_j / \alpha_j^d & \text{TM polarization} \end{cases}. \tag{8.32}$$

$\eta_v$ is the impedance of vacuum and $\mu_r$ is the relative magnetic permeability common to the two media ($\mu_r = 1$). Substituting expression (8.31) into the first of the two boundary equations (8.29), we get

$$\vec{z} \wedge [\tilde{n}_1^d(\vec{z} \wedge \vec{\mathbb{A}}_{1,d}^{T+}) + \tilde{n}_0^d(\vec{z} \wedge \vec{\mathbb{A}}_{0,d}^{T-})] = \vec{\mathbb{J}}(\vec{\nu}_d). \tag{8.33}$$

Using the properties of the vector triple product, we obtain

$$\tilde{n}_1^d \vec{\mathbb{A}}_{1,d}^{T+} + \tilde{n}_0^d \vec{\mathbb{A}}_{0,d}^{T-} = -\vec{\mathbb{J}}(\vec{\nu}_d). \tag{8.34}$$

Now we calculate the quantity $\vec{z} \wedge \vec{\mathbb{M}}$ from equation (8.28), again using the properties of the vector triple product:

$$\vec{z} \wedge \vec{\mathbb{M}}(\vec{\nu}_d) = \vec{z} \wedge \left( \vec{z} \wedge [\vec{\mathbb{A}}_{1,d}^{+} - \vec{\mathbb{A}}_{0,d}^{-}] \right) = -\vec{\mathbb{A}}_{1,d}^{T+} + \vec{\mathbb{A}}_{0,d}^{T-}. \tag{8.35}$$

To determine the amplitudes of the tangential components of the elementary scattered fields, it is necessary only to solve the following system of equations:

$$\begin{cases} \tilde{n}_1^d \vec{\mathbb{A}}_{1,d}^{T+} + \tilde{n}_0^d \vec{\mathbb{A}}_{0,d}^{T-} = -\vec{\mathbb{J}}(\vec{\nu}_d) \\ \vec{\mathbb{A}}_{1,d}^{T+} - \vec{\mathbb{A}}_{0,d}^{T-} = -\vec{z} \wedge \vec{\mathbb{M}}(\vec{\nu}_d) \end{cases} \tag{8.36}$$

leading in turn to the desired expressions

$$\begin{cases} \vec{\mathbb{A}}_{0,d}^{T-} = -\dfrac{\vec{\mathbb{J}}(\vec{\nu}_d) - \tilde{n}_1^d \, \vec{z} \wedge \vec{\mathbb{M}}(\vec{\nu}_d)}{\tilde{n}_0^d + \tilde{n}_1^d} \\ \vec{\mathbb{A}}_{1,d}^{T+} = -\dfrac{\vec{\mathbb{J}}(\vec{\nu}_d) + \tilde{n}_0^d \, \vec{z} \wedge \vec{\mathbb{M}}(\vec{\nu}_d)}{\tilde{n}_0^d + \tilde{n}_1^d} \end{cases} \tag{8.37}$$

This last pair of equations expresses the relationship between the fictitious currents $\vec{\mathbb{M}}$ and $\vec{\mathbb{J}}$ associated with the roughness of the interface and the amplitudes of the elementary components of the fields that they generate on either side of this interface.

### 8.2.5 Source Terms

We now have to calculate the source terms $\vec{\mathbb{J}}$ and $\vec{z} \wedge \vec{\mathbb{M}}$ that appear in relations (8.37). Note at this point that the ideal field is also made up of a spatial wave packet, since it is not restricted to a single illuminating incidence.

### The Fictitious Electric Current $\vec{\mathbb{J}}$

Making use of relations (8.21) and (8.30), the electric current is written in the Fourier plane as

$$\vec{\mathbb{J}}(\vec{\nu}_d) = -\vec{z} \wedge \int_{\vec{r}} h(\vec{r}) \left[ \frac{\partial \vec{\mathcal{H}}_{1,e}}{\partial z} - \frac{\partial \vec{\mathcal{H}}_{0,e}}{\partial z} \right]_{\vec{r},0} e^{-2i\pi \vec{\nu}_d \cdot \vec{r}} \, d^2 \vec{r}. \tag{8.38}$$

In this expression, the exciting magnetic fields, i.e., those corresponding to a perfectly plane interface, are spatial wave packets made up

- in the incident medium, by superposing progressive and retrograde waves, these being generated by reflection at the interface between incident medium and substrate;
- and in the substrate, by progressive waves only, corresponding to those transmitted by this same interface.

Consequently, and using the notation $\vec{\nu}_e$ for the spatial frequency of the exciting field, we have

$$\begin{cases} \vec{\mathcal{H}}_{0,e}(\vec{r}, z) = \displaystyle\int_{\vec{\nu}_e} \left[ \vec{\mathbb{B}}_{0,e}^+(\vec{\nu}_e)e^{i\alpha_0^e z} + \vec{\mathbb{B}}_{0,e}^-(\vec{\nu}_e)e^{-i\alpha_0^e z} \right] e^{2i\pi\vec{\nu}_e.\vec{r}} \, d^2\vec{\nu}_e \\[2mm] \vec{\mathcal{H}}_{1,e}(\vec{r}, z) = \displaystyle\int_{\vec{\nu}_e} \vec{\mathbb{B}}_{1,e}^+(\vec{\nu}_e)e^{i\alpha_1^e z} e^{2i\pi\vec{\nu}_e.\vec{r}} \, d^2\vec{\nu}_e \end{cases} . \quad (8.39)$$

We immediately deduce that

$$\begin{cases} \left[ \dfrac{\partial\vec{\mathcal{H}}_{0,e}}{\partial z} \right]_{\vec{r},0} = i \displaystyle\int_{\vec{\nu}_e} \alpha_0^e \left[ \vec{\mathbb{B}}_{0,e}^+ - \vec{\mathbb{B}}_{0,e}^- \right] e^{2i\pi\vec{\nu}_e.\vec{r}} \, d^2\vec{\nu}_e \\[4mm] \left[ \dfrac{\partial\vec{\mathcal{H}}_{1,e}}{\partial z} \right]_{\vec{r},0} = i \displaystyle\int_{\vec{\nu}_e} \alpha_1^e \vec{\mathbb{B}}_{1,e}^+ e^{2i\pi\vec{\nu}_e.\vec{r}} \, d^2\vec{\nu}_e \end{cases} . \quad (8.40)$$

The elementary component of current $\vec{\mathbb{J}}$ is therefore expressed as

$$\vec{\mathbb{J}}(\vec{\nu}_d) = i \int_{\vec{r}} \int_{\vec{\nu}_e} h(\vec{r})\alpha_0^e \left[ \vec{z} \wedge \vec{\mathbb{B}}_{0,e}^+ - \vec{z} \wedge \vec{\mathbb{B}}_{0,e}^- \right] e^{-2i\pi(\vec{\nu}_d-\vec{\nu}_e).\vec{r}} \, d^2\vec{\nu}_e d^2\vec{r}$$

$$- i \int_{\vec{r}} \int_{\vec{\nu}_e} h(\vec{r})\alpha_1^e \left[ \vec{z} \wedge \vec{\mathbb{B}}_{1,e}^+ \right] e^{-2i\pi(\vec{\nu}_d-\vec{\nu}_e).\vec{r}} \, d^2\vec{\nu}_e d^2\vec{r}. \quad (8.41)$$

Introducing the two-dimensional Fourier transform $\widehat{h}$ of the profile of the rough surface and integrating over $\vec{r}$, we get

$$\vec{\mathbb{J}}(\vec{\nu}_d) = i \int_{\vec{\nu}_e} \widehat{h}(\vec{\nu}_d - \vec{\nu}_e)\alpha_0^e \left[ \vec{z} \wedge \vec{\mathbb{B}}_{0,e}^+ - \vec{z} \wedge \vec{\mathbb{B}}_{0,e}^- \right] d^2\vec{\nu}_e$$

$$- i \int_{\vec{\nu}_e} \widehat{h}(\vec{\nu}_d - \vec{\nu}_e)\alpha_1^e \left[ \vec{z} \wedge \vec{\mathbb{B}}_{1,e}^+ \right] d^2\vec{\nu}_e. \quad (8.42)$$

Furthermore, using the definition of effective refractive indices and the properties of the vector product, we can write

$$\vec{z} \wedge \vec{\mathbb{B}}_{0,e}^+ = \vec{z} \wedge \vec{\mathbb{B}}_{0,e}^{T+} = \vec{z} \wedge [\tilde{n}_0^e(\vec{z} \wedge \vec{\mathbb{A}}_{0,e}^{T+})] = -\tilde{n}_0^e \vec{\mathbb{A}}_{0,e}^{T+} \quad (8.43)$$

$$\vec{z} \wedge \vec{\mathbb{B}}_{0,e}^- = \vec{z} \wedge \vec{\mathbb{B}}_{0,e}^{T-} = -\vec{z} \wedge [\tilde{n}_0^e(\vec{z} \wedge \vec{\mathbb{A}}_{0,e}^{T-})] = \tilde{n}_0^e \vec{\mathbb{A}}_{0,e}^{T-} \quad (8.44)$$

$$\vec{z} \wedge \vec{\mathbb{B}}_{1,e}^+ = \vec{z} \wedge \vec{\mathbb{B}}_{1,e}^{T+} = \vec{z} \wedge [\tilde{n}_1^e(\vec{z} \wedge \vec{\mathbb{A}}_{1,e}^{T+})] = -\tilde{n}_1^e \vec{\mathbb{A}}_{1,e}^{T+}, \quad (8.45)$$

while continuity of the tangential components of the ideal field at $z = 0$ means that

$$\vec{z} \wedge \vec{\mathbb{E}}_{0,e}(\vec{\nu}, 0) = \vec{z} \wedge \vec{\mathbb{E}}_{1,e}(\vec{\nu}, 0)$$

$$\Rightarrow \quad \vec{\mathbb{E}}_{0,e}^T(0) = \vec{\mathbb{A}}_{0,e}^{T+} + \vec{\mathbb{A}}_{0,e}^{T-} = \vec{\mathbb{A}}_{1,e}^{T+} = \vec{\mathbb{E}}_{1,e}^T(0). \quad (8.46)$$

The elementary component of the surface current is therefore finally expressed as

$$\vec{\mathbb{J}}(\vec{\nu}_d) = i \int_{\vec{\nu}_e} \widehat{h}(\vec{\nu}_d - \vec{\nu}_e) \left[ (\alpha_1^e \tilde{n}_1^e - \alpha_0^e \tilde{n}_0^e) \vec{\mathbb{E}}_{0,e}^T \right] d^2 \vec{\nu}_e \qquad (8.47)$$

and corresponds to the **convolution** of the Fourier transform of the surface profile with the elementary component of the ideal field at the surface, weighted by a coefficient that depends on the effective refractive indices of the two media.

Recall at this point that we also have the relations

$$\vec{\mathbb{E}}_{0,e}^T(0) = (1 + r^e)\vec{\mathbb{A}}_{0,e}^{T+} = \vec{\mathbb{A}}_{1,e}^{T+} = t^e \vec{\mathbb{A}}_{0,e}^{T+} , \qquad (8.48)$$

where

$$r^e = \frac{\tilde{n}_0^e - \tilde{n}_1^e}{\tilde{n}_0^e + \tilde{n}_1^e} \text{ and } t^e = \frac{2\tilde{n}_0^e}{\tilde{n}_0^e + \tilde{n}_1^e} . \qquad (8.49)$$

### The Fictitious Magnetic Current $\vec{z} \wedge \vec{\mathbb{M}}$

Consider the quantity $\vec{z} \wedge \vec{\mathbb{M}}$. Recall that this is expressed as

$$\vec{z} \wedge \vec{\mathbb{M}}(\vec{\nu}_d) = \vec{z} \wedge \int_{\vec{r}} \mathbf{grad}\, h(\vec{r}) \wedge \left[ \vec{\mathcal{E}}_{1,e}(\vec{r},0) - \vec{\mathcal{E}}_{0,e}(\vec{r},0) \right] e^{-2i\pi\vec{\nu}_d \cdot \vec{r}} d^2\vec{r}$$

$$- \vec{z} \wedge \int_{\vec{r}} h(\vec{r})\, \vec{z} \wedge \left[ \frac{\partial \vec{\mathcal{E}}_{1,e}}{\partial z} - \frac{\partial \vec{\mathcal{E}}_{0,e}}{\partial z} \right]_{\vec{r},0} e^{-2i\pi\vec{\nu}_d \cdot \vec{r}} d^2\vec{r}. \qquad (8.50)$$

Making use of the vector triple product and expressing the vector $\vec{\mathcal{E}}_{j,e}(\vec{r},0)$ as the sum of its tangential and normal components, i.e.,

$$\vec{\mathcal{E}}_{j,e}(\vec{r},0) = \vec{\mathcal{E}}_{j,e}^T(\vec{r},0) + [\vec{z} \cdot \vec{\mathcal{E}}_{j,e}(\vec{r},0)]\, \vec{z},$$

we get

$$\vec{z} \wedge \vec{\mathbb{M}}(\vec{\nu}_d) = \mathbb{I}_1 + \mathbb{I}_2 = \int_{\vec{r}} \mathbf{grad}\, h(\vec{r}) \left\{ \vec{z} \cdot [\vec{\mathcal{E}}_{1,e}(\vec{r},0) - \vec{\mathcal{E}}_{0,e}(\vec{r},0)] \right\} e^{-2i\pi\vec{\nu}_d \cdot \vec{r}} d^2\vec{r}$$

$$+ \int_{\vec{r}} h(\vec{r}) \left[ \frac{\partial \vec{\mathcal{E}}_{1,e}^T}{\partial z} - \frac{\partial \vec{\mathcal{E}}_{0,e}^T}{\partial z} \right]_{\vec{r},0} e^{-2i\pi\vec{\nu}_d \cdot \vec{r}} d^2\vec{r}. \qquad (8.51)$$

The calculation of $\mathbb{I}_2$, which therefore corresponds to the second integral in equation (8.51), is similar to the calculation we have just carried out for the elementary current $\vec{\mathbb{J}}$, and leads to

$$\mathbb{I}_2 = i \int_{\vec{\nu}_e} \widehat{h}(\vec{\nu}_d - \vec{\nu}_e) \left[ \frac{\alpha_1^e \tilde{n}_0^e - \alpha_0^e \tilde{n}_1^e}{\tilde{n}_0^e} \vec{\mathbb{E}}_{0,e}^T(\vec{\nu}_e,0) \right] d^2\vec{\nu}_e . \qquad (8.52)$$

On the other hand, if we denote the quantity $\vec{z} \cdot [\vec{\mathcal{E}}_{1,e}(\vec{r},0) - \vec{\mathcal{E}}_{0,e}(\vec{r},0)]$ as $g(\vec{r})$, the integral $\mathbb{I}_1$ can be written

$$\mathbb{I}_1 = \int_{\vec{r}} g(\vec{r}) \, \mathbf{grad} \, h(\vec{r}) \, e^{-2i\pi\vec{\nu}_d \cdot \vec{r}} \, d^2\vec{r} = \text{TF} \, \{g.\mathbf{grad} \, h\} \, (\vec{\nu}_d) \, . \tag{8.53}$$

Using the properties of the Fourier transform, we obtain

$$\text{TF} \, \{g.\mathbf{grad} \, h\} = \text{TF} \, \{g\} \star \text{TF} \, \{\mathbf{grad} \, h\} = \widehat{g} \star \left[2i\pi\vec{\nu}\,\widehat{h}\right], \tag{8.54}$$

and, spelling out the convolution product,

$$\mathbb{I}_1 = \text{TF} \, \{g.\mathbf{grad} \, h\} \, (\vec{\nu}_d) = 2i\pi \int_{\vec{\nu}} \widehat{g}(\vec{\nu}) \times (\vec{\nu}_d - \vec{\nu})\widehat{h}(\vec{\nu}_d - \vec{\nu}) \, d^2\vec{\nu} \, . \tag{8.55}$$

Furthermore, the boundary conditions (8.14) relating to the normal components of the exciting electric field enable us to write

$$\tilde{\epsilon}_0 \, \vec{z} \cdot [\vec{\mathcal{E}}_{1,e}(\vec{r},0) - \vec{\mathcal{E}}_{0,e}(\vec{r},0)] = (\tilde{\epsilon}_0 - \tilde{\epsilon}_1)\vec{z} \cdot \vec{\mathcal{E}}_{1,e}(\vec{r},0) \, . \tag{8.56}$$

This allows us to calculate the Fourier transform of the quantity $g$, namely

$$\begin{aligned}
\widehat{g}(\vec{\nu}) &= \frac{\tilde{\epsilon}_0 - \tilde{\epsilon}_1}{\tilde{\epsilon}_0} \int_{\vec{r}} \vec{z} \cdot \vec{\mathcal{E}}_{1,e}(\vec{r},0) \, e^{-2i\pi\vec{\nu}.\vec{r}} \, d^2\vec{r} \\
&= \frac{\tilde{\epsilon}_0 - \tilde{\epsilon}_1}{\tilde{\epsilon}_0} \int_{\vec{r}} \int_{\vec{\nu}_e} \vec{z} \cdot \vec{A}_{1,e}^+(\vec{\nu}_e) \, e^{-2i\pi(\vec{\nu} - \vec{\nu}_e).\vec{r}} \, d^2\vec{\nu}_e d^2\vec{r} \\
&= \frac{\tilde{\epsilon}_0 - \tilde{\epsilon}_1}{\tilde{\epsilon}_0} \int_{\vec{\nu}_e} \vec{z} \cdot \vec{A}_{1,e}^+(\vec{\nu}_e) \, \delta(\vec{\nu} - \vec{\nu}_e) \, d^2\vec{\nu}_e \\
&= \frac{\tilde{\epsilon}_0 - \tilde{\epsilon}_1}{\tilde{\epsilon}_0} \left[\vec{z} \cdot \vec{A}_{1,e}^+(\vec{\nu})\right] \, .
\end{aligned} \tag{8.57}$$

In order to appeal to the simple relationship between the normal and tangential components of the magnetic field (see Section 6.3.3), we can make use of the fact that the emerging elementary wave in medium 1 is a progressive monochromatic one-dimensional wave, namely

$$\vec{z} \cdot \vec{A}_{1,e}^+(\vec{\nu}) = -\frac{2\pi}{\alpha_1}\,\vec{\nu} \cdot \vec{A}_{1,e}^+(\vec{\nu}) = -\frac{2\pi}{\alpha_1(\vec{\nu})}\,\vec{\nu} \cdot \vec{\mathbb{E}}_{0,e}^T(\vec{\nu},0) \, . \tag{8.58}$$

From this an expression for the integral $\mathbb{I}_1$ can be deduced:

$$\mathbb{I}_1 = -4\pi^2 i \frac{\tilde{\epsilon}_0 - \tilde{\epsilon}_1}{\tilde{\epsilon}_0} \int_{\vec{\nu}} (\vec{\nu}_d - \vec{\nu}) \, \widehat{h}(\vec{\nu}_d - \vec{\nu}) \left[\frac{\vec{\nu} \cdot \vec{\mathbb{E}}_{0,e}^T(\vec{\nu},0)}{\alpha_1(\vec{\nu})}\right] d^2\vec{\nu} \, . \tag{8.59}$$

### 8.2.6 Scattered Fields

Combining all the theoretical results established in Sections 8.2.4 and 8.2.5, we are now in a position to provide an analytical expression for the tangential components of the fields scattered in the two media by this single interface of profile $h(x, y)$, namely

$$
\vec{\mathcal{E}}_{0,d}^{T}(\vec{r}, z) = -\int_{\vec{\nu}_d} \frac{\vec{\mathbb{J}}(\vec{\nu}_d) - \tilde{n}_1^d [\vec{z} \wedge \vec{\mathbb{M}}](\vec{\nu}_d)}{\tilde{n}_0^d + \tilde{n}_1^d} e^{2i\pi\vec{\nu}_d.\vec{r}} e^{-i\alpha_0^d z} d^2\vec{\nu}_d
$$

$$
\vec{\mathcal{E}}_{1,d}^{T}(\vec{r}, z) = -\int_{\vec{\nu}_d} \frac{\vec{\mathbb{J}}(\vec{\nu}_d) + \tilde{n}_0^d [\vec{z} \wedge \vec{\mathbb{M}}](\vec{\nu}_d)}{\tilde{n}_0^d + \tilde{n}_1^d} e^{2i\pi\vec{\nu}_d.\vec{r}} e^{i\alpha_1^d z} d^2\vec{\nu}_d \,,
$$

(8.60)

where

$$
\vec{\mathbb{J}}(\vec{\nu}_d) = i\int_{\vec{\nu}_e} \widehat{h}(\vec{\nu}_d - \vec{\nu}_e) \left[ (\alpha_1^e \tilde{n}_1^e - \alpha_0^e \tilde{n}_0^e) \vec{\mathbb{E}}_{0,e}^{T}(\vec{\nu}_e, 0) \right] d^2\vec{\nu}_e
$$

(8.61)

$$
\vec{z} \wedge \vec{\mathbb{M}}(\vec{\nu}_d) = i\int_{\vec{\nu}_e} \widehat{h}(\vec{\nu}_d - \vec{\nu}_e) \left[ \frac{\alpha_1^e \tilde{n}_0^e - \alpha_0^e \tilde{n}_1^e}{\tilde{n}_0^e} \vec{\mathbb{E}}_{0,e}^{T}(\vec{\nu}_e, 0) \right] d^2\vec{\nu}_e
$$

$$
- 4\pi^2 i \frac{\tilde{\epsilon}_0 - \tilde{\epsilon}_1}{\tilde{\epsilon}_0} \int_{\vec{\nu}_e} (\vec{\nu}_d - \vec{\nu}_e) \widehat{h}(\vec{\nu}_d - \vec{\nu}_e) \left[ \frac{\vec{\nu}_e}{\alpha_1^e} \cdot \vec{\mathbb{E}}_{0,e}^{T}(\vec{\nu}_e, 0) \right] d^2\vec{\nu}_e \,.
$$

(8.62)

Note here that the currents are expressed as a function of the tangential component of the ideal electric field.

Recall that these relations were established in the general case where the ideal field is described by a spatial wave packet characterized by multiple incidences of illumination. We shall now see how these formulae can be simplified in the case where this ideal field is the field associated with a weakly diverging beam of light; in practice, this will nearly always be the configuration used. Notice here that weakly diverging does not mean a plane wave; indeed working with a plane wave would lead to difficulties in the normalization of the roughness spectrum (see further), since the illumination area would be infinite (see Chapter 4).

### 8.2.7 The Particular Case of a Weakly Divergent Incident Beam

We shall now assume that the ideal field at the surface of the plane interface is that created by a weakly divergent monochromatic beam of light whose mean direction of propagation $w$ makes an angle $\theta_0^i$ with the normal to this interface, as shown schematically in Figure 8.2.

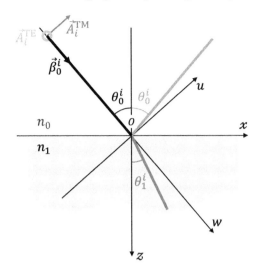

Figure 8.2 Schematic representation of the configuration corresponding to a weakly divergent beam illuminating a plane interface at an angle of incidence $\theta_0^i$.

We shall assume that the transverse distribution of the field in the plane $uOv$ perpendicular to the direction of propagation $w$ is (for example) Gaussian and can therefore be described by an expression of the type

$$\vec{\mathcal{E}}_i(u, v, 0) = \vec{\mathcal{E}}_i \, e^{-\frac{u^2 + v^2}{w_0^2}}, \tag{8.63}$$

where $w_0$ corresponds to the waist of this Gaussian beam and $(u, v)$ are the coordinates in the $Ouv$ plane. Recall that $\vec{v}$ and $\vec{y}$ are parallel vectors with $\vec{w} = \vec{u} \wedge \vec{v}$ and $\vec{z} = \vec{x} \wedge \vec{y}$.

We then break this spatial distribution down into unidirectional elementary components:

$$\vec{\mathbb{E}}_i(\nu_u, \nu_v, 0) = \iint\limits_{u,v} \vec{\mathcal{E}}(u, v, 0) \, e^{-2i\pi(u\nu_u + v\nu_v)} \, du \, dv, \tag{8.64}$$

i.e.,

$$\vec{\mathbb{E}}_i(\nu_u, \nu_v, 0) = \pi w_0^2 \, \vec{\mathcal{E}}_i \, e^{-\pi^2 w_0^2(\nu_u^2 + \nu_v^2)}. \tag{8.65}$$

We immediately deduce the transverse distribution of the incident field at any point in the half-space $z < 0$ in the reference frame $uvw$:

$$\vec{\mathcal{E}}_i(u, v, w) = \iint\limits_{\nu_u, \nu_v} \pi w_0^2 \, \vec{\mathcal{E}}_i \, e^{-\pi^2 w_0^2(\nu_u^2 + \nu_v^2)} \, e^{2i\pi(\nu_u u + \nu_v v)} \, e^{i\alpha_{uv} w} \, d\nu_u \, d\nu_v, \tag{8.66}$$

where

$$\alpha_{uv}^2 = k_0^2 - 4\pi^2(\nu_u^2 + \nu_v^2).$$ (8.67)

To a first approximation, the assumption of weak divergence of the incident beam allows us to neglect the contribution to this expression of the transverse components of the wave vector, and hence we can write

$$\alpha_{uv} \approx k_0.$$ (8.68)

In order to obtain the spatial dependence of this field on the interface, we must first pass from the $uvw$ coordinate system to the $xyz$ coordinate system:

$$\begin{cases} u = x\cos\theta_0^i - z\sin\theta_0^i \\ v = y \\ w = x\sin\theta_0^i + z\cos\theta_0^i \end{cases}.$$ (8.69)

From this we deduce the general expression for the incident field in this new reference frame:

$$\vec{\mathcal{E}}_i(x, y, z) = \iint_{\nu_u, \nu_v} \pi w_0^2 \, \vec{\mathcal{E}}_i \, e^{-\pi^2 w_0^2(\nu_u^2 + \nu_v^2)} \, e^{2i\pi[\nu_u(x\cos\theta_0^i - z\sin\theta_0^i) + \nu_v y]}$$

$$\times \, e^{ik_0(x\sin\theta_0^i + z\cos\theta_0^i)} \, d\nu_u \, d\nu_v,$$ (8.70)

and finally we consider the special case $z = 0$, i.e.,

$$\vec{\mathcal{E}}_i(x, y, 0) = \pi w_0^2 \, \vec{\mathcal{E}}_i \, e^{ik_0 x \sin\theta_0^i} \times$$

$$\times \iint_{\nu_u, \nu_v} e^{-\pi^2 w_0^2(\nu_u^2 + \nu_v^2)} \, e^{2i\pi[\nu_u x \cos\theta_0^i + \nu_v y]} \, d\nu_u \, d\nu_v.$$ (8.71)

Since the Fourier transform of a Gaussian function is also Gaussian, we get

$$\vec{\mathcal{E}}_i(x, y, 0) = \vec{\mathcal{E}}_i \, e^{ik_0 x \sin\theta_0^i} \, e^{-\frac{(x\cos\theta_0^i)^2 + y^2}{w_0^2}}.$$ (8.72)

Henceforth we shall put

$$s_e(x, y) = e^{ik_0 x \sin\theta_0^i} \, e^{-\frac{(x\cos\theta_0^i)^2 + y^2}{w_0^2}}.$$ (8.73)

It only remains to decompose this field distribution on the interface into elementary waves, i.e.,

$$\vec{\mathbb{E}}_i(\vec{\nu}, 0) = \vec{\mathbb{A}}_i^+(\vec{\nu}) = \int_{\vec{r}} \vec{\mathcal{E}}_i(\vec{r}, 0) \, e^{-2i\pi\vec{\nu}.\vec{r}} \, d^2\vec{r} = \hat{s}_e(\vec{\nu}) \, \vec{\mathcal{E}}_i,$$ (8.74)

where

$$\hat{s}_e(\nu_x, \nu_y) = \frac{\pi w_0^2}{\cos \theta_0^i} e^{-\pi^2 \frac{w_0^2}{\cos^2 \theta_0^i}(\nu_x - \nu_i)^2} e^{-\pi^2 w_0^2 \nu_y^2}, \tag{8.75}$$

where

$$\nu_i = \frac{n_0}{\lambda} \sin \theta_0^i. \tag{8.76}$$

Equation (8.74) enables an analytical expression for $\vec{\mathbb{E}}_{0,e}^T(\vec{\nu}_e, 0)$ to be obtained, namely

$$\vec{\mathbb{E}}_{0,e}^T(\vec{\nu}_e, 0) = (1 + r^e)\vec{\mathbb{A}}_i^{T+}(\vec{\nu}_e) = \frac{2\tilde{n}_0^e}{\tilde{n}_0^e + \tilde{n}_1^e} \hat{s}_e(\vec{\nu}_e)\,\vec{\mathcal{E}}_i^T. \tag{8.77}$$

We are now interested in the expression for the current $\vec{\mathbb{J}}(\vec{\nu}_d)$ given by equation (8.62) and we replace the quantity $\vec{\mathbb{E}}_{0,e}^T(\vec{\nu}_e, 0)$ by its analytical expression. We get

$$\vec{\mathbb{J}}(\vec{\nu}_d) = i\vec{\mathcal{E}}_i^T \int_{\vec{\nu}_e} \frac{2\tilde{n}_0^e}{\tilde{n}_0^e + \tilde{n}_1^e}(\alpha_1^e \tilde{n}_1^e - \alpha_0^e \tilde{n}_0^e)\,\hat{h}(\vec{\nu}_d - \vec{\nu}_e)\hat{s}_e(\vec{\nu}_e)\,d^2\vec{\nu}_e. \tag{8.78}$$

Note that the two Gaussian terms centered respectively on $\nu_x^e = \nu_i$ and $\nu_y^e = 0$ in the expression for $\hat{s}_e(\vec{\nu}_e)$ demonstrate a very narrow width at all angles $\theta_0^i$, on account of the assumption of weak divergence of the incident beam. Furthermore, the variation (as a function of $\nu_x^e$ and $\nu_y^e$) of the effective refractive indices and normal components of the wave vectors appearing under the integral sign is sufficiently slow that these quantities can be replaced by constants corresponding to their values at $(\nu_i, 0)$; this leads to

$$\vec{\mathbb{J}}(\vec{\nu}_d) = \frac{2i\tilde{n}_0^i(\alpha_1^i \tilde{n}_1^i - \alpha_0^i \tilde{n}_0^i)}{\tilde{n}_0^i + \tilde{n}_1^i}\,\vec{\mathcal{E}}_i^T \int_{\vec{\nu}_e} \hat{h}(\vec{\nu}_d - \vec{\nu}_e)\,\hat{s}_e(\vec{\nu}_e)\,d^2\vec{\nu}_e; \tag{8.79}$$

that is, in the form of convolution product,

$$\vec{\mathbb{J}}(\vec{\nu}_d) = \frac{2i\tilde{n}_0^i(\alpha_1^i \tilde{n}_1^i - \alpha_0^i \tilde{n}_0^i)}{\tilde{n}_0^i + \tilde{n}_1^i}\,[\hat{h} \star \hat{s}_e]_{\vec{\nu}_d}\,\vec{\mathcal{E}}_i^T. \tag{8.80}$$

Similarly, for the current $\vec{z} \wedge \vec{\mathbb{M}}(\vec{\nu}_d)$ we get

$$\vec{z} \wedge \vec{\mathbb{M}}(\vec{\nu}_d) = \frac{2i(\alpha_1^i \tilde{n}_0^i - \alpha_0^i \tilde{n}_1^i)}{\tilde{n}_0^i + \tilde{n}_1^i}\,[\hat{h} \star \hat{s}_e]_{\vec{\nu}_d}\,\vec{\mathcal{E}}_i^T$$
$$- \frac{8i\pi^2 \tilde{n}_0^i}{\tilde{n}_0^i + \tilde{n}_1^i}\frac{\tilde{\epsilon}_0 - \tilde{\epsilon}_1}{\tilde{\epsilon}_0}(\vec{\nu}_d - \vec{\nu}_i)\left\{\frac{\vec{\nu}_i}{\alpha_1^i} \cdot [\hat{h} \star \hat{s}_e]_{\vec{\nu}_d}\,\vec{\mathcal{E}}_i^T\right\}. \tag{8.81}$$

It remains only to make use of relations (8.37) to deduce expressions for the amplitudes of the elementary components of the scattered fields on either side of the interface, i.e.,

$$
\vec{\mathbb{A}}_{0,d}^{T-} = -\frac{2i}{(\tilde{n}_0^i + \tilde{n}_1^i)(\tilde{n}_0^d + \tilde{n}_1^d)} \left\{ \left[ \tilde{n}_0^i(\alpha_1^i \tilde{n}_1^i - \alpha_0^i \tilde{n}_0^i) - \tilde{n}_1^d(\alpha_1^i \tilde{n}_0^i - \alpha_0^i \tilde{n}_1^i) \right] \right.
$$

$$
\left. [\widehat{h} \star \widehat{s}_e]_{\vec{v}_d} \vec{\mathcal{E}}_i^T + 4\pi^2(\vec{v}_d - \vec{v}_i) \frac{\tilde{n}_0^i \tilde{n}_1^d}{\alpha_1^i} \frac{\tilde{\epsilon}_0 - \tilde{\epsilon}_1}{\tilde{\epsilon}_0} [\widehat{h} \star \widehat{s}_e]_{\vec{v}_d} \{\vec{v}_i \cdot \vec{\mathcal{E}}_i^T\} \right\}, \quad (8.82)
$$

$$
\vec{\mathbb{A}}_{1,d}^{T+} = -\frac{2i}{(\tilde{n}_0^i + \tilde{n}_1^i)(\tilde{n}_0^d + \tilde{n}_1^d)} \left\{ \left[ \tilde{n}_0^i(\alpha_1^i \tilde{n}_1^i - \alpha_0^i \tilde{n}_0^i) + \tilde{n}_0^d(\alpha_1^i \tilde{n}_0^i - \alpha_0^i \tilde{n}_1^i) \right] \right.
$$

$$
\left. [\widehat{h} \star \widehat{s}_e]_{\vec{v}_d} \vec{\mathcal{E}}_i^T - 4\pi^2(\vec{v}_d - \vec{v}_i) \frac{\tilde{n}_0^i \tilde{n}_0^d}{\alpha_1^i} \frac{\tilde{\epsilon}_0 - \tilde{\epsilon}_1}{\tilde{\epsilon}_0} [\widehat{h} \star \widehat{s}_e]_{\vec{v}_d} \{\vec{v}_i \cdot \vec{\mathcal{E}}_i^T\} \right\}. \quad (8.83)
$$

To summarize, the assumption of a weakly divergent beam has allowed the equations of Section 8.2.6 to be reduced to a single product of convolution between the Fourier transform of the profile and that of the shape of the exciting beam.

### 8.2.8 Including the Polarization States

In order to express the effective refractive indices associated with the incident and scattered fields that appear in the relations we have just established, we need to specify the polarization states of these two fields.

#### Description of the Method

The weak divergence assumption that we adopted in Section 8.2.7 allows us qualify the polarization state of the incident light in a simple manner, since this is expressed in relation to the plane of incidence; this incidence plane is defined (see Figure 8.2) by the mean direction of the incident wave vector $\vec{\beta}_0^i$ (or the associated spatial frequency $\vec{v}_i$) and the normal $\vec{z}$ to the surface of the sample. Here, this plane of incidence is therefore the plane $y = 0$, and the tangential components of the incident electric field of polarization TE (or S) and TM (or P) are respectively parallel to the vectors $Oy$ and $Ox$.

Similarly, the polarization of an elementary component of the scattered field can be expressed in relation to the scattering plane defined by the scattered wave vector (or the scattered spatial frequency $\vec{v}_d$) and the normal $\vec{z}$ to the surface of the plane optical interface.

As we saw in Chapter 4, we can associate a normal direction ($\theta$) and a polar direction ($\phi$) with each spatial frequency, so in each medium,

$$\vec{\nu}_d = n_j \sin \theta_j^d \begin{bmatrix} \cos \phi_d \\ \sin \phi_d \end{bmatrix} \quad \text{where} \quad j = 0, 1 \,. \tag{8.84}$$

Note that the polar (or azimuthal) angle is independent of the medium; it allows the polar scattering plane, i.e., the vertical plane making an angle $\phi$ with the plane $y = 0$ containing the direction of illumination, to be defined.

We must therefore express the tangential components of the scattered electric field with respect to this polar plane: the TE (or S) component will be normal to this plane, while the TM (or P) component will be contained within it. In addition, each of these components may be associated with a particular polarization of the incident wave. An incident polarization S will therefore give rise to two scattered components (SS and SP), while an incident polarization P will similarly give rise to two other scattered components (PP and PS). All these situations are shown schematically in Figure 8.3.

Equations (8.74), (8.82), and (8.83) established in, Section 8.2.7 show that the amplitude of the tangential component of the scattered field in the incident medium can be put into the following general form:

$$\vec{A}_{0,d}^{T-} = S_{0,d}^i \vec{\mathcal{E}}_i^T + P_{0,d}^i (\vec{\nu}_d - \vec{\nu}_i) \{ \vec{\nu}_i \cdot \vec{\mathcal{E}}_i^T \} \,, \tag{8.85}$$

where $S_{0,d}^i$ and $P_{0,d}^i$ are coefficients that are functions of the refractive indices of the two media, of the scattering angle and angle of illumination, and of the polarization states of the incident and scattered waves.

Similarly, in the case of the amplitude of the tangential component of the scattered field within the substrate, we can write

$$\vec{A}_{1,d}^{T+} = S_{1,d}^i \vec{\mathcal{E}}_i^T + P_{1,d}^i (\vec{\nu}_d - \vec{\nu}_i) \{ \vec{\nu}_i \cdot \vec{\mathcal{E}}_i^T \} \,. \tag{8.86}$$

One simple solution for calculating the effective components of the scattered field as a function of their states of polarization involves associating a direct unit orthogonal system $(\vec{a}, \vec{b}, \vec{c})$ with each of these fields; vector $\vec{c}$ is collinear with the scattered wave vector $\vec{\beta}^d$, while vector $\vec{b}$ is perpendicular to the scattering plane defined by $(\vec{c}, \vec{z})$; vector $\vec{a}$ can be deduced from the other two using the vector product $(\vec{a} = \vec{b} \wedge \vec{c})$.

As we shall be considering only the tangential components of the incident and scattered components, we can restrict our approach to the plane $xOy$ and to the two vectors $\vec{a}$ and $\vec{b}$, i.e.,

$$\vec{a}_0 = \begin{bmatrix} -\cos \phi_d \\ -\sin \phi_d \end{bmatrix} \quad ; \quad \vec{b}_0 = \begin{bmatrix} -\sin \phi_d \\ \cos \phi_d \end{bmatrix} \tag{8.87}$$

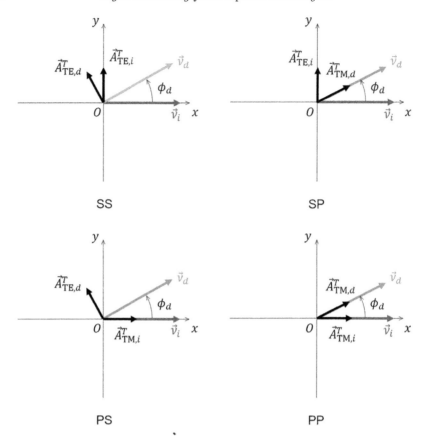

Figure 8.3 Schematic representation of the relative orientation of the vectors that are characteristic of incident and scattered waves in the tangential plane, for different polarization states (SS, SP, PS, and PP).

for the orthogonal system associated with the scattered wave in the incident medium, and in the case of that associated with the scattered wave within the substrate, we have

$$\vec{a}_1 = \begin{bmatrix} \cos\phi_d \\ \sin\phi_d \end{bmatrix} \quad ; \quad \vec{b}_1 = \begin{bmatrix} -\sin\phi_d \\ \cos\phi_d \end{bmatrix}. \tag{8.88}$$

We can immediately deduce the effective amplitudes (in S polarization and P polarization) of the tangential components of the scattered field in each of the two media:

$$[\mathbb{A}_{j,d}^T]_S = S_{j,d}^i\, [\vec{b}_j \cdot \vec{\mathcal{E}}_i^T] + P_{j,d}^i\, [\vec{b}_j \cdot (\vec{v}_d - \vec{v}_i)]\{\vec{v}_i \cdot \vec{\mathcal{E}}_i^T\} \quad j = 0,1 \tag{8.89}$$

$$[\mathbb{A}_{j,d}^T]_P = S_{j,d}^i\, [\vec{a}_j \cdot \vec{\mathcal{E}}_i^T] + P_{j,d}^i\, [\vec{a}_j \cdot (\vec{v}_d - \vec{v}_i)]\{\vec{v}_i \cdot \vec{\mathcal{E}}_i^T\} \quad j = 0,1. \tag{8.90}$$

Furthermore, as the diagrams in Figure 8.3 show, regardless of the polarization states of the incident and scattered waves, we can write

$$\vec{\nu}_d - \vec{\nu}_i = \begin{bmatrix} (n_j/\lambda) \sin \theta_j^d \cos \phi_d \\ (n_j/\lambda) \sin \theta_j^d \sin \phi_d \end{bmatrix} - \begin{bmatrix} (n_j/\lambda) \sin \theta_j^i \\ 0 \end{bmatrix} \quad \text{where} \quad j = 0, 1 . \quad (8.91)$$

When the incident field is S-polarized, we have

$$\vec{\mathcal{E}}_i^T = \mathcal{E}_i \, \vec{y} = \begin{bmatrix} 0 \\ \mathcal{E}_i \end{bmatrix} . \quad (8.92)$$

The scalar product $\{\vec{\nu}_i \cdot \vec{\mathcal{E}}_i^T\}$ is 0, so

$$\begin{cases} \mathbb{A}_{0,d}^{SS} = S_{0,d}^i \mathcal{E}_i \cos \phi_d \\ \mathbb{A}_{0,d}^{SP} = -S_{0,d}^i \mathcal{E}_i \sin \phi_d \\ \mathbb{A}_{1,d}^{SS} = S_{s,d}^i \mathcal{E}_i \cos \phi_d \\ \mathbb{A}_{1,d}^{SP} = S_{s,d}^i \mathcal{E}_i \sin \phi_d \end{cases} \quad (8.93)$$

This equation gives the polar dependence of the scattered field in the case of a S-polarized illumination. Notice the single law given by a product with a cosine or sine function.

For an incident field of polarization P, still in the tangential plane, we have

$$\vec{\mathcal{E}}_i^T = \mathcal{E}_i \cos \theta_0^i \, \vec{x} = \begin{bmatrix} \mathcal{E}_i \cos \theta_0^i \\ 0 \end{bmatrix} , \quad (8.94)$$

which leads us to include the terms that depend on the coefficients $P_{j,d}^i$ since the scalar product $\{\vec{\nu}_i \cdot \vec{\mathcal{E}}_i^T\}$ is no longer identically zero.

We get:

$$\begin{cases} \mathbb{A}_{0,d}^{PS} = -\left\{ S_{0,d}^i - \frac{\nu_i P_{0,d}^i}{\lambda} \, n_0 \sin \theta_0^i \right\} \sin \phi_d \mathcal{E}_i \cos \theta_0^i \\ \mathbb{A}_{0,d}^{PP} = -\left\{ S_{0,d}^i \cos \phi_d + \frac{\nu_i P_{0,d}^i}{\lambda} \left[ n_0 \sin \theta_0^d - n_0 \sin \theta_0^i \cos \phi_d \right] \right\} \mathcal{E}_i \cos \theta_0^i \\ \mathbb{A}_{1,d}^{PS} = -\left\{ S_{s,d}^i - \frac{\nu_i P_{s,d}^i}{\lambda} \, n_s \sin \theta_s^i \right\} \sin \phi_d \mathcal{E}_i \cos \theta_0^i \\ \mathbb{A}_{1,d}^{PP} = \left\{ S_{s,d}^i \cos \phi_d + \frac{\nu_i P_{s,d}^i}{\lambda} \left[ n_s \sin \theta_s^d - n_s \sin \theta_s^i \cos \phi_d \right] \right\} \mathcal{E}_i \cos \theta_0^i \end{cases} .$$

Note that these expressions obtained for incident P-polarization are more complex than those obtained for S-polarization. In particular, in the PP case, the $\cos \phi_d$ polar dependence is affine, though it is linear by nature in all the other cases.

Note also that, generally speaking, there is no change in polarization modes when scattering takes place in the plane of incidence. If $\phi_d$ is 0, the cross-components vanish ($\mathbb{A}_{j,d}^{SP} = \mathbb{A}_{j,d}^{PS} = 0$ for $j = 0, 1$). This is a general result analogous to that observed for diffraction by a Mie sphere, and enables second-order effects of scattering (not included here) to be observed experimentally.

### The SS Case

The coefficients $S_{0,d}^i$ and $S_{1,d}^i$ introduced in the previous section are expressed generally as

$$S_{0,d}^i = -\frac{2i}{(\tilde{n}_0^d + \tilde{n}_1^d)(\tilde{n}_0^i + \tilde{n}_1^i)} \left\{ \tilde{n}_0^i(\alpha_1^i \tilde{n}_1^i - \alpha_0^i \tilde{n}_0^i) - \tilde{n}_1^d(\alpha_1^i \tilde{n}_0^i - \alpha_0^i \tilde{n}_1^i) \right\} [\widehat{h} \star \widehat{s}_e]_{\bar{\nu}_d}$$

$$S_{1,d}^i = -\frac{2i}{(\tilde{n}_0^d + \tilde{n}_1^d)(\tilde{n}_0^i + \tilde{n}_1^i)} \left\{ \tilde{n}_0^i(\alpha_1^i \tilde{n}_1^i - \alpha_0^i \tilde{n}_0^i) + \tilde{n}_0^d(\alpha_1^i \tilde{n}_0^i - \alpha_0^i \tilde{n}_1^i) \right\} [\widehat{h} \star \widehat{s}_e]_{\bar{\nu}_d} .$$

where

$$\alpha_0^i = k_0 \cos \theta_0^i \quad ; \quad \alpha_1^i = k_1 \cos \theta_1^i \quad ; \quad \alpha_0^d = k_0 \cos \theta_0^d \quad ; \quad \alpha_1^d = k_1 \cos \theta_1^d$$

$$\tilde{n}_0^i = \frac{n_0}{\eta_v} \cos \theta_0^i \quad ; \quad \tilde{n}_1^i = \frac{n_1}{\eta_v} \cos \theta_1^i \quad ; \quad \tilde{n}_0^d = \frac{n_0}{\eta_v} \cos \theta_0^d \quad ; \quad \tilde{n}_1^d = \frac{n_1}{\eta_v} \cos \theta_1^d .$$

Note that in this case,

$$\alpha_1^i \tilde{n}_0^i - \alpha_0^i \tilde{n}_1^i = 0$$

$$\tilde{n}_0^i(\alpha_1^i \tilde{n}_1^i - \alpha_0^i \tilde{n}_0^i) = -k_0 \cos \theta_0^i (\tilde{n}_0^i - \tilde{n}_1^i)(\tilde{n}_0^i + \tilde{n}_1^i) .$$

The first equation shows that the two coefficients $S_{0,d}^i$ and $S_{1,d}^i$ are equal; using the second, they can be simplified and we can write the amplitudes of the tangential components of the scattered elementary waves in the two media in the following form:

$$\mathbb{A}_{0,d}^{SS} = \mathbb{A}_{1,d}^{SS} = \frac{2ik_0 \cos \theta_0^i (n_0 \cos \theta_0^i - n_1 \cos \theta_1^i)}{n_0 \cos \theta_0^d + n_1 \cos \theta_1^d} \cos \phi_d \, [\widehat{h} \star \widehat{s}_e]_{\bar{\nu}_d} \, \mathcal{E}_i . \quad (8.95)$$

### The SP Case

The polarization states of the incident and scattered waves are different here (S and P respectively), leading to the following relations:

$$\tilde{n}_0^i = \frac{n_0}{\eta_v} \cos \theta_0^i \quad ; \quad \tilde{n}_1^i = \frac{n_1}{\eta_v} \cos \theta_1^i \quad ; \quad \tilde{n}_0^d = \frac{n_0}{\eta_v \cos \theta_0^d} \quad ; \quad \tilde{n}_1^d = \frac{n_1}{\eta_v \cos \theta_1^d} .$$

Since the quantity $\alpha_1^i \tilde{n}_0^i - \alpha_0^i \tilde{n}_1^i$ is still 0, the expressions for the tangential scattered components in the two media are analogous to those obtained for the SS case, and may be written

$$\mathbb{A}_{0,d}^{SP} = -\mathbb{A}_{1,d}^{SP} = -\frac{2ik_0\cos\theta_0^i(n_0\cos\theta_0^i - n_1\cos\theta_1^i)}{(n_0/\cos\theta_0^d + n_1/\cos\theta_1^d)}\sin\phi_d\,[\widehat{h}\!\star\!\widehat{s}_e]_{\bar{v}_d}\,\mathcal{E}_e\,. \quad (8.96)$$

### The PS Case

This time we have

$$\tilde{n}_0^i = \frac{n_0}{\eta_v\cos\theta_0^i} \quad;\quad \tilde{n}_1^i = \frac{n_1}{\eta_v\cos\theta_1^i} \quad;\quad \tilde{n}_0^d = \frac{n_0}{\eta_v}\cos\theta_0^d \quad;\quad \tilde{n}_1^d = \frac{n_1}{\eta_v}\cos\theta_1^d\,.$$

Let

$$K = -\frac{2i\,[\widehat{h}\star\widehat{s}_e]_{\bar{v}_d}}{(\tilde{n}_0^d + \tilde{n}_1^d)(\tilde{n}_0^i + \tilde{n}_1^i)} = \frac{-2i\eta_v^2\,[\widehat{h}\star\widehat{s}_e]_{\bar{v}_d}}{(\frac{n_0}{\cos\theta_0^i} + \frac{n_1}{\cos\theta_1^i})(n_0\cos\theta_0^d + n_1\cos\theta_1^d)}\,. \quad (8.97)$$

Under these conditions,

$$S_{0,d}^i = \frac{k_0}{\cos\theta_0^i}\frac{K}{\eta_v^2}\left\{(n_1^2 - n_0^2) - \frac{n_1^2\cos\theta_1^d}{\cos\theta_1^i}(\cos^2\theta_1^i - \cos^2\theta_0^i)\right\}\,. \quad (8.98)$$

Note that

$$n_1^2(\cos^2\theta_1^i - \cos^2\theta_0^i) = n_1^2(\sin^2\theta_0^i - \sin^2\theta_1^i) = \frac{n_1^2}{n_0^2}\sin^2\theta_1^i(n_1^2 - n_0^2)\,. \quad (8.99)$$

On the other hand, the coefficient $P_{0,d}^i$ can be expressed generally as

$$P_{0,d}^i = 4\pi^2\,\frac{\tilde{n}_0^i\tilde{n}_1^d}{\alpha_1^i}\frac{\tilde{\epsilon}_0 - \tilde{\epsilon}_1}{\tilde{\epsilon}_0}\,K\,, \quad (8.100)$$

leading to the following equation:

$$\frac{\nu_i P_{0,d}^i}{\lambda}\,n_0\sin\theta_0^i = \frac{k_0}{\cos\theta_0^i}\frac{K}{\eta_v^2}(n_0^2 - n_1^2)\frac{\cos\theta_1^d}{\cos\theta_1^i}\sin^2\theta_0^i\,. \quad (8.101)$$

Consequently,

$$S_0^i - \frac{\nu_i P_0^i}{\lambda}\,n_0\sin\theta_0^i = \frac{k_0}{\cos\theta_0^i}\frac{K}{\eta_v^2}(n_1^2 - n_0^2)\left\{1 - \frac{\cos\theta_1^d}{\cos\theta_1^i}\left[\frac{n_1^2}{n_0^2}\sin^2\theta_1^i - \sin^2\theta_0^i\right]\right\}\,, \quad (8.102)$$

where, by applying Snell's law, the last term in square brackets is 0.

A calculation at any similar point can be carried out in the case of a scattered wave in the substrate, leading ultimately to the following expression:

$$\mathbb{A}_{0,d}^{PS} = \mathbb{A}_{1,d}^{PS} = \frac{-2ik_0(n_0^2 - n_1^2)\sin\phi_d}{(\frac{n_0}{\cos\theta_0^i} + \frac{n_1}{\cos\theta_1^i})(n_0\cos\theta_0^d + n_1\cos\theta_1^d)}[\widehat{h}\star\widehat{s}_e]_{\bar{v}_d}\,\mathcal{E}_i\,. \quad (8.103)$$

## The PP Case

The polarization states of the incident and scattered waves are both of type P, such that

$$\tilde{n}_0^i = \frac{n_0}{\eta_v \cos\theta_0^i} \quad ; \quad \tilde{n}_1^i = \frac{n_1}{\eta_v \cos\theta_1^i} \quad ; \quad \tilde{n}_0^d = \frac{n_0}{\eta_v \cos\theta_0^d} \quad ; \quad \tilde{n}_1^d = \frac{n_1}{\eta_v \cos\theta_1^d} \,.$$

The calculation process is very similar to that described for the PS case, and leads to the following expressions:

$$\begin{cases} \mathbb{A}_{0,d}^{PP} = -\dfrac{2ik_0(n_0^2 - n_1^2)}{\left(\frac{n_0}{\cos\theta_0^d} + \frac{n_1}{\cos\theta_1^d}\right)\left(\frac{n_0}{\cos\theta_1^i} + \frac{n_1}{\cos\theta_1^i}\right)}\left[\cos\phi_d - \dfrac{\sin\theta_0^i \sin\theta_0^d}{\cos\theta_1^i \cos\theta_1^d}\right][\hat{h} \star \hat{s}_e]_{\bar{\nu}_d}\,\mathcal{E}_i \\[3mm] \mathbb{A}_{1,d}^{PP} = \dfrac{2ik_0(n_0^2 - n_1^2)}{\left(\frac{n_0}{\cos\theta_0^d} + \frac{n_1}{\cos\theta_1^d}\right)\left(\frac{n_0}{\cos\theta_0^i} + \frac{n_1}{\cos\theta_1^i}\right)}\left[\cos\phi_d + \dfrac{\sin\theta_0^i \sin\theta_1^d}{\cos\theta_1^i \cos\theta_0^d}\right][\hat{h} \star \hat{s}_e]_{\bar{\nu}_d}\,\mathcal{E}_i \end{cases}$$

$$(8.104)$$

### 8.2.9 Scattered Flux

In each of the two media, the scattered flux is given as a function of the tangential component of the corresponding field by the following general formula (see Chapters 4 and 6):

$$\Phi_j^d = \iint\limits_{\nu_x^d,\nu_y^d} \frac{1}{2}\Re\{\tilde{n}_j^d\}\left|\vec{\mathbb{A}}_{j,d}^T(\nu_x^d,\nu_y^d)\right|^2 d\nu_x^d \, d\nu_y^d \quad \text{where} \quad j = 0,1. \qquad (8.105)$$

However, we wish to determine not the total value of this scattered flux but how it varies as a function of the direction of observation. To achieve this, we first change to polar coordinates in the expression for scattered flux, using the following change of variables:

$$(\nu_x^d,\nu_y^d) \to (\nu_d,\phi_d) \quad \text{where} \quad \nu_d = \sqrt{(\nu_x^d)^2 + (\nu_y^d)^2}\,,$$

which leads to the following new formulation:

$$\Phi_j^d = \iint\limits_{\nu_d,\phi_d} \frac{1}{2}\tilde{n}_j^d\left|\vec{\mathbb{A}}_{j,d}^T(\nu_d,\phi_d)\right|^2 \nu_d d\nu_d d\phi_d\,, \qquad (8.106)$$

where $\Re\{\tilde{n}_j^d\}$ has been replaced by $\tilde{n}_j^d$ (the two media are assumed nonabsorbing).

We also recall that the spatial frequency $\nu_d$ is, quite generally, defined by the relation $2\pi\nu_d = k_j \sin\theta_j^d$, so we get

$$d\nu_d = \frac{k_j}{2\pi}\cos\theta_j d\theta_j^d\,. \qquad (8.107)$$

The spectral density of the scattered flux in the direction $(\theta_j^d, \phi_d)$ can therefore be defined as

$$d\Phi_j^d = \frac{1}{2}\tilde{n}_j^d \left|\vec{\mathbb{A}}_{j,d}^T(\nu_d, \phi_d)\right|^2 \left(\frac{k_j}{2\pi}\right)^2 \cos\theta_j^d \sin\theta_j^d d\theta_j^d d\phi^d . \qquad (8.108)$$

We point out again that going from the integral equation (8.106) to the infinitesimal (8.108) needs to be justified with greater mathematical rigor, especially for problems of uniqueness. This result can be retrieved by studying the asymptotic (far field) behavior of the wave packet expressing the scattered field before extracting the flux from the associated Poynting vector (see Chapter 7, Section 7.10). However, difficulties can arise in special cases connected with high-frequency variations in the argument of the amplitude of the wave packet: in these specific cases found, for example, in the event of giant enhancements in the electromagnetic field (see Chapter 7), equation (8.108) no longer works for describing the fine angular variations of the scattered field around a resonance, even though its integral continues to describe perfectly the loss of energy associated with scattering. Hence for these specific situations the exact calculation should be performed numerically considering the wave packets of the electric and magnetic fields at each space location, so as to calculate the Poynting vector.

Now, provided that these specific cases are avoided, one can work with equation (8.108) and the spectral density of the scattered flux per unit solid angle in each of the two media can be deduced:

$$\frac{d\Phi_j^d}{d\Omega_j} = \frac{1}{2}\left(\frac{k_j}{2\pi}\right)^2 \tilde{n}_j^d \cos\theta_j^d \left|\vec{\mathbb{A}}_{j,d}^T(\nu_d, \phi_d)\right|^2 \quad \text{with} \quad d\Omega_j = \sin\theta_j^d d\theta_j^d d\phi^d . \qquad (8.109)$$

We now need to normalize these expressions by dividing them by the incident flux, which is expressed here, taking into account equation (8.74), as

$$\Phi_0^i = \int_{\vec{\nu}_e} \frac{1}{2}\Re\{\tilde{n}_0^e\}\left|\vec{\mathbb{A}}_i^{T+}(\vec{\nu}_e)\right|^2 d^2\vec{\nu}_e = \frac{1}{2}\left\{\int_{\vec{\nu}_e} \tilde{n}_0^e |\hat{s}_i(\vec{\nu}_e)|^2 d^2\vec{\nu}_e\right\}|\vec{\mathcal{E}}_i^T|^2 . \qquad (8.110)$$

As with calculating the currents $\vec{\mathbb{J}}(\vec{\nu}_d)$ and $\vec{z} \wedge \vec{\mathbb{M}}(\vec{\nu}_d)$, we at once make use of the weak divergence assumption (see Section 8.2.7) to replace $\tilde{n}_0^e$ by $\tilde{n}_0^i$, thereby making it possible to calculate the integral analytically. Applying Parseval's theorem, we have

$$\int_{\vec{\nu}_e} |\hat{s}_e(\vec{\nu}_e)|^2 d^2\vec{\nu}_e = \int_{\vec{r}} |s_e(\vec{r})|^2 d^2\vec{r} = S \quad \text{where} \quad S = \frac{1}{2}\frac{\pi w_0^2}{\cos\theta_0^i} . \qquad (8.111)$$

These two remarks allow us to give the following generic form to the expression for the incident flux:

$$\Phi_0^i = \frac{1}{2}\tilde{n}_0^i \, S |\vec{\mathcal{E}}_i^T|^2 \,, \tag{8.112}$$

where $S$ is the effective area of the sample, defined by the illumination conditions.

We deduce an expression for the spectral density of the scattered flux in each of the two media, normalized with respect to the incident flux:

$$\frac{1}{\Phi_0^i}\frac{d\Phi_j^d}{d\Omega_j} = \left(\frac{n_j}{\lambda}\right)^2 \frac{\tilde{n}_j^d}{\tilde{n}_0^i} \frac{|\vec{\mathbb{A}}_{j,d}^T(\vec{\nu}_d)|^2}{S|\vec{\mathcal{E}}_i^T|^2} \cos\theta_j^d \quad \text{where} \quad j = 0,1\,, \tag{8.113}$$

and where $\vec{\mathbb{A}}_{j,d}^T(\vec{\nu}_d)$ is defined in a general sense by equations (8.82) and (8.83).

Note that this entire calculation process has been carried out in the particular case of a weakly divergent incident beam, whose spatial amplitude distribution in the plane $w = 0$ was Gaussian. However, it can be transposed without difficulty to any other amplitude distribution $\vec{\mathcal{E}}_i(u,v,0)$, provided the Fourier transform $\vec{\mathbb{E}}_i(\nu_u,\nu_v,0)$ of the chosen spatial distribution corresponds to a weakly divergent illuminating beam, so that the approximation $\alpha_{uv} \approx k_0$ can be justified.

The modification of this amplitude distribution will simply have an effect on the analytical form of the function $\hat{s}_e(\vec{\nu})$, and by the same token on the expression for the effective area $S$; the resulting effect will be small because the relevant quantity is here a ratio involving these two terms, i.e., $(1/S)|\hat{h} \star \hat{s}_e|^2$. Note, however, that this could modify the fine structure (at a speckle size) of the scattering pattern.

### 8.2.10 Angle Resolved Scattering

At a given wavelength, equations (8.113) express the scattered flux per unit solid angle, normalized with respect to the incident flux. These quantities are commonly termed ARS (Angle Resolved Scattering), their units being that of the reciprocal of a solid angle ($\text{sr}^{-1}$).

Also appearing in the literature is the term BSDF (Bidirectional Scattering Distribution Function), which is related to ARS by the following definition:

$$\text{ARS} = \text{BSDF}\,\cos\theta_d\,. \tag{8.114}$$

Hence ARS is also called BSDF $\cos\theta$.

If we are interested only in the reflected part of the scattered light we speak of the BRDF (Bidirectional Reflectance Distribution Function). Similarly, if only the transmitted part is taken into account, we speak of the BTDF (Bidirectional Transmittance Distribution Function).

Generally speaking, all these variables describe the angular distribution of the scattered flux and are therefore collectively known as scattering distributions.

For example, consider the particular case in which these scattering distributions are registered in the plane of incidence ($\phi_d = 0$); we give the literal expression of BRDF $\cos\theta$ for the only two polarization configurations to be considered here, namely SS and PP:

$$\mathrm{ARS}_0^{SS}(\theta_i; \theta_d, 0) = \left(\frac{n_0}{\lambda}\right)^2 \frac{\cos^2\theta_0^d}{\cos\theta_0^i} \left[\frac{2k_0\cos\theta_0^i(n_0\cos\theta_0^i - n_1\cos\theta_1^i)}{n_0\cos\theta_0^d + n_1\cos\theta_1^d}\right]^2$$

$$\times \frac{1}{S}\left|[\hat{h} \star \hat{s}_e]\right|^2 , \tag{8.115}$$

$$\mathrm{ARS}_0^{PP}(\theta_i; \theta_d, 0) = \left(\frac{n_0}{\lambda}\right)^2 \frac{1}{\cos\theta_0^i} \left[\frac{2k_0(n_0^2 - n_1^2)}{\left(\frac{n_0}{\cos\theta_0^d} + \frac{n_1}{\cos\theta_1^d}\right)\left(\frac{n_0}{\cos\theta_0^i} + \frac{n_1}{\cos\theta_1^i}\right)}\right]^2$$

$$\times \left[1 - \frac{\sin\theta_0^i \sin\theta_0^d}{\cos\theta_1^i \cos\theta_1^d}\right]^2 \frac{1}{S}\left|[\hat{h} \star \hat{s}_e]\right|^2 . \tag{8.116}$$

### 8.2.11 Roughness Spectrum of the Interface

#### Definitions

We saw in the previous section that the scattering distributions from an interface are proportional to the modulus squared of $\hat{h}$ convolved with $\hat{s}_e$. At this stage we introduce the function $h_e(x, y)$, defined by

$$h_e(x, y) = h(x, y)\, s_e(x, y). \tag{8.117}$$

This function $h_e(x, y)$ describes the surface topography as seen by the illuminating beam, and it is therefore its Fourier transform that appears logically in expressions for the scattering distributions. Recall at this point that the Fourier transform of a product of two functions is equal to the convolution of the two Fourier transforms, i.e.,

$$h_e(\vec{r}) = h(\vec{r})\, s_e(\vec{r}) \xrightarrow{\mathrm{TF}} [\hat{h} \star \hat{s}_e](\vec{\nu}). \tag{8.118}$$

For completeness, note that the function $\hat{s}_e$ related to the structure of the illumination tends toward a Dirac distribution centered on $\vec{\nu}_i$ when the parameter characterizing the footprint of the function $s_e$ on the surface (e.g., the waist $w_0$ for the Gaussian profile introduced in Section 8.2.7) becomes very large. Consequently, in the limit of large illumination areas, the convolution vanishes:

$$[\widehat{h} \star \widehat{s}_e](\vec{\nu}_d) \longrightarrow \int_{\vec{\nu}} \widehat{h}(\vec{\nu}_d - \vec{\nu})\delta(\vec{\nu} - \vec{\nu}_i)\, d^2\vec{\nu} = \widehat{h}(\vec{\nu}_d - \vec{\nu}_i)\,. \qquad (8.119)$$

In the most general case, if we introduce the *apparent* roughness spectrum for the interface (given in m$^4$) as

$$\gamma_e(\vec{\nu}) = \frac{1}{S}|\widehat{h}_e(\vec{\nu})|^2\,, \qquad (8.120)$$

then it is possible to represent the angular dependence of the scattering distributions for this interface by the following general equation:

$$\text{ARS}(\vec{\nu}_d, \vec{\nu}_i) = C(\vec{\nu}_d, \vec{\nu}_i)\gamma_e(\vec{\nu}_d)\,. \qquad (8.121)$$

Hence we see that the angle resolved scattering appears as the product of two terms containing information that is independent and different in nature. The first term $(C)$ is a coefficient that can be calculated theoretically if we know the angles of illumination and observation as well as the refractive indices of the two media at the illuminating wavelength. The incident polarization should also be known. The second term $(\gamma_e)$ contains all the information related to the topography of the interface and that can be extracted from experiment with the beam of light used. This can be determined easily from a scattering measurement by dividing the measured ARS by the value of $C$ that results from a theoretical calculation. In addition, if the illuminated area becomes very large, this second term $(\gamma_e)$ can be identified with the value taken at point $\vec{\nu}_d - \vec{\nu}_i$ of the function $\gamma$ defined as

$$\gamma_e(\vec{\nu}) = \gamma(\vec{\nu} - \vec{\nu}_i)\,, \qquad (8.122)$$

and that only depends on the surface topography, meaning that, in the limit of increasing illumination area, the convolution vanishes, as per (8.119).

Generally speaking, the roughness spectrum of a surface is used to characterize its topography. This variable is also known as the **roughness power spectral density**, and corresponds to the Fourier transform of the autocorrelation function $\Gamma$ of the spatial profile of this surface, defined by the following expression:

$$\Gamma(\vec{\tau}) = \frac{1}{S}\int_{\vec{r}} h(\vec{r})\, h(\vec{r} - \vec{\tau})\, d^2\vec{r} \quad \overset{\text{TF}}{\longrightarrow} \quad \gamma(\vec{\nu}) = \frac{1}{S}|\widehat{h}(\vec{\nu})|^2\,. \qquad (8.123)$$

Using the properties of the Fourier transform, the root mean square $\delta$ of the roughness of a surface can be extracted by measuring its roughness spectrum. We have

$$\delta^2 = \Gamma(\vec{0}) = \frac{1}{S} \int_{\vec{r}} h^2(\vec{r}) \, d^2\vec{r} = \int_{\vec{\nu}} \gamma(\vec{\nu}) \, d^2\vec{\nu} \,. \tag{8.124}$$

Thus, the integral of the measured spectrum is equal to the variance of the fluctuations in the surface profile (recall that the mean of the function $h(\vec{r})$ is 0). It may be noted that, since the roughness spectrum has been derived from optical measurement, it is obtained with the relative precision that commonly characterizes such measurements, i.e., around the percent. For this reason, roughness values of the order of a fraction of a nanometer can be commonly displayed with a relative precision around 1%. For its part, absolute precision depends entirely on the calibration accuracy of the scatterometer used for the measurement.

Note that this variance $\delta^2$ can also be expressed as a function of the angular dependence of the roughness spectrum, since

$$\begin{aligned}\delta^2 &= \int_{\vec{\nu}} \gamma(\vec{\nu}) \, d^2\vec{\nu} = \iint_{\nu,\phi} \gamma(\nu,\phi) \, \nu d\nu d\phi \\ &= \left(\frac{n}{\lambda}\right)^2 \iint_{\theta,\phi} \gamma(\theta,\phi) \cos\theta \sin\theta d\theta d\phi \,. \end{aligned} \tag{8.125}$$

Subject to the introduction of the polar mean $\bar{\gamma}(\theta)$ of this spectrum, viz.

$$\bar{\gamma}(\theta) = \frac{1}{2\pi} \int_{\phi} \gamma(\theta,\phi) \, d\phi \,, \tag{8.126}$$

this mean square roughness can be expressed via a single integration over the angle $\theta$, i.e.,

$$\delta^2 = 2\pi \left(\frac{n}{\lambda}\right)^2 \int_{\theta} \bar{\gamma}(\theta) \cos\theta \sin\theta d\theta \,. \tag{8.127}$$

For completeness, note finally that the variance of the fluctuations in slope $p$ of the surface can be calculated from a measurement of its roughness spectrum by using the following equation:

$$p^2 = (2\pi)^3 \int_{\nu} \nu^3 \bar{\gamma}(\nu) \, d\nu \,. \tag{8.128}$$

Indeed:

$$p^2 = \frac{1}{S} \int_{\vec{r}} |\mathbf{grad} \, h(\vec{r})|^2 \, d^2\vec{r} = \frac{1}{S} \int_{\vec{r}} \left\{ \left|\frac{\partial h}{\partial x}\right|^2 + \left|\frac{\partial h}{\partial y}\right|^2 \right\} d^2\vec{r}. \tag{8.129}$$

Using Parseval's theorem and the Fourier transform of a derivative, we get

$$\int_{\vec{r}} \left\{ \left| \frac{\partial h}{\partial x} \right|^2 + \left| \frac{\partial h}{\partial y} \right|^2 \right\} d^2\vec{r} = \int_{\vec{\nu}} \left\{ \left| (2i\pi\nu_x \widehat{h}(\vec{\nu}) \right|^2 + \left| (2i\pi\nu_y \widehat{h}(\vec{\nu}) \right|^2 \right\} d^2\vec{\nu}$$

$$= (2\pi)^2 \int_{\vec{\nu}} \nu^2 |\widehat{h}(\vec{\nu})|^2 \, d^2\vec{\nu} . \qquad (8.130)$$

Consequently,

$$p^2 = (2\pi)^2 \int_{\vec{\nu}} \nu^2 \gamma(\vec{\nu}) \, d^2\vec{\nu} = (2\pi)^2 \iint_{\nu,\phi} \nu^2 \gamma(\nu, \phi) \, \nu d\nu d\phi$$

$$= (2\pi)^3 \int_{\nu} \nu^3 \bar{\gamma}(\nu) \, d\nu . \qquad (8.131)$$

### Bandwidth

As often with measurable physical quantities, roughness has little meaning unless it is associated with an experimentally feasible frequency window, known as the measurement bandwidth. We know, for example, that in the far field light cannot see spatial periods less than the wavelength (for normal incidence illumination), as we have already shown in Chapter 4. This observation implies a variation with wavelength of the measured roughness, and this roughness dispersion will be pronounced to a greater or lesser degree depending on the particular statistical topography of the surface. Furthermore, the notion of bandwidth is essential in making a pertinent comparison of the roughness values obtained using different measurement techniques, whether optical or nonoptical (see Section 8.2.11).

We just saw in the previous section that, in the limiting case of a large illuminated area, the argument of the roughness spectrum is written in the plane $\nu_x O \nu_y$ as

$$\vec{\nu}_d - \vec{\nu}_i = \vec{\nu}_{i,d} = \left( \frac{n}{\lambda} \right) \begin{bmatrix} \sin\theta_d \cos\phi_d - \sin\theta_i \\ \sin\theta_d \sin\phi_d \end{bmatrix} , \qquad (8.132)$$

whose modulus squared is therefore

$$|\vec{\nu}_{i,d}|^2 = \left( \frac{n}{\lambda} \right)^2 \left\{ \sin^2\theta_i + \sin^2\theta_d - 2\sin\theta_i \sin\theta_d \cos\phi_d \right\} . \qquad (8.133)$$

Consequently, the highest spatial frequency that a measurement of the roughness spectrum can achieve is obtained for $\phi_d = \pi$ and $\theta_d = \theta_i = \pi/2$, i.e., when illumination is under grazing incidence and when the scattering measurement is also at a grazing angle and in the half-plane containing the incident beam. This highest spatial frequency therefore has the value $\nu_{\max} = 2n/\lambda$, an expression reminiscent of far-field optical resolution.

The lowest spatial frequency $\nu_{\min}$ at which a measurement can be made is obtained for $\phi_d = 0$ and $\theta_d = \theta_i$, i.e., when the measurement is made in the half-plane of the reflected beam. Theoretically, it might be possible to go down to zero frequency, but that is not feasible due to the intrinsic divergence of the illuminating beam, which we have assumed to be weakly divergent, but obviously not zero. In the case of a beam with a Gaussian spatial profile characterized by a waist ray $w_0$ (see Section 8.2.7), this minimum frequency can be estimated by requiring that it be in the order of twice the effective width of the associated wave packet, i.e.,

$$\nu_{\min} \approx 2 \frac{\cos \theta_i}{\pi w_0} .$$

To summarize, in the most common case of normal incident illumination ($\theta_i = 0$), the range of theoretically accessible frequencies is defined as

$$\frac{2}{\pi w_0} \leqslant \nu_d \leqslant \frac{n}{\lambda} . \tag{8.134}$$

To extend this range, we can therefore change the wavelength and/or reduce the divergence of the beam. Note also that, strictly speaking, the minimum frequency depends on the level of scattering compared to that of the reflected Gaussian beam. Indeed we used a severe assumption to establish (8.134), that is, the receiver cannot enter the reflected beam at low angles (in order to make sure that we measure scattering and not reflection). But this condition may vanish in case where low-angle scattering dominates reflection.

### Polar Variations and Isotropic Surfaces

Up to this point we have dealt specifically in terms of the modulus of the spatial frequency and hence the modulus of the vector $\vec{\nu}_{id}$. However, since the surfaces are two-dimensional, we must also take account of this vector's direction. Figure 8.4 shows the relative orientation of the vectors $\vec{\nu}_i$, $\vec{\nu}_d$ and $\vec{\nu}_{i,d}$, on the left in the case of some arbitrary incidence, and on the right in the case of zero incidence.

If the scattering measurement angle varies, the point D representing the end of the vector $\vec{\nu}_{i,d}$ moves inside the light gray disk. As described by equation (8.133), we observe that the modulus of the bandwidth varies with the polar angle of vector $\vec{\nu}_{i,d}$ and that the result of this construction depends on the angle of incidence.

To reduce this complexity we often prefer to work with a normally incident beam: the frequencies involved are then distributed uniformly within a disk centered on the origin (see the right-hand graph in Figure 8.4).

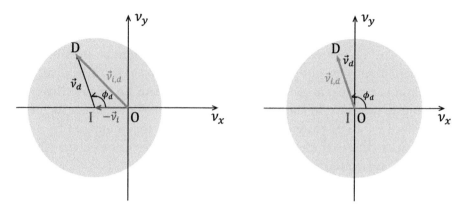

Figure 8.4 Domain of spatial frequencies covered by an angular scattering measurement (zero angle of incidence on the right, and the general case on the left).

For this particular case of illumination, the expressions for the scattering distributions simplify, and can be expressed in the following form:

$$\mathrm{ARS}_j^{SS}(\nu_d) = C_j^S \cos^2\phi_d \, \gamma(\nu_d) \quad ; \quad \mathrm{ARS}_j^{PS}(\nu_d) = C_j^S \sin^2\phi_d \, \gamma(\nu_d)$$

$$\text{where} \quad C_j^S = \left(\frac{n_j}{\lambda}\right)^2 \frac{n_j \cos^2\theta_j^d}{n_0} \left[\frac{2k_0(n_0^2 - n_1^2)}{n_0 \cos\theta_0^d + n_1 \cos\theta_1^d}\right]^2 . \tag{8.135}$$

and

$$\mathrm{ARS}_j^{SP}(\nu_d) = C_j^P \sin^2\phi_d \, \gamma(\nu_d) \quad ; \quad \mathrm{ARS}_j^{PP}(\nu_d) = C_j^P \cos^2\phi_d \, \gamma(\nu_d) \,,$$

$$\text{where} \quad C_j^P = \left(\frac{n_j}{\lambda}\right)^2 \frac{n_j}{n_0} \left[\frac{2k_0(n_0^2 - n_1^2)}{n_0/\cos\theta_0^d + n_1/\cos\theta_1^d}\right]^2 .$$

$$\tag{8.136}$$

In the case of a scattering measurement in unpolarized light, both for illumination and detection, this leads to

$$\mathrm{ARS}(0; \theta_d, \phi_d) = \frac{1}{2}\left[\mathrm{ARS}_j^{SS} + \mathrm{ARS}_j^{SP} + \mathrm{ARS}_j^{PS} + \mathrm{ARS}_j^{PP}\right]$$

$$= \frac{1}{2}\left[C_j^S + C_j^P\right]\gamma(\nu_d), \tag{8.137}$$

where the coefficients $C_j^S$ and $C_j^P$ have no polar dependence. Under these conditions, only the roughness spectrum can break the scattering symmetry of revolution about the normal to the sample.

Polishing techniques these days allow almost perfectly isotropic surfaces to be made, with the possible exception of substrates intended for infrared. For these isotropic surfaces, the autocorrelation function is radial by definition, likewise its Fourier transform (the spectrum $\gamma$). The scattering distribution is then one of revolution, so we need only measure it in the plane of incidence ($\phi_d = 0$); this saves considerably on time and in simplicity, and it is not necessary to allow for moving the detector outside the plane of incidence.

For anisotropic surfaces, it is also possible to make use of the two-dimensional spectrum $\gamma(\nu_d, \phi_d)$ without leaving the plane of incidence. Using non-polarized light under normal incidence, it is sufficient to turn the sample about its normal for each scattering direction $\theta_d$. If we associate each angular position of the sample in its plane with an angle $\alpha$, an $\mathrm{ARS}_\alpha(\theta_d, \phi_d = 0)$ indicatrix is thus registered. However, the only physical variable modified by this rotation is the surface topography $h(\vec{r})$ which must properly be written $h[R_\alpha(\vec{r})]$, where $R_\alpha$ denotes the rotation through angle $\alpha$. Since the Fourier transform preserves rotations, this rotation through angle $\alpha$ can be applied to the roughness spectrum, i.e.,

$$\gamma_\alpha(\nu, \phi_d = 0) = \gamma[R_\alpha(\nu\vec{x})] = \gamma(\nu, \phi_d = \alpha). \qquad (8.138)$$

We see that the angle of rotation $\alpha$ plays the same role as the polar angle $\phi_d$, so the 2D measurement of the spectrum can effectively be obtained while remaining constantly in the plane of incidence.

### Experimental Determination of the Roughness Spectrum

Measurement of the roughness spectrum of a substrate provides an essential result, since when this substrate is coated with a stack of thin layers, the trace of this information can be recovered in the scattering properties of each interface of this multilayer; this is as a result of a topographical replication mechanism that can vary from mediocre to perfect (see Section 8.3.8).

Knowledge of the roughness spectrum of the substrate can then be used in this way to predict the spectral and angular dependence of the scattering distribution of a stack before it is manufactured. Making use of the (optical) character of the final application, the direct measurement of an angle resolved scattering is therefore the most appropriate and reliable method, but it does require a scatterometer working at the wavelength of interest. Note that in case of broadband applications, multiwavelength scatterometers should be used in order to take account of the roughness dispersion (connected with the bandpass variation versus wavelength).

However, other measurement techniques can be used and provide helpful alternatives to the scattering characterization of transparent substrates (in this case the scattering technique must be coupled with a confocal system allowing the two substrate surfaces to be separated). Among these alternative techniques are white light interferometry and atomic force microscopy.

With these techniques the measurements are sampled along $x$ and $y$ with a regular pitch $\Delta$. The result of this measurement is therefore of the form $h(n\Delta, m\Delta)$ where $n$ and $m$ vary between 1 and $N$, $N$ being a power of 2 (e.g. 1024). The transverse dimension of the sampled surface is therefore $L = N\Delta$.

A Fourier transform of this table of values is then carried out numerically (FFT, Fast Fourier Transform), and after recentering on the origin we obtain a new table of $N \times N$ values that correspond with the Fourier transform $\widehat{h}(\nu_x, \nu_y)$ of the recorded topography, sampled at $(p\,\Delta\nu, q\,\Delta\nu)$, where $p$ and $q$ vary between $-N/2$ and $N/2$; the step length in spatial frequency $\Delta\nu$ is $1/L$. Note that one can also compute a classical Fourier transform as

$$\gamma(p\Delta\nu, q\Delta\nu) = \frac{1}{N^2} \left| \sum h(n\Delta, m\Delta)\, e^{-2i\pi \frac{np+mq}{N}} \right|^2. \tag{8.139}$$

Using a direct application of Shannon's criterion, the domain of spatial frequencies for which such a measurement can be made therefore lies between $\nu_{\min} = 1/L$ and $\nu_{\max} = N/2L = 1/2\Delta$; it is clearly essential that this bandwidth be consistent with the measurement of a scattering distribution for this surface. If this condition is not met, the measured spectrum may be unconnected with the topography that generates the scattered flux at the wavelength of interest. In case of normal illumination (for the scattering technique), the maximum frequency will be the same if the pitch satisfies the following condition:

$$\frac{n_0}{\lambda} = \frac{1}{2\Delta} \quad \Rightarrow \quad \Delta = \frac{\lambda}{2n_0}. \tag{8.140}$$

For white light interferometry, this condition might be best approached by optimizing the ratio of the magnification $M$ of the optical system used for the measurement to the pitch $P$ of the photodiode matrix used in the detector, i.e.,

$$\Delta = \frac{P}{M} \approx \frac{\lambda}{2}. \tag{8.141}$$

Finally, we can also perform a **multiscale analysis** of this spectrum by modifying the wavelength or the length of the sampling step; in all cases, the partial spectra must be superposed in the overlapping areas of the frequency windows: this allows the quality of the measurements to be tested, while analyzing the roughness dispersion at the same time. The literature contains many results relating to scattering measurement at several wavelengths, as well as implementations of techniques of atomic force microscopy and white light interferometry.

### The Statistical Nature of Roughness

For the most part, surfaces of optical quality are isotropic and random; consequently, their roughness spectra are radial in nature and decrease with frequency.

It is common to approximate these spectra with exponential or Gaussian statistical distributions (or with frequency to a negative power). These approximations are not essential since the spectrum is already known (and usable) point by point, though this does highlight the statistical nature of roughness; moreover, this statistical nature can vary with the bandwidth used for measuring the topography.

That said, it is often useful to have analytical expressions for roughness spectra; for example, surfaces that scatter very weakly in the visible are correctly described by an autocorrelation function that is the sum of a Gaussian and an exponential function:

$$\Gamma(\tau) = \Gamma_g(\tau) + \Gamma_e(\tau) = \delta_g^2\, e^{-(\tau/L_g)^2} + \delta_e^2\, e^{-|\tau|/L_e} \,. \tag{8.142}$$

The associated roughness spectrum then takes the following form:

$$\gamma(\nu) = \pi(\delta_g L_g)^2\, e^{-(\pi L_g \nu)^2} + 2\pi(\delta_e L_e)^2\, \left[1 + (2\pi L_e \nu)^2\right]^{-3/2} \,. \tag{8.143}$$

Note that the root mean square $\delta = \sqrt{\delta_g^2 + \delta_e^2}$ does not define roughness in practice since it is relative to an infinite bandwidth. Note also that depending on the ratio of correlation length to wavelength, several solutions can be found. Hence we have to integrate this spectrum within the measurement bandwidth to obtain a value of roughness that has any physical significance.

For example, in the frequency window associated with visible light, superpolished substrates are characterized by correlation lengths in the order of 2 µm for the exponential contribution (which describes small-angle scattering), and in the order of 0.1 µm for the Gaussian contribution (which defines wide-angle scattering). On a logarithmic scale, Figure 8.5 shows the behavior of each of these two contributors (Gaussian and exponential), plus their sum, when the quantities $\delta_g$ and $\delta_e$ are both equal to 1 nm.

### Influence of the Profile of the Illuminating Beam

In preceding three sections we deliberately focused our analysis on the asymptotic function $\gamma$, which is the limit of the effective spectrum $\gamma_e$ when the illumination surface tends to infinity (that is, when the convolution effect disappears). This asymptotic spectrum is associated uniquely with the roughness of the substrate, and not with the optogeometric properties of the illumination beam.

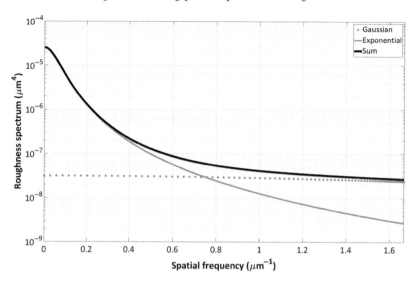

Figure 8.5 Roughness spectrum of an isotropic surface (Gaussian component: gray dots, exponential component: gray line; the result of their combination is in black).

The expression for $\gamma_e$ introduced in Section 8.2.11, i.e.,

$$\gamma_e(\vec{\nu}) = \frac{1}{S}|\widehat{h}_e(\vec{\nu})|^2 = \frac{1}{S}|[\widehat{h} \star \widehat{s}_e](\vec{\nu})|^2\,,$$

shows that the effectively measured spectrum will be related to the mean of the asymptotic profile spectrum over a frequency band corresponding to the width of the Fourier transform of the illumination profile. The frequency resolution of an optical spectral measurement will therefore be completely under the influence of this mean effect (that is, there is no information about the scattering distribution below this spectral resolution that corresponds to a speckle size).

This mean spectral effect also provides the justification as to why it is possible to describe the roughness spectrum of a surface using statistical distributions that vary slowly in frequency (see Section 8.2.11), while the Fourier transforms of the topographies generally show extremely rapid variations with this same spatial frequency.

Finally, this convolution mechanism explains why the scattering distribution does not vanish at zero frequency, i.e., in the direction of the specular beam ($\vec{\nu}_d = \vec{\nu}_i$). At zero frequency we get in the asymptotic case

$$\gamma(\vec{0}) = \frac{1}{S}\left|\int_{\vec{r}} h(\vec{r})\right|^2 d^2\vec{r} = 0\,, \tag{8.144}$$

since the function $h$ has 0 mean (see Section 8.2.1); this result is not valid when the existence of a convolution is included. We will come back to more details in Section 8.3.8 devoted to cross-correlation laws.

We should also stress on the fact that, in addition to the smoothing effect connected with the illumination beam [see relation (8.118)], we should also consider the intensity integration over the detector area (or solid angle $\Delta\Omega$). Actually since the scattering coefficients (not roughness-related) of single surfaces are quasi-constant over the angular aperture of receivers, any receiver integration will be reduced to a frequency integration of the apparent spectrum over $\Delta\Omega$, that is:

$$\gamma_e' = \int \gamma_e \, d\Omega \tag{8.145}$$

Hence strictly speaking, the apparent spectrum must be modified following equation (8.145) to take account of both averaging effects (incident beam size and detector area). The two spectra ($\gamma_e$ and $\gamma_e'$) become identical when the detector area is decreased.

### Example

To illustrate all the concepts introduced in Section 8.2, Figure 8.6 shows the result of a theoretical calculation of a few scattering distributions for an air/silica interface registered in the plane of incidence ($\phi_d = 0$) at a wavelength of 600 nm. The roughness spectrum used for these calculations corresponds to that in Figure 8.5. The angles of incidence chosen correspond to zero incidence (top), 30-degree incidence (middle) and finally, 60-degree incidence (bottom).

All these figures show that the scattering level is much lower than unity, for which reason we use logarithmic coordinates to describe its angular behaviour. We first notice a strong level decrease when the angles get away from the specular direction (that of reflection). Also, for the larger illumination angle (bottom figure), we observe that light scattering vanishes for PP polarization, at a *pseudo-Brewster* angle given by

$$\frac{\sin\theta_0^i \sin\theta_0^d}{\cos\theta_1^i \cos\theta_1^d} = 1 \, ; \tag{8.146}$$

that is,

$$\sin\theta_0^d = \sqrt{\frac{n_1^4 - n_1^2 n_0^2 \sin^2\theta_0^i}{n_1 n_0^2 + (n_1^4 - n_0^4)\sin^2\theta_0^i}} \quad \text{for} \quad n_1 > n_0 . \tag{8.147}$$

This angle is found to be 52 degrees when the illumination angle is 60 degrees. However, it does not exist for low illumination angles (0 degrees and 30 degrees).

To conclude, we keep in mind that for oblique illumination, severe polarization effects may appear in light scattering from a single interface.

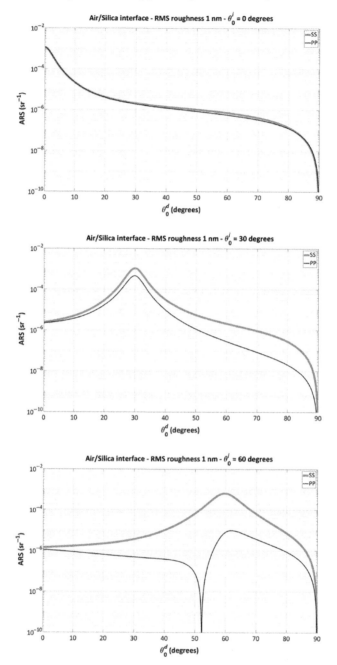

Figure 8.6 Angle resolved scattering (ARS) of an air/silica interface of roughness 1 nm RMS, registered in the plane of incidence ($\phi_d = 0$) for various angles of illumination incidence $\theta_1^i$ (top, $\theta_1^i = 0$ – middle, $\theta_1^i = 30$ degree – bottom, $\theta_1^i = 60$ degrees) and two states of polarization (SS, gray curves – PP, black curves).

## 8.3 Scattering by a Stack of Thin Layers

### *8.3.1 Introduction*

We now look at the case in which the element being considered is no longer a simple rough interface, but rather a stack of $p$ thin layers deposited on a rough substrate, as shown in Figure 8.7. Hence each interface $j$ is rough, with a profile $h_j(\vec{r})$ of mean centered on height $z_j$.

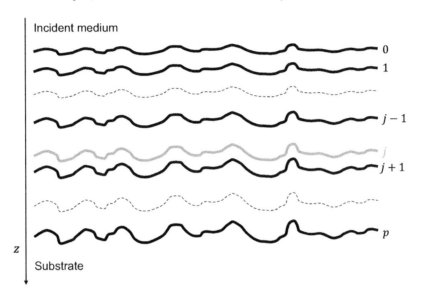

Figure 8.7 Stack comprising $p$ rough layers.

As we saw in the case of a single interface, these rough interfaces are equivalent to plane optical interfaces supporting fictitious currents, both electric $\vec{\mathcal{J}}_j$ and magnetic $\vec{\mathcal{M}}_j$, analogous to those introduced in Section 8.2.3; these fictitious currents are proportional to the ideal fields $\vec{\mathcal{E}}_{j,e}$ and $\vec{\mathcal{H}}_{j,e}$ found at the $j$th interface.

For a stack of $p$ layers there are therefore $p+1$ pairs of fictitious sources $(\vec{\mathcal{J}}_j, \vec{\mathcal{M}}_j)$. The field scattered by interface $j$ can be calculated assuming only the pair of sources $(\vec{\mathcal{J}}_j, \vec{\mathcal{M}}_j)$ to be nonzero, the total field being obtained by summing these different contributions, in accordance with the principle of superposition. Obviously, this approach is valid only because the interfaces are slightly rough, so that they do not impact the ideal field that excites the currents. Furthermore, our calculation is still only first-order so that we ignore any *scattering of scattering*.

We shall see that our calculations are somewhat similar to the previous case (the single interface), except that the (ideal) exciting field is predominantly stationary (not progressive or retrograde) within the stack, in which case the effective index must be replaced by the complex admittance.

### 8.3.2 Elementary Components of the Field Scattered by a Single Interface Within the Stack

Consider the interface $j$ separating the two media of refractive indices $n_j$ and $n_{j+1}$, and supporting the fictitious currents $(\vec{\mathcal{J}}_j, \vec{\mathcal{M}}_j)$. These currents are responsible for stationary scattered fields on either side of this interface whose elementary components are given by the following relations:

- For $z_{j-1} \leqslant z \leqslant z_j$

$$
\begin{aligned}
\vec{\mathbb{E}}_{j,d}(\vec{\nu}_d, z) &= \vec{\mathbb{A}}^+_{j,d}(\vec{\nu}_d)e^{i\alpha^d_j z} + \vec{\mathbb{A}}^-_{j,d}(\vec{\nu}_d)e^{-i\alpha^d_j z} \\
\vec{\mathbb{H}}_{j,d}(\vec{\nu}_d, z) &= \vec{\mathbb{B}}^+_{j,d}(\vec{\nu}_d)e^{i\alpha^d_j z} + \vec{\mathbb{B}}^-_{j,d}(\vec{\nu}_d)e^{-i\alpha^d_j z} \,.
\end{aligned}
\tag{8.148}
$$

- For $z_j \leqslant z \leqslant z_{j+1}$

$$
\begin{aligned}
\vec{\mathbb{E}}_{j+1,d}(\vec{\nu}_d, z) &= \vec{\mathbb{A}}^+_{j+1,d}(\vec{\nu}_d)e^{i\alpha^d_{j+1} z} + \vec{\mathbb{A}}^-_{j+1,d}(\vec{\nu}_d)e^{-i\alpha^d_{j+1} z} \\
\vec{\mathbb{H}}_{j+1,d}(\vec{\nu}_d, z) &= \vec{\mathbb{B}}^+_{j+1,d}(\vec{\nu}_d)e^{i\alpha^d_{j+1} z} + \vec{\mathbb{B}}^-_{j+1,d}(\vec{\nu}_d)e^{-i\alpha^d_{j+1} z} \,.
\end{aligned}
\tag{8.149}
$$

The amplitudes of the scattered fields on this $j$th interface correspond to the previous expressions at $z = z_j$ in the medium concerned ($n_j$ or $n_{j+1}$ refractive index). Recall that these fields are discontinuous at interface $j$.

The boundary conditions between fields scattered at interface $j$ are expressed as

$$
\begin{aligned}
\vec{z} \wedge [\vec{\mathbb{E}}_{j+1,d}(\vec{\nu}_d, z_j) - \vec{\mathbb{E}}_{j,d}(\vec{\nu}_d, z_j)] &= \vec{\mathbb{M}}_j(\vec{\nu}_d) \\
\vec{z} \wedge [\vec{\mathbb{H}}_{j+1,d}(\vec{\nu}_d, z_j) - \vec{\mathbb{H}}_{j,d}(\vec{\nu}_d, z_j)] &= \vec{\mathbb{J}}_j(\vec{\nu}_d) \,,
\end{aligned}
\tag{8.150}
$$

with, in addition, the following relations between the tangential components on this interface:

$$
\begin{aligned}
\vec{\mathbb{H}}^T_{j+1,d}(\vec{\nu}_d, z_j) &= Y_{j,d}\, \vec{z} \wedge \vec{\mathbb{E}}^T_{j+1,d}(\vec{\nu}_d, z_j) \\
\vec{\mathbb{H}}^T_{j,d}(\vec{\nu}_d, z_j) &= Y'_{j,d}\, \vec{z} \wedge \vec{\mathbb{E}}^T_{j,d}(\vec{\nu}_d, z_j) \,.
\end{aligned}
\tag{8.151}
$$

It is important to point out again that, including the scattering sources at interface $j$, the scattered field (electric or magnetic, normal or tangential) is, in the general case, no longer continuous – this differs from the context studied in Chapter 6; similarly for the admittance, connected with the ratio of these fields and which becomes a discontinuous function at interface $j$. Consequently, admittance $Y_{j,d}$ must be calculated by recurrence (see Chapter 6) in its region of continuity ($z_j < z < z_p$), that is, from the admittance of the optical interface $p$ (given by the effective refractive index $\tilde{n}_s^d$ of the substrate), while admittance $Y'_{j,d}$ must be calculated by recurrence in its region of continuity $(0 < z < z_j)$, from the admittance of optical interface 0 (given by the effective refractive index $-\tilde{n}_0^d$ of air).

Substituting equations (8.151) into the second of equations (8.150), we get immediately

$$- Y_{j,d} \vec{\mathbb{E}}^T_{j+1,d}(\vec{\nu}_d, z_j) + Y'_{j,d} \vec{\mathbb{E}}^T_{j,d}(\vec{\nu}_d, z_j) = \vec{\mathbb{J}}_j(\vec{\nu}_d). \tag{8.152}$$

We now calculate the quantity $\vec{z} \wedge \vec{\mathbb{M}}_j$ using the first of equations (8.150):

$$\vec{z} \wedge \vec{\mathbb{M}}_j(\vec{\nu}_d) = -\vec{\mathbb{E}}^T_{j+1,d}(\vec{\nu}_d, z_j) + \vec{\mathbb{E}}^T_{j,d}(\vec{\nu}_d, z_j). \tag{8.153}$$

The system of equations arising from (8.152) and (8.153) has the solution

$$\begin{cases} \vec{\mathbb{E}}^T_{j,d}(\vec{\nu}_d, z_j) = -\dfrac{\vec{\mathbb{J}}_j(\vec{\nu}_d) - Y_{j,d}\, \vec{z} \wedge \vec{\mathbb{M}}_j(\vec{\nu}_d)}{Y_{j,d} - Y'_{j,d}} \\[3mm] \vec{\mathbb{E}}^T_{j+1,d}(\vec{\nu}_d, z_j) = -\dfrac{\vec{\mathbb{J}}_j(\vec{\nu}_d) - Y'_{j,d}\, \vec{z} \wedge \vec{\mathbb{M}}_j(\vec{\nu}_d)}{Y_{j,d} - Y'_{j,d}} \end{cases}. \tag{8.154}$$

Note that these relations are analogous to those obtained for the single optical interface [see equations (8.37)], except that the effective refractive indices (used for progressive and retrograde waves) are replaced by admittances (used for stationary waves).

It now remains to calculate the fictitious currents associated with each interface; to this end we will first need to define an expression for the stationary exciting fields.

### 8.3.3 Exciting Field on Interface $j$

Just as for the single interface, we shall assume here that the exciting field on interface 0 corresponds to that created by a weakly divergent monochromatic beam of light whose mean direction of propagation subtends an angle $\theta_0^i$ with the normal to the multilayer component. The elementary Fourier components of the field incident on interface 0 are written (see Section 8.2.7.)

$$\vec{\mathbb{A}}^+_i(\vec{\nu}_e) = \widehat{s}_e(\vec{\nu}_e)\, \vec{\mathcal{E}}_i, \tag{8.155}$$

where

$$\widehat{s}_e(\nu_x, \nu_y) = \frac{\pi w_0^2}{\cos \theta_0^i}\, e^{-\pi^2 \frac{w_0^2}{\cos^2 \theta_0^i}(\nu_x - \nu_i)^2}\, e^{-\pi^2 w_0^2 \nu_y^2} \quad \text{and} \quad \nu_i = \frac{n_0}{\lambda}\sin\theta_0^i. \tag{8.156}$$

Only the tangential component of the exciting field on the $j$th interface will be necessary to calculate the field scattered by this same interface $j$. The expression for this tangential component follows directly from the relations established in Section 6.3.8, and we have therefore

$$\vec{\mathbb{E}}_{j,e}^{T}(\vec{v}_e, z_j) = \frac{1 + r_0^e}{\prod\limits_{l=1}^{j} \left[\cos\delta_l^e - i(Y_{l,e}/\tilde{n}_l^e)\sin\delta_l^e\right]}\,\hat{s}_e(\vec{v}_e)\,\vec{\mathcal{E}}_i^T, \qquad (8.157)$$

where

$$\delta_l^e = \alpha_l^e d_l \quad ; \quad r_0^e = \frac{\tilde{n}_0^e - Y_{0,e}}{\tilde{n}_0^e + Y_{0,e}}. \qquad (8.158)$$

This can also be written as

$$\vec{\mathbb{E}}_{j,e}^{T}(\vec{v}_e, z_j) = \left[\vec{\mathbb{A}}_{j,e}^{+} e^{i\alpha_j^e z_j} + \vec{\mathbb{A}}_{j,e}^{-} e^{-i\alpha_j^e z_j}\right] = t_{0j}^e\,\hat{s}_e(\vec{v}_e)\,\vec{\mathcal{E}}_i^T, \qquad (8.159)$$

where

$$t_{0j}^e = \frac{1 + r_0^e}{\prod\limits_{l=1}^{j} \left[\cos\delta_l^e - i(Y_{l,e}/\tilde{n}_l^e)\sin\delta_l^e\right]} \qquad \text{for } j = 1,\dots,p. \qquad (8.160)$$

In the case of a single interface (i.e., no stack), these relations remain true provided it is assumed that the product appearing in the denominator of the last expression is equal to 1, i.e.,

$$t_{00}^e = 1 + r_0^e. \qquad (8.161)$$

### 8.3.4 Source Terms Associated with Interface $j$

#### Fictitious Electric Currents

The elementary Fourier component of the fictitious electric current associated with the roughness of interface $j$ is defined by the following general expression:

$$\vec{\mathbb{J}}_j(\vec{v}_d) = -\vec{z} \wedge \int_{\vec{r}} h_j(\vec{r}) \left[\frac{\partial \vec{\mathcal{H}}_{j+1,e}}{\partial z} - \frac{\partial \vec{\mathcal{H}}_{j,e}}{\partial z}\right]_{\vec{r},z_j} e^{-2i\pi\vec{v}_d\cdot\vec{r}}\,d^2\vec{r}, \qquad (8.162)$$

where

$$\vec{\mathcal{H}}_{j,e}(\vec{r}, z) = \int_{\vec{v}_e} \left[\vec{\mathbb{B}}_{j,e}^{+} e^{i\alpha_j^e z} + \vec{\mathbb{B}}_{j,e}^{-} e^{-i\alpha_j^e z}\right] e^{2i\pi\vec{v}_e\cdot\vec{r}}\,d^2\vec{v}_e. \qquad (8.163)$$

After differentiation and application of the vector product, we get

$$\vec{z} \wedge \left[ \frac{\partial \vec{\mathcal{H}}_{j,e}}{\partial z} \right]_{\vec{r},z_j} = i \int_{\vec{\nu}_e} \alpha_j^e \left[ \vec{z} \wedge \vec{\mathbb{B}}_{j,e}^+ e^{i\alpha_j^e z_j} - \vec{z} \wedge \vec{\mathbb{B}}_{j,e}^- e^{-i\alpha_j^e z_j} \right] e^{2i\pi \vec{\nu}_e.\vec{r}} \, d^2\vec{\nu}_e \, .$$

(8.164)

Furthermore,

$$\begin{cases} \vec{z} \wedge \vec{\mathbb{B}}_{j,e}^+ = \vec{z} \wedge \vec{\mathbb{B}}_{j,e}^{T+} = \vec{z} \wedge [\tilde{n}_j^e \vec{z} \wedge \vec{\mathbb{A}}_{j,e}^{T+}] = -\tilde{n}_j^e \vec{\mathbb{A}}_{j,e}^{T+} \\ \vec{z} \wedge \vec{\mathbb{B}}_{j,e}^- = \vec{z} \wedge \vec{\mathbb{B}}_{j,e}^{T-} = -\vec{z} \wedge [\tilde{n}_j^e \vec{z} \wedge \vec{\mathbb{A}}_{j,e}^{T-}] = \tilde{n}_j^e \vec{\mathbb{A}}_{j,e}^{T-} \end{cases} \, .$$

(8.165)

Substituting these results into (8.164), we get

$$\vec{z} \wedge \left[ \frac{\partial \vec{\mathcal{H}}_{j,e}}{\partial z} \right]_{\vec{r},z_j} = -i \int_{\vec{\nu}_e} \alpha_j^e \tilde{n}_j^e \left[ \vec{\mathbb{A}}_{j,e}^+ e^{i\alpha_j^e z_j} + \vec{\mathbb{A}}_{j,e}^- e^{-i\alpha_j^e z_j} \right] e^{2i\pi \vec{\nu}_e.\vec{r}} \, d^2\vec{\nu}_e \, ,$$

(8.166)

and also, using equation (8.159),

$$\vec{z} \wedge \left[ \frac{\partial \vec{\mathcal{H}}_{j,e}}{\partial z} \right]_{\vec{r},z_j} = -i \int_{\vec{\nu}_e} \alpha_j^e \tilde{n}_j^e t_{0j}^e \, \hat{s}_e(\vec{\nu}_e) \, \vec{\mathcal{E}}_i^T \, e^{2i\pi \vec{\nu}_e.\vec{r}} \, d^2\vec{\nu}_e \, .$$

(8.167)

This is where we make use of our assumption that the incident beam is weakly divergent, identifying the slowly varying quantities $\alpha_j^e$, $\tilde{n}_j^e$ and $t_{0j}^e$ with their values in $\vec{\nu}_e = \vec{\nu}_i$, namely

$$\vec{z} \wedge \left[ \frac{\partial \vec{\mathcal{H}}_{j,e}}{\partial z} \right]_{\vec{r},z_j} = -i\alpha_j^i \tilde{n}_j^i t_{0j}^i \, \vec{\mathcal{E}}_i^T \int_{\vec{\nu}_e} \hat{s}_e(\vec{\nu}_e) \, e^{2i\pi \vec{\nu}_e.\vec{r}} \, d^2\vec{\nu}_e$$

$$= -i\alpha_j^i \tilde{n}_j^i t_{0j}^i \, s_e(\vec{r}) \, \vec{\mathcal{E}}_i^T \, .$$

(8.168)

Similarly,

$$\vec{z} \wedge \left[ \frac{\partial \vec{\mathcal{H}}_{j+1,e}}{\partial z} \right]_{\vec{r},z_j} = -i \int_{\vec{\nu}_e} \alpha_{j+1}^e \tilde{n}_{j+1}^e \vec{\mathbb{E}}_{j+1,i}^T(\vec{\nu}_i, z_j) \, e^{2i\pi \vec{\nu}_e.\vec{r}} \, d^2\vec{\nu}_e \, .$$

(8.169)

Furthermore, the continuity of the tangential component of the exciting field, i.e.,

$$\vec{\mathbb{E}}_{j+1,i}^T(\vec{\nu}_i, z_j) = \vec{\mathbb{E}}_{j,i}^T(\vec{\nu}_i, z_j) \, ,$$

(8.170)

leads to

$$\vec{z} \wedge \left[ \frac{\partial \vec{\mathcal{H}}_{j+1,e}}{\partial z} \right]_{\vec{r},z_j} = -i\alpha_{j+1}^i \tilde{n}_{j+1}^i t_{0j}^i \, s_e(\vec{r}) \, \vec{\mathcal{E}}_i^T \, .$$

(8.171)

Substituting expressions (8.168) and (8.171) into (8.162), we get finally

$$\vec{J}_j(\vec{v}_d) = i \left[ \alpha^i_{j+1} \tilde{n}^i_{j+1} - \alpha^i_j \tilde{n}^i_j \right] t^i_{0j} \, \vec{\mathcal{E}}^T_i \int_{\vec{r}} h_j(\vec{r}) \, s_e(\vec{r}) e^{-2i\pi \vec{v}_d \cdot \vec{r}} \, d^2\vec{r}, \quad (8.172)$$

and also, using the fact that the Fourier transform of a product of two functions is equal to the convolution of their Fourier transforms,

$$\vec{J}_j(\vec{v}_d) = i \, \{ \alpha^i_{j+1} \tilde{n}^i_{j+1} - \alpha^i_j \tilde{n}^i_j \} \, t^i_{0j} \, [\hat{h}_j \star \hat{s}_e]_{\vec{v}_d} \, \vec{\mathcal{E}}^T_i. \quad (8.173)$$

### Fictitious Magnetic Currents

The fictitious magnetic current introduced by the roughness of interface $j$ is defined by the following equation:

$$\vec{z} \wedge \vec{M}_j(\vec{v}_d) = \vec{z} \wedge \int_{\vec{r}} \mathbf{grad} \, h_j(\vec{r}) \wedge \left[ \vec{\mathcal{E}}_{j+1,e}(\vec{r}, z_j) - \vec{\mathcal{E}}_{j,e}(\vec{r}, z_j) \right] e^{-2i\pi \vec{v}_d \cdot \vec{r}} \, d^2\vec{r}$$

$$- \vec{z} \wedge \int_{\vec{r}} h_j(\vec{r}) \, \vec{z} \wedge \left[ \frac{\partial \vec{\mathcal{E}}_{j+1,e}}{\partial z} - \frac{\partial \vec{\mathcal{E}}_{j,e}}{\partial z} \right]_{\vec{r}, z_j} e^{-2i\pi \vec{v}_d \cdot \vec{r}} \, d^2\vec{r}, \quad (8.174)$$

where

$$\vec{\mathcal{E}}_{j,e}(\vec{r}, z) = \int_{\vec{v}_e} \left[ \vec{A}^+_{j,e} e^{i\alpha^e_j z} + \vec{A}^-_{j,e} e^{-i\alpha^e_j z} \right] e^{2i\pi \vec{v}_e \cdot \vec{r}} \, d^2\vec{v}_e. \quad (8.175)$$

Just as for the single interface, we put

$$\mathbb{I}_{1,j} = \vec{z} \wedge \int_{\vec{r}} \mathbf{grad} \, h_j(\vec{r}) \wedge \left[ \vec{\mathcal{E}}_{j+1,e}(\vec{r}, z_j) - \vec{\mathcal{E}}_{j,e}(\vec{r}, z_j) \right] e^{-2i\pi \vec{v}_d \cdot \vec{r}} \, d^2\vec{r}, \quad (8.176)$$

and

$$\mathbb{I}_{2,j} = -\vec{z} \wedge \int_{\vec{r}} h_j(\vec{r}) \, \vec{z} \wedge \left[ \frac{\partial \vec{\mathcal{E}}_{j+1,e}}{\partial z} - \frac{\partial \vec{\mathcal{E}}_{j,e}}{\partial z} \right]_{\vec{r}, z_j} e^{-2i\pi \vec{v}_d \cdot \vec{r}} \, d^2\vec{r}. \quad (8.177)$$

In the first instance, we are interested in the second of these two integrals. Differentiating (8.175) with respect to $z$, we get

$$\vec{z} \wedge \left[ \frac{\partial \vec{\mathcal{E}}_{j,e}}{\partial z} \right]_{\vec{r}, z_j} = i \int_{\vec{v}_e} \alpha^e_j \, \vec{z} \wedge \left[ \vec{A}^+_{j,e} e^{i\alpha^e_j z_j} - \vec{A}^-_{j,e} e^{-i\alpha^e_j z_j} \right] e^{2i\pi \vec{v}_e \cdot \vec{r}} \, d^2\vec{v}_e. \quad (8.178)$$

Moreover,

$$\vec{\mathbb{H}}^T_{j,e}(\vec{v}_e, z_j) = \left[ \vec{\mathbb{B}}^{T+}_{j,e} e^{i\alpha^e_j z_j} + \vec{\mathbb{B}}^{T-}_{j,e} e^{-i\alpha^e_j z_j} \right]$$

$$= \tilde{n}^e_j \, \vec{z} \wedge \left[ \vec{A}^+_{j,e} e^{i\alpha^e_j z_j} - \vec{A}^-_{j,e} e^{-i\alpha^e_j z_j} \right]. \quad (8.179)$$

Recall at this stage the definition of the complex admittance of interface $j$, namely

$$\vec{\mathbb{H}}_{j,e}^T(\vec{\nu}_e, z_j) = Y_{j,e}\,\vec{z} \wedge \vec{\mathbb{E}}_{j,e}^T(\vec{\nu}_e, z_j)\,. \tag{8.180}$$

Combining (8.159), (8.178), (8.179), and (8.180), we obtain

$$\vec{z} \wedge \left[\frac{\partial \vec{\mathcal{E}}_{j,e}}{\partial z}\right]_{\vec{r},z_j} = i \int_{\vec{\nu}_e} \frac{\alpha_j^e}{\tilde{n}_j^e} Y_{j,e}\, t_{0j}^e\, \widehat{s}_e(\vec{\nu}_e)\,[\vec{z} \wedge \vec{\mathcal{E}}_i^T]\, e^{2i\pi\vec{\nu}_e\cdot\vec{r}}\, d^2\vec{\nu}_e\,, \tag{8.181}$$

and, using as before the assumption of weak beam divergence,

$$\vec{z} \wedge \left[\frac{\partial \vec{\mathcal{E}}_{j,e}}{\partial z}\right]_{\vec{r},z_j} = i\frac{\alpha_j^i}{\tilde{n}_j^i} Y_{j,i}\, t_{0j}^i\, s_e(\vec{r})\,[\vec{z} \wedge \vec{\mathcal{E}}_i^T]\,. \tag{8.182}$$

Following a procedure that is everywhere analogous to the procedure just described and also making use of the continuity of the tangential component of the exciting magnetic field $\vec{\mathbb{H}}_{j+1,e}^T$ on interface $j$, we can similarly show that

$$\vec{z} \wedge \left[\frac{\partial \vec{\mathcal{E}}_{j+1,e}}{\partial z}\right]_{\vec{r},z_j} = i\frac{\alpha_{j+1}^i}{\tilde{n}_{j+1}^i} Y_{j,i}\, t_{0j}^i\, s_e(\vec{r})\,[\vec{z} \wedge \vec{\mathcal{E}}_i^T]\,. \tag{8.183}$$

Using relations (8.182) and (8.183), combined with the properties of the vector triple product, we can put the integral $\mathbb{I}_{2,j}$ in the following form:

$$\mathbb{I}_{2,j} = i\left[\frac{\alpha_{j+1}^i}{\tilde{n}_{j+1}^i} - \frac{\alpha_j^i}{\tilde{n}_j^i}\right] Y_{j,i}\, t_{0j}^i\, \vec{\mathcal{E}}_i^T \int_{\vec{r}} h_j(\vec{r})\, s_e(\vec{r})\, e^{-2i\pi\vec{\nu}_d\cdot\vec{r}}\, d^2\vec{r}\,, \tag{8.184}$$

or,

$$\mathbb{I}_{2,j} = i\left[\frac{\alpha_{j+1}^i}{\tilde{n}_{j+1}^i} - \frac{\alpha_j^i}{\tilde{n}_j^i}\right] Y_{j,i}\, t_{0j}^i\, [\widehat{h}_j \star \widehat{s}_e]_{\vec{\nu}_d}\, \vec{\mathcal{E}}_i^T\,. \tag{8.185}$$

We are now interested in calculating the integral $\mathbb{I}_{1,j}$. The properties of the vector triple product enable us to write

$$\mathbb{I}_{1,j} = \int_{\vec{r}} \mathbf{grad}\, h_j(\vec{r})\,\{\vec{z}\cdot[\vec{\mathcal{E}}_{j+1,e}(\vec{r}, z_j) - \vec{\mathcal{E}}_{j,e}(\vec{r}, z_j)]\}\, e^{-2i\pi\vec{\nu}_d\cdot\vec{r}}\, d^2\vec{r}\,. \tag{8.186}$$

As with the single interface, we put

$$g_j(\vec{r}) = \vec{z}\cdot\vec{\mathcal{E}}_{j+1,e}(\vec{r}, z_j) - \vec{z}\cdot\vec{\mathcal{E}}_{j,e}(\vec{r}, z_j)\,, \tag{8.187}$$

which allows us to express the integral $\mathbb{I}_{1,j}$ as a convolution:

$$\mathbb{I}_{1,j} = \mathrm{TF}\{g_j\} \star \left[2i\pi\widehat{h}_j\vec{\nu}\right] = 2i\pi \int_{\vec{\nu}_e} [\vec{\nu}_d - \vec{\nu}_e]\,\widehat{g}_j(\vec{\nu}_e)\,\widehat{h}_j(\vec{\nu}_d - \vec{\nu}_e)\, d^2\vec{\nu}_e\,. \tag{8.188}$$

We first use the boundary condition between the normal components of the exciting field at interface $j$, i.e.,

$$\tilde{\epsilon}_{j+1}\vec{z}\cdot\vec{\mathcal{E}}_{j+1,e}(\vec{r},z_j) = \tilde{\epsilon}_j\vec{z}\cdot\vec{\mathcal{E}}_{j,e}(\vec{r},z_j)$$

$$\Rightarrow\quad \vec{z}\cdot\vec{\mathcal{E}}_{j+1,e}(\vec{r},z_j) = \frac{\tilde{\epsilon}_j}{\tilde{\epsilon}_{j+1}}\vec{z}\cdot\vec{\mathcal{E}}_{j,e}(\vec{r},z_j)\,, \qquad (8.189)$$

to reformulate the quantity $g_j(\vec{r})$

$$g_j(\vec{r}) = \left[\frac{\tilde{\epsilon}_j}{\tilde{\epsilon}_{j+1}} - 1\right]\vec{z}\cdot\vec{\mathcal{E}}_{j,e}(\vec{r},z_j)\,, \qquad (8.190)$$

and finally give an expression for its Fourier transform:

$$\hat{g}_j(\vec{\nu}_e) = \int_{\vec{r}} g_j(\vec{r})\,e^{-2i\pi\vec{\nu}_e\cdot\vec{r}}\,d^2\vec{r} = \left[\frac{\tilde{\epsilon}_j}{\tilde{\epsilon}_{j+1}} - 1\right]\vec{z}\cdot\vec{\mathbb{E}}_{j,e}(\vec{\nu}_e,z_j)\,. \qquad (8.191)$$

If the exciting field is TE polarized, the vectors $\vec{z}$ and $\vec{\mathbb{E}}_{j,e}(\vec{\nu},z_j)$ are orthogonal and the quantity $\hat{g}_j(\vec{\nu})$ is therefore identically zero, as is also the integral $\mathbb{I}_{1,j}$. If the field is TM polarized, this is clearly no longer the case, but using the relations established in Section 6.3.8 we can express the normal component of the exciting field as a function of its tangential component, i.e.,

$$\vec{z}.\vec{\mathbb{E}}_{j,e}(\vec{\nu}_e,z_j) = \mathbb{E}_{j,e}^N(\vec{\nu}_e,z_j) = -\frac{2\pi\nu_e}{\omega\tilde{\epsilon}_j}Y_{j,e}\mathbb{E}_{j,e}^T(\vec{\nu}_e,z_j)\,. \qquad (8.192)$$

Putting expressions (8.159), (8.191), and (8.192) into equation (8.188), we then obtain

$$\mathbb{I}_{1,j,P} = i\frac{4\pi^2}{\omega}\left[\frac{\tilde{\epsilon}_{j+1}-\tilde{\epsilon}_j}{\tilde{\epsilon}_j\tilde{\epsilon}_{j+1}}\right]\mathcal{E}_i^T\int_{\vec{\nu}_e}(\vec{\nu}_d-\vec{\nu}_e)\,\nu_e Y_{j,e}\,t_{0j}^e\,\hat{h}_j(\vec{\nu}_d-\vec{\nu}_e)\,\hat{s}_e(\vec{\nu}_e)\,d^2\vec{\nu}_e\,, \qquad (8.193)$$

and so, after invoking the assumption that the beam is weakly divergent,

$$\mathbb{I}_{1,j,P} = i\frac{4\pi^2}{\omega}\left(\frac{\tilde{\epsilon}_{j+1}-\tilde{\epsilon}_j}{\tilde{\epsilon}_j\tilde{\epsilon}_{j+1}}\right)\{\vec{\nu}_d-\vec{\nu}_i\}\,\nu_i Y_{j,i}\,t_{0j}^i\,[\hat{h}_j\star\hat{s}_e]_{\vec{\nu}_d}\,\mathcal{E}_i^T\,. \qquad (8.194)$$

So as to better emphasize the homogeneous nature of this last equation, we shall use the definition of the effective refractive index of medium $j+1$ under polarization P, namely

$$\tilde{n}_{j+1}^i = \frac{\omega\tilde{\epsilon}_{j+1}}{\alpha_{j+1}^i}\,, \qquad (8.195)$$

to write

$$\mathbb{I}_{1,j,P} = 4i\pi^2\left(\frac{\tilde{\epsilon}_{j+1}-\tilde{\epsilon}_j}{\tilde{\epsilon}_j}\right)\{\vec{\nu}_d-\vec{\nu}_i\}\,\frac{Y_{j,i}}{\alpha_{j+1}^i\tilde{n}_{j+1}^i}\,t_{0j}^i\,[\hat{h}_j\star\hat{s}_e]_{\vec{\nu}_d}\,\nu_i\,\mathcal{E}_i^T\,. \qquad (8.196)$$

Provided the quantity $\nu_i \mathcal{E}_i^T$ is replaced by the scalar product $\vec{\nu}_i \cdot \vec{\mathcal{E}}_i^T$ in this last equation, we then have a completely general formulation for the value of integral $\mathbb{I}_{1,j}$ since this scalar product is 0 for the S case, and is identified with $\nu_i \mathcal{E}_i^T$ in the P case.

In conclusion, the fictitious magnetic current associated with the roughness of the $j$th interface is defined by the following general expression, and is valid whatever the polarization state of the exciting incident wave:

$$\vec{z} \wedge \vec{\mathbb{M}}_j(\vec{\nu}_d) = iY_{j,i}\, t_{0j}^i\, [\hat{h} \star \hat{s}_e]_{\vec{\nu}_d}$$

$$\times \left\{ \left[ \frac{\alpha_{j+1}^i}{\tilde{n}_{j+1}^i} - \frac{\alpha_j^i}{\tilde{n}_j^i} \right] \vec{\mathcal{E}}_i^T + \frac{4\pi^2}{\alpha_{j+1}^i \tilde{n}_{j+1}^i} \left[ \frac{\tilde{\epsilon}_{j+1} - \tilde{\epsilon}_j}{\tilde{\epsilon}_j} \right] (\vec{\nu}_d - \vec{\nu}_i)\{\vec{\nu}_i \cdot \vec{\mathcal{E}}_i^T\} \right\}. \tag{8.197}$$

### 8.3.5 Fields Scattered at Interface $j$

To calculate the scattered fields either side of interface $j$, we now have to substitute the expressions for the fictitious currents established in the previous section into equation (8.154), i.e.,

$$\vec{E}_{j,d}^T(\vec{\nu}_d, z_j) = -\frac{i\,t_{0j}^i}{Y_{j,d} - Y_{j,d}'} \left\{ \left[ (\alpha_{j+1}^i \tilde{n}_{j+1}^i - \alpha_j^i \tilde{n}_j^i) - Y_{j,d}\, Y_{j,i} \left( \frac{\alpha_{j+1}^i}{\tilde{n}_{j+1}^i} - \frac{\alpha_j^i}{\tilde{n}_j^i} \right) \right] \vec{\mathcal{E}}_i^T \right.$$

$$\left. -4\pi^2(\vec{\nu}_d - \vec{\nu}_i)\frac{Y_{j,d}\, Y_{j,i}}{\alpha_{j+1}^i \tilde{n}_{j+1}^i} \left[ \frac{\tilde{\epsilon}_{j+1} - \tilde{\epsilon}_j}{\tilde{\epsilon}_j} \right] \{\vec{\nu}_i \cdot \vec{\mathcal{E}}_i^T\} \right\} [\hat{h}_j \star \hat{s}_e]_{\vec{\nu}_d}, \tag{8.198}$$

$$\vec{E}_{j+1,d}^T(\vec{\nu}_d, z_j) = -\frac{i\,t_{0j}^i}{Y_{j,d} - Y_{j,d}'} \left\{ \left[ (\alpha_{j+1}^i \tilde{n}_{j+1}^i - \alpha_j^i \tilde{n}_j^i) - Y_{j,d}'\, Y_{j,i} \left( \frac{\alpha_{j+1}^i}{\tilde{n}_{j+1}^i} - \frac{\alpha_j^i}{\tilde{n}_j^i} \right) \right] \vec{\mathcal{E}}_i^T \right.$$

$$\left. -4\pi^2(\vec{\nu}_d - \vec{\nu}_i)\frac{Y_{j,d}'\, Y_{j,i}}{\alpha_{j+1}^i \tilde{n}_{j+1}^i} \left[ \frac{\tilde{\epsilon}_{j+1} - \tilde{\epsilon}_j}{\tilde{\epsilon}_j} \right] \{\vec{\nu}_i \cdot \vec{\mathcal{E}}_i^T\} \right\} [\hat{h}_j \star \hat{s}_e]_{\vec{\nu}_d}. \tag{8.199}$$

Note that these general formulas can be identified with those established for a single interface, provided the following correspondences are used:

$$j = 0 \quad ; \quad Y_{0,i} = \tilde{n}_1^i \quad ; \quad Y_{0,d} = \tilde{n}_1^d \quad ; \quad Y_{0,d}' = -\tilde{n}_0^d \quad ; \quad t_{00}^i = \frac{2\tilde{n}_0^i}{\tilde{n}_0^i + \tilde{n}_1^i}.$$

### 8.3.6 Outgoing Scattered Fields

These *outgoing* scattered fields correspond to the sum of the fields scattered by each of the $p+1$ interfaces, expressed in the incident medium and the substrate.

*Scattered Field Within the Substrate*

The tangential component of this progressive field is expressed by

$$\vec{\mathcal{E}}_{s,d}^T(\vec{r}, z) = \int_{\vec{\nu}_d} \vec{\mathbb{A}}_{s,d}^{T+}(\vec{\nu}_d)\, e^{2i\pi \vec{\nu}_d \cdot \vec{r}} e^{i\alpha_s^d(z - z_p)}\, d^2\vec{\nu}_d \quad z \geqslant z_p, \qquad (8.200)$$

where

$$\vec{\mathbb{A}}_{s,d}^{T+}(\vec{\nu}_d) = \vec{\mathbb{E}}_{p+1,d}^T(\vec{\nu}_d, z_p) + \sum_{j=1}^{p} t_{jp}^d \vec{\mathbb{E}}_{j,d}^T(\vec{\nu}_d, z_{j-1}), \qquad (8.201)$$

and

$$t_{jp}^d = \frac{1}{\prod_{m=j}^{p} [\cos \delta_m^d - i(Y_{m,d}/\tilde{n}_m^d)\sin \delta_m^d]} \quad \text{avec} \quad \delta_m^d = \alpha_m^d d_m. \qquad (8.202)$$

Recall at this point the method used to calculate complex admittances, which are continuous between the scattering optical interface and the substrate:

$$Y_{p,d} = \tilde{n}_s^d \quad ; \quad Y_{j-1,d} = \frac{Y_{j,d}\cos \delta_j^d - i\tilde{n}_j^d \sin \delta_j^d}{\cos \delta_j^d - i(Y_{j,d}/\tilde{n}_j^d)\sin \delta_j^d} \quad j = 1, \ldots, p. \qquad (8.203)$$

*Scattered Field Within the Incident Medium*

The tangential component of this retrograde field is expressed by

$$\vec{\mathcal{E}}_{0,d}^T(\vec{r}, z) = \int_{\vec{\nu}_d} \vec{\mathbb{A}}_{0,d}^{T-}(\vec{\nu}_d)\, e^{2i\pi \vec{\nu}_d \cdot \vec{r}} e^{-i\alpha_0^d z}\, d^2\vec{\nu}_d \quad z \leqslant 0, \qquad (8.204)$$

where

$$\vec{\mathbb{A}}_{0,d}^{T-}(\vec{\nu}_d) = \vec{\mathbb{E}}_{0,d}^T(\vec{\nu}_d, 0) + \sum_{j=1}^{p} t_{j0,d}' \vec{\mathbb{E}}_{j,d}^T(\vec{\nu}_d, z_j), \qquad (8.205)$$

and

$$t_{j0,d}' = \frac{1}{\prod_{m=1}^{j} \left[\cos \delta_m^d + i(Y_{m-1,d}'/\tilde{n}_m^d)\sin \delta_m^d\right]}. \qquad (8.206)$$

Now recall the method used to calculate complex admittances, which are continuous between the scattering optical interface and the incident medium:

$$Y_{0,d}' = -\tilde{n}_0^d \quad ; \quad Y_{j,d}' = \frac{Y_{j-1,d}'\cos \delta_j^d + i\tilde{n}_j^d \sin \delta_j^d}{\cos \delta_j^d + i(Y_{j-1,d}'/\tilde{n}_j^d)\sin \delta_j^d} \quad j = 1, \ldots, p. \qquad (8.207)$$

### *8.3.7 Angle Resolved Scattering of a Multilayer*

In section 8.3.4 we saw that all fictitious currents are proportional to the Fourier transform $[\widehat{h}_j \star \widehat{s}_e]$ of the effective profile $h_e$. Therefore the results established in Sections 8.3.5 and 8.3.6 show that the tangential components of the elementary fields scattered by the stack in one of the two extreme media (incident medium or substrate) may be described by a general expression of the type

$$\vec{\mathbb{A}}_d^{T\pm} = \sum_{j=0}^{p} C_j^{\pm} [\widehat{h}_j \star \widehat{s}_e] \vec{\mathcal{E}}_i^T . \tag{8.208}$$

This expression is valid for each of the four polarization cases (SS, SP, PS and PP) of which only the quantities $[\widehat{h}_j \star \widehat{s}_e]$ are independent.

In particular, remember that the coefficients $C_j^{\pm}$ include not only the sources at the origins of scattering, but also the crossing of the layers between the extreme medium and the $j$th scattering interface. The $+$ superscript is reserved for the emerging scattered field in the substrate (the associated wave is progressive), while the $-$ superscript denotes the scattered field that emerges on the side of the incident medium (the associated wave is retrograde).

The scattering distributions embodied in each of the merging transparent media are, as in the case of the single interface, defined by equation (8.113), i.e.,

$$\text{ARS}^{\pm} = \left(\frac{n_{\pm}}{\lambda}\right)^2 \frac{\widetilde{n}_{\pm}^d}{\widetilde{n}_0^i} \frac{|\vec{\mathbb{A}}_d^{T\pm}|^2}{S|\vec{\mathcal{E}}_i^T|^2} \cos\theta_{\pm}^d , \tag{8.209}$$

where the various quantities identified with the $\pm$ indices use the same logic as that adopted for the scattered fields ($+$ for the substrate, $-$ for the incident medium).

Putting

$$D_j^{\pm} = \frac{n_{\pm}}{\lambda} C_j^{\pm} \sqrt{\frac{\widetilde{n}_{\pm}^d}{\widetilde{n}_0^i}} \cos\theta_{\pm}^d , \tag{8.210}$$

and using equation (8.208), we can write equation (8.209) in the following form:

$$\text{ARS}^{\pm} = \frac{1}{S} \left| \sum_{j=0}^{p} D_j^{\pm} [\widehat{h}_j \star \widehat{s}_e] \right|^2 . \tag{8.211}$$

Also, replacing the convolution $\widehat{h}_j \star \widehat{s}_e$ by the Fourier transform $\widehat{h}_{e,j}$ of the product $h_j s_e$ (see Section 8.2.11), and expanding, we get

$$\text{ARS}^{\pm} = \frac{1}{S} \sum_{j=0}^{p} |D_j^{\pm} \widehat{h}_{e,j}|^2 + \frac{1}{S} \sum_{j=0}^{p} \sum_{k \neq j} D_j^{\pm} [D_k^{\pm}]^* \, \widehat{h}_{e,j} \, \widehat{h}_{e,k}^* \tag{8.212}$$

$$= \text{ARS}_{\text{inc}}^{\pm} + \text{ARS}_{\text{coh}}^{\pm} \,.$$

The first term, $\text{ARS}_{\text{inc}}^{\pm}$, appearing in equation (8.212) is an always-positive quadratic term often referred to as the **incoherent scattering component**. This term generally increases with the number of layers in the stack.

Conversely, the second term, $\text{ARS}_{\text{coh}}^{\pm}$, often referred to as the **coherent scattering component**, can be either positive or negative and consequently can reduce or increase the resultant $\text{ARS}^{\pm}$ scattering in certain directions in space. This will be emphasized later on in the chapter.

It is important at this stage to stress that equation (8.212) shows that the topography of rough interfaces is decoupled from the optical characteristics of the problem (wavelength, thicknesses and refractive indices of the layers that make up the stack, angles of illumination and scattering, polarization states of the incident and scattered waves). We now turn our attention to those terms specifically related to the microstructure of the interfaces.

### 8.3.8 Correlations between Interfaces

#### General Observations

For the case of a single interface, Section 8.2.11 introduced the effective roughness spectrum of this interface using equation (8.120), namely

$$\gamma_e(\vec{\nu}) = \frac{1}{S} |\widehat{h}_e(\vec{\nu})|^2 \quad \text{where} \quad h_e(\vec{r}) = h(\vec{r}) \, s_e(\vec{r}) \,,$$

and we also showed that, when the divergence of the incident beam tends to 0, this spectrum tends to an asymptotic expression $\gamma(\vec{\nu})$ that is characteristic only of the surface topography.

Consequently, the incoherent scattering component can be written

$$\text{ARS}_{\text{inc}}^{\pm} = \frac{1}{S} \sum_{j=0}^{p} |D_j^{\pm} \widehat{h}_{e,j}|^2 = \sum_{j=0}^{p} |D_j^{\pm}|^2 \, \gamma_{e,j} \,, \tag{8.213}$$

where $\gamma_{e,j} = (1/S) |\widehat{h}_{e,j}|^2$ is the effective roughness spectrum of interface $j$.

Also, the coherent scattering component can be written in the form

$$\text{ARS}_{\text{coh}}^{\pm} = \frac{1}{S} \sum_{j=0}^{p} \sum_{k \neq j} D_j^{\pm} [D_k^{\pm}]^* \, \widehat{h}_{e,j} \, \widehat{h}_{e,k}^* = 2\Re \left\{ \sum_{j=0}^{p-1} \sum_{k>j} D_j^{\pm} [D_k^{\pm}]^* \, \gamma_{e,jk} \right\}, \quad (8.214)$$

provided we introduce the variable

$$\gamma_{e,jk} = \frac{1}{S} \widehat{h}_{e,j} \, \widehat{h}_{e,k}^* . \quad (8.215)$$

This quantity $\gamma_{e,jk}$ denotes the intercorrelation spectrum for interfaces $j$ and $k$, which is the Fourier transform of the function $\Gamma_{e,jk}$ giving the intercorrelation function between the effective topographies of these interfaces. We have

$$\Gamma_{e,jk} = \frac{1}{S} [h_{e,j} \star h'_{e,k}] \xrightarrow{\text{TF}} \frac{1}{S} \widehat{h}_{e,j} \, \widehat{h}_{e,k}^* , \quad (8.216)$$

where $h'_{e,k}(\vec{r}) = h_{e,k}(-\vec{r})$. This function $\Gamma_{e,jk}$ quantifies the similarity between the effective topographies of the two interfaces $j$ and $k$.

It is common at this stage to introduce a normalized correlation coefficient $\alpha_{e,jk}$ defined as

$$\gamma_{e,jk} = \alpha_{e,jk} \sqrt{\gamma_{e,j} \gamma_{e,k}} , \quad (8.217)$$

which can be expressed in detail as

$$\alpha_{e,jk} = \frac{\frac{1}{S} \widehat{h}_{e,j} \, \widehat{h}_{e,k}^*}{\sqrt{\frac{1}{S} |\widehat{h}_{e,j}|^2 \frac{1}{S} |\widehat{h}_{e,k}|^2}} = \frac{[\widehat{h}_j \star \widehat{s}_i] [\widehat{h}_k \star \widehat{s}_i]^*}{\sqrt{|\widehat{h}_j \star \widehat{s}_i|^2 |\widehat{h}_k \star \widehat{s}_i|^2}} . \quad (8.218)$$

Under these conditions, the scattered flux may finally be written as

$$\text{ARS}^{\pm} = \sum_{j=0}^{p} |D_j^{\pm}|^2 \gamma_{e,j} + 2\Re \left\{ \sum_{j=0}^{p-1} \sum_{k>j} D_j^{\pm} [D_k^{\pm}]^* \, \alpha_{e,jk} \sqrt{\gamma_{e,j} \gamma_{e,k}} \right\}. \quad (8.219)$$

Note at this step that the normalized correlation coefficients defined by equation equation (8.218) are unity in modulus, so that they introduce phase terms via complex exponentials. Hence one may wonder why low correlation values are introduced in the literature when analyzing experimental data. Actually this is due to an additional intensity integration performed by the receiver, as already discussed in Section 8.2.11.

Indeed though the impact of this averaging was said to be minor on the apparent roughness spectrum [see relation (8.145)], it may have significant effects on the cross-correlation values. To clarify this point, we again consider slight receiver apertures in which the scattering coefficients $D_j$ of multilayers do not vary. The integration of the ARS in equations (8.213) and (8.214) lead to replace all spectra and cross-correlation spectra by their integration over the receiver solid angle, that is, respectively:

$$\gamma_{e,j} \longrightarrow \gamma'_{e,j} = \int \gamma_{e,j} \, d\Omega \tag{8.220}$$

$$\gamma_{e,jk} \longrightarrow \gamma'_{e,jk} = \int \gamma_{e,jk} \, d\Omega \tag{8.221}$$

Then a cross-correlation coefficient must be defined and corrected in a way similar to equation (8.217):

$$\gamma'_{e,jk} = \alpha'_{e,jk} \sqrt{\gamma'_{e,j} \gamma'_{e,k}} \tag{8.222}$$

where

$$\alpha'_{e,jk} = \frac{\displaystyle\int \gamma_{e,jk} \, d\Omega}{\sqrt{\displaystyle\int \gamma_{e,j} \, d\Omega \int \gamma_{e,k} \, d\Omega}} \tag{8.223}$$

and

$$\int \gamma_{e,jk} \, d\Omega = \frac{1}{S} \int \widehat{h}_{e,j} \widehat{h}^*_{e,k} \, d\Omega \tag{8.224}$$

Eventually, due to the additional integration of complex functions which may be independent in equation (8.224), the modified cross-correlation $\alpha'_{e,jk}$ coefficient may take zero values. We keep in mind that this is a receiver effect; in other words, the modulus of cross-correlation is around unity at the speckle size, but its value decreases with the receiver aperture and may approach zero in case of independent functions $\widehat{h}_{e,j}$ and $\widehat{h}^*_{e,k}$.

To summarize, if the effective topographies of the various interfaces are totally uncorrelated, i.e., if all the correlation coefficients are 0, the angle resolved scattering reduces simply to the incoherent component:

$$\mathrm{ARS}^{\pm} = \mathrm{ARS}^{\pm}_{\mathrm{inc}} = \sum_{j=0}^{p} |D_j^{\pm}|^2 \gamma'_{e,j}, \tag{8.225}$$

while for totally correlated interfaces ($\alpha'_{e,jk} = 1$), the coherent component is added to the previous one in the expression for the angle resolved scattering, i.e.,

$$\alpha'_{e,jk} = 1 \ \forall j, k \quad \Rightarrow \quad \gamma'_{e,j} = \gamma'_{e,k} = \gamma'_e$$

$$\Rightarrow \quad \text{ARS}^\pm = \left( \sum_{j=0}^{p} |D_j^\pm|^2 + 2\Re \left\{ \sum_{j=0}^{p-1} \sum_{k>j} D_j^\pm [D_k^\pm]^* \right\} \right) \gamma'_e = D \gamma'_e .$$

$$(8.226)$$

Note that, rigorously, the correlated case corresponds to $|\alpha'_{e,jk}| = 1$, i.e., to $\alpha'_{e,jk} = e^{i\eta_{j,k}}$, which adds overall phase terms to the coherent component.

Finally, we can conclude that the correlations between interfaces play the role of mutual coherence between the scattering sources since these correlations allow (or otherwise) interference to occur.

### *The Influence of Technology on the Correlations between Interfaces*

From experience, the correlation coefficients are often close to one. This observation is related to the frequency window responsible for far-field scattering, which extends from $\sin\theta^d_{\min}/\lambda$ to $1/\lambda$ at normal illumination. Hence the spatial periods concerned go from $\lambda$ to $\lambda/\sin\theta^d_{\min}$, and are therefore generally much greater than the thickness of each layer, at least if scattering at very large angles ($\sin\theta^d$ close to 1) can be avoided.

Consequently, starting from a substrate roughness described by the profile $h_s = h_p$, the first thin layer deposited tends to reproduce this first topography. For greater generality, this replication can be described by a convolution, i.e.,

$$h_{p-1} = h_s \star b_{\text{mat},p} , \qquad (8.227)$$

where $b_{\text{mat},p}$ is a replication function characteristic of the material used to deposit the layer $p$. If there were only these replication effects, the topography of interface 0 would then be described by a relation of the form

$$h_0 = h_s \star b_{\text{mat},p} \star \cdots \star b_{\text{mat},1} \quad \Rightarrow \quad \widehat{h}_0 = \widehat{h}_s \prod_{k=1}^{p} \widehat{b}_{\text{mat},k} . \qquad (8.228)$$

Taking account of the observations we made on bandwidth, these replication effects are almost perfect, in such a way that, determined by the initial roughness of the substrate, we obtain a threshold effect for the roughness of each interface in the stack.

However, other effects can intervene with this replication and arise, in particular, from the intrinsic microstructure of the materials used in depositing the thin layers. To include this contribution, a phenomenological approach is appropriate, so for each interface we write

$$h_{k-1} = h_k \star b_{\text{mat},k} + g_k . \qquad (8.229)$$

This equation shows that, added to the replication effect, there is an additional topographical contribution associated with the material, often inaccurately referred to as the *grain of the material*. Finally, in the most general case, equation (8.229) can be used to describe the change in topography at each interface in the stack, given that there are normally only two materials to be deposited. Hence the mutual coherence of the scattering sources given in (8.218) can be related to the effects of replication and grain.

With modern highly energetic deposition technologies that give rise to very dense thin layers, there is a competition between the threshold effect (near-perfect replication of the substrate) and the effect of the materials intrinsic grain. Consequently, the effect of the material can be seen only if we start with extremely well polished substrates. If the threshold effect dominates, the correlation is close to one; on the other hand, the correlation is almost zero if the grain of the material dominates.

Finally, it is important to point out that at very low scattering levels the existence of localized defects must be taken into account. These can be on the substrate or they might appear during the course of manufacture; from a phenomenological point of view, they are sufficiently well described by the grain of the materials.

### 8.3.9 Application to a Few Simple Cases

We have at this point all the theoretical relations we need to calculate the angular and spectral dependence of the flux scattered by a multilayer stack. However, these theoretical expressions are relatively complex, and at first sight it is difficult to see the structure and properties of the scattering distributions associated with such and such a stack. Also, it seemed useful to detail these calculations in a few simple cases corresponding to some accessible basic functions already encountered in dedicated sections of Chapter 6, namely antireflective coatings, the Fabry–Perot filter and the quarter-wave mirror.

#### Antiscattering Coatings

We showed in Section 6.4.1 that a single quarter-wave layer (at wavelength $\lambda_0$ and whose refractive index is equal to the square root of that of the substrate) deposited onto a substrate caused, at normal incidence and at wavelength $\lambda_0$, the reflection coefficient of a wave impinging on that substrate to vanish. Hence such a layer was called an *anti-reflection* layer.

In the previous section we highlighted certain phenomena of interference between the waves scattered by the various interfaces within a multilayer stack, and we can make good use of these to reduce, even to zero, the resultant scattering at certain wavelengths or in certain directions in space. To achieve this, this interference must be destructive and the result will depend on the structure of the multilayer being considered. In our first example, we shall consider the simple case of a single layer and will seek, analytically, the corresponding conditions under which scattering vanishes. Hence we shall refer generically to such a layer as *antiscattering*.

Suppose that the topographies of the two interfaces comprising this layer (substrate and upper interface) are perfectly correlated, such as to guarantee full interference.

To make the analytical calculation slightly easier, we shall assume that illumination is at normal incidence ($\theta_0^i = 0$) and we will restrict ourselves to calculating the back-scattered field ($\theta_0^d = \theta_0^i = 0$) and in the plane of incidence ($\phi_d = 0$). This will allow us to address both polarizations (SS and PP) in the same way. Besides, this will also be a good choice if we include the expression in $D\gamma$ [see equation (8.226)] for the scattered intensity: since coefficient $D$ has a slowly varying angular dependence, it is the roughness spectrum $\gamma$ common to both interfaces which most often imposes a maximum value at small angles, and a rapid decay at large angles. Consequently, if we succeed in getting the coefficient $D$ to vanish at $\theta_0^d = 0$, it is probable that scattering will remain very weak over the entire angular domain as a result of the decay of the spectrum.

Consider equation (8.230) that describes the general case of the amplitude of the tangential component of the scattered field in the incident medium, and apply it to the particular case of a single layer:

$$\vec{A}_{0,d}^{T-}(\vec{\nu}_d) = \vec{\mathbb{E}}_{0,d}^{T}(\vec{\nu}_d, z_0) + t'_{10,d}\vec{\mathbb{E}}_{1,d}^{T}(\vec{\nu}_d, z_1). \qquad (8.230)$$

We recognize here the sum of the fields back-scattered by interfaces 0 and 1 in the thin layer. The retrograde transmission coefficient $t'_{10,d}$ can be calculated using equation (8.206), i.e.,

$$t'_{10,d} = \frac{1}{\cos\delta_1^d + i(Y'_{0,d}/\tilde{n}_1^d)\sin\delta_1^d} = \frac{1}{\cos\delta_1^d - i(\tilde{n}_0^d/\tilde{n}_1^d)\sin\delta_1^d}. \qquad (8.231)$$

Now we express the fields $\vec{\mathbb{E}}_{j,d}^{T}$ using equation (8.198), taking account of the particular values adopted for the angles of incidence and scattering; this leads to the following equalities (still for nonmagnetic media):

$$\alpha_j^i = \alpha_j^d = k_j = \frac{2\pi}{\lambda} n_j = k_v n_j \quad ; \quad \tilde{n}_j^i = \tilde{n}_j^d = n_j/\eta_v \, .$$

Hence

$$\vec{\mathbb{E}}_{0,d}^T(\vec{0}, z_0) = -\frac{i \, t_{00}^i}{Y_{0,d} - Y_{0,d}'} (k_1 n_1/\eta_v - k_0 n_0/\eta_v) \, [\widehat{h}_0 \star \widehat{s}_e]_{\vec{0}} \, \vec{\mathcal{E}}_i^T \, , \quad (8.232)$$

$$\vec{\mathbb{E}}_{1,d}^T(\vec{0}_d, z_1) = -\frac{i \, t_{01}^i}{Y_{1,d} - Y_{1,d}'} (k_s n_s/\eta_v - k_1 n_1/\eta_v) \, [\widehat{h}_1 \star \widehat{s}_e]_{\vec{0}} \, \vec{\mathcal{E}}_i^T \, . \quad (8.233)$$

The scattered component in the incident medium $\vec{\mathbb{A}}_{0,d}^{T-}$ therefore vanishes when

$$\vec{\mathbb{E}}_{0,d}^T(\vec{0}, z_0) + t_{10,d}' \vec{\mathbb{E}}_{1,d}^T(\vec{0}, z_1) = \vec{0} \, ; \quad (8.234)$$

i.e.,

$$\frac{n_1^2 - n_0^2}{Y_0 - Y_0'} t_{00} \, [\widehat{h}_0 \star \widehat{s}_e]_{\vec{0}} + \frac{n_s^2 - n_1^2}{Y_1 - Y_1'} t_{01} \, t_{10}' \, [\widehat{h}_1 \star \widehat{s}_e]_{\vec{0}} = 0 \, . \quad (8.235)$$

We need to express the values taken by the admittances that appear in the equation above; to do this, we need to specify the optical and geometric characteristics of the layer. Consider the quarter-wave layer at the illuminating wavelength, which amounts to saying that the phase shift $\delta_1$ introduced by this layer is an odd multiple of $\pi/2$.

We have therefore

$$Y_1 = \tilde{n}_s = n_s/\eta_v \quad \Rightarrow \quad Y_0 = \frac{Y_1 \cos\delta_1 - i\tilde{n}_1 \sin\delta_1}{\cos\delta_1 - i(Y_1/\tilde{n}_1)\sin\delta_1} = \frac{n_1^2}{n_s \eta_v}$$

$$Y_0' = -\tilde{n}_0 = -n_0/\eta_v \quad \Rightarrow \quad Y_1' = \frac{Y_0' \cos\delta_1 + i\tilde{n}_1 \sin\delta_1}{\cos\delta_1 + i(Y_0'/\tilde{n}_1)\sin\delta_1} = -\frac{n_1^2}{n_0 \eta_v}$$

$$t_{00} = 1 + r_0 = \frac{2\tilde{n}_0}{\tilde{n}_0 + Y_0} = \frac{2 n_0 n_s}{n_0 n_s + n_1^2} \quad (8.236)$$

$$t_{01} = \frac{1 + r_0}{\cos\delta_1 - i(Y_1/\tilde{n}_1)\sin\delta_1} = \frac{2 i n_0 n_1}{n_0 n_s + n_1^2}$$

$$t_{10}' = i\frac{n_1}{n_0} \, .$$

Substituting all these results into (8.235) we finally get

$$n_s^2(n_1^2 - n_0^2) \, [\widehat{h}_0 \star \widehat{s}_i]_{\vec{0}} - n_1^2(n_s^2 - n_1^2) \, [\widehat{h}_1 \star \widehat{s}_e]_{\vec{0}} = 0 \, . \quad (8.237)$$

In the case of identical effective roughnesses ($[\widehat{h}_0 \star \widehat{s}_e]_{\vec{0}} = [\widehat{h}_1 \star \widehat{s}_e]_{\vec{0}}$), the condition for scattering to vanish becomes

$$n_s^2(n_1^2 - n_0^2) - n_1^2(n_s^2 - n_1^2) = 0 \quad \Rightarrow \quad n_1 = \sqrt{n_0 n_s} \quad (8.238)$$

and is therefore identical to that of the antireflection layer (see section 6.4.1).

This result shows that the antireflection layer is also antiscattering, at least in the direction of back-scatter. It now remains for the scattering behavior over the whole angular domain to be analyzed numerically. In SS configuration, this behavior is illustrated in Figure 8.8. Observe that scattering is effectively zero in the backscattering direction, and considerably reduced over the entire reflected half-space (0 degrees $< \theta <$ 90 degrees), at least when compared to scattering from the bare substrate. Finally, this result should be compared with that obtained for identical, but uncorrelated, roughnesses (see Figure 8.8), which shows much higher scattering levels. In this Figure, calculation is performed at wavelength $\lambda_0 = 700$ nm for SS polarization, with refractive indices $n_0 = 1$, $n_s = 1.50$, $n_1 = \sqrt{n_0 n_s} = 1.225$, and $n_1 d_1 = \lambda_0/4$. The roughness spectrum is that used in Section 8.2.11.

More detail is given on the antiscattering process in Figure 8.9. In this figure we have plotted the ARS level in the case of a constant roughness spectrum (equal to unity), which in fact gives the angular variations of the scattering coefficient $D$. The exact roughness spectrum is given in the same figure. These results show how the scattering coefficient reduces ARS at low angles, while the spectrum reduces ARS at high angles.

Finally, if we depart from the antireflection condition, the backscatter is no longer zero but the roughness ratio $(\delta_0/\delta_1)$ could be calculated so as to recover the zero-scattering condition; however such roughness ratio would be very difficult to control experimentally. On the other hand, given that we are interested in reducing (rather than canceling) scattering as compared with the substrate, such a reduction is nevertheless allowable over a significant range of roughness values.

This result is important, since it provides a simple way to improve the performance of light absorbers. One way to improve the performance of a specular absorber at one wavelength consists in covering an opaque polished glass with a quarter-wave low-index layer; at the same time, reflection and scattering are reduced by more than a factor of 10 (depending on the layer refractive index). A second type (not specular) of absorber uses black paint, a substrate whose reflection is totally diffuse with quasi-lambertian scattering carrying around 4% of the incident light; if this paint is covered with a low refractive index quarter-wave layer, the antiscattering phenomenon appears again and reduces the scattered flux to a value lower than one percent, causing the absorption to exceed 0.99. Although this change in absorption may appear insignificant, the visual effect, on the other hand, is especially striking since it results in a strong reduction in scattering. Note that these coatings of black paint are very rough (their scattering distributions are quasi-Lambertian) and are not covered by the first-order theory described in

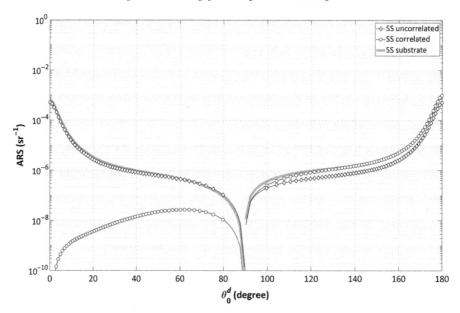

Figure 8.8 Angle resolved scattering from an antiscattering layer in SS polarization (open circles, correlated interfaces – open diamonds, uncorrelated interfaces – gray line, bare substrate) [see text for more details].

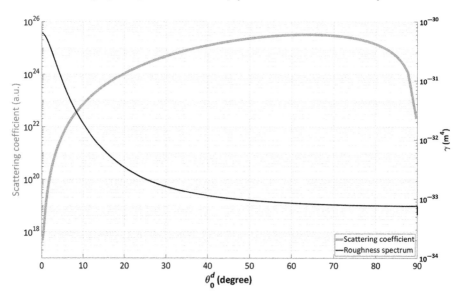

Figure 8.9 Angular variations of scattering coefficient $D$ and roughness spectrum $\gamma$, for the antireflective layer of Figure 8.8. The two surfaces are fully correlated [See text for more details].

this chapter. Despite that, however, the shallow roughness slopes allow this antiscattering effect to be maintained; this is consistently observed (using an exact theory and validation by experiment). Finally, note that in both cases absorption takes place within the volume of the substrate; in order to miniaturize broad-band light absorbers, metal-dielectric stacks need to be used and will confine absorption within their bulk.

Before concluding this section it would be helpful to comment on maintaining the antiscattering effect when the ideal refractive index material (here $n_1 = 1.225$) is not available. For that we use a double layer antireflective coating made with current materials. We therefore consider the two analytical solutions highlighted in Section 6.4.2, characterized by the following formulas:

- Solution 1 : Air / 0.672L 1.645H / Silica
- Solution 2 : Air / 1.328L 0.355H / Silica

for a wavelength of 700 nm and a material pair whose high and low refractive indices at this wavelength are 2.130 ($n_H$) and 1.489 ($n_L$) respectively.

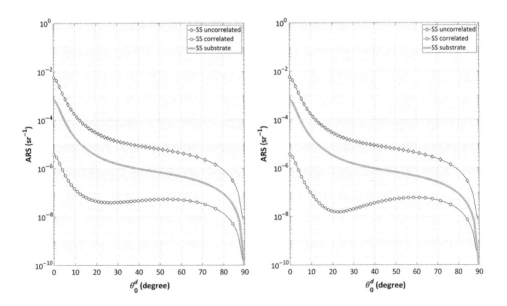

Figure 8.10 Angle resolved scattering (reflected half-space) for two different bilayer antireflective coatings in SS polarization (open circles, correlated interfaces – open diamonds, uncorrelated interfaces – gray line, bare substrate) [see text for more details].

Figure 8.10 shows the scattering distributions corresponding to these two double-layer antireflection coating configurations with the first (resp. second) solution on the left (resp. right) graph. In both cases we observe a strong reduction of scattering in the reflected half-space, and this allows to think of broadband antiscattering, antireflective multilayers.

### *Fabry–Perot Filter*

We have already seen the advantage in using deposition technologies that replicate the topology of the interfaces, since this guarantees a strong mutual coherence (i.e., unity correlation) between scattering sources. Such coherence is required to achieve full interference, but it must be destructive in nature in order to lead to a reduction in scattering levels.

There are other stacking arrangements that lead to effects similar to those highlighted in the antiscattering case: the Fabry–Perot configuration (see Chapter 6) is an especially important example.

Consider a Fabry–Perot filter defined by the formula HLHLH 6L HLHLH, centered on wavelength $\lambda_0 = 700$ nm, this being also the illuminating wavelength. The materials of upper and lower refractive index are those of the previous section. It will help if we start by assuming that the substrate and superstrate are identical (air), which makes this Fabry–Perot structure perfectly symmetrical. We shall restrict ourselves to SS polarization in the incidence plane and illumination at normal incidence. Figure 8.11 shows the

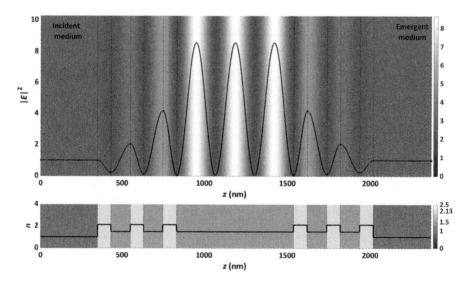

Figure 8.11 Distribution of the stationary field within the volume of a symmetrical Fabry–Perot filter arranged as Air / HLHLH 6L HLHLH / Air

distribution of the stationary field within the filter. Recall that this field is responsible for exciting the scattering sources.

We therefore observe four major scattering sources on this figure at the LH interfaces (in addition to the extreme surfaces), the field being, conversely, virtually zero at the HL interfaces. However, it can be shown analytically that the fields scattered back into the air by the two scattering surfaces to the left of the spacer are opposite in phase (though with the same modulus) to the fields radiated into the air by the two other surfaces to the right of the spacer. Consequently, the field back-scattered by the Fabry–Perot filter vanishes.

This result is illustrated in Figures 8.12 and 8.13, where we considered the case of correlated and uncorrelated interfaces. There is zero backscatter in Figure 8.12, but, in contrast to the antiscatter monolayer, we observe that scattering is relatively pronounced as we move away from the back-scattering direction; this result is related to high values of the correlated scattering coefficient $D$, which are no longer fully compensated by the decay of the roughness spectrum.

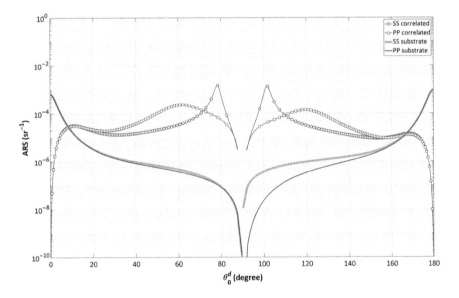

Figure 8.12 Angle resolved scattering of a symmetrical Fabry–Perot filter for correlated ($\alpha_{jk} = 1$) interfaces.

We also notice in these figures the difference in the polarization behaviors of the ARS curves of a Fabry–Perot filter. In particular large scattering rings with narrow widths appear at large angles for PP polarization; such rings are often responsible for cross-talk limitation in optical communication systems.

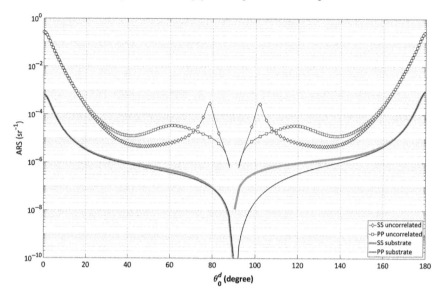

Figure 8.13 Angle resolved scattering of a symmetrical Fabry–Perot filter for uncorrelated ($\alpha_{jk} = 0$) interfaces.

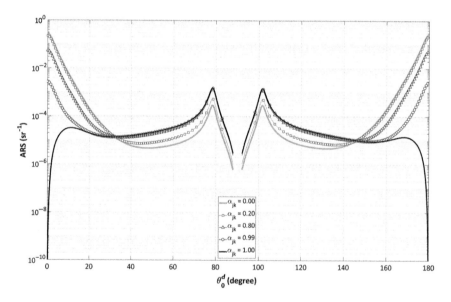

Figure 8.14 Sensitivity of angle resolved scattering for a symmetrical Fabry–Perot filter versus the cross-correlations within the stack (PP polarization).

This comparison of the scattering distributions obtained in the two extreme correlation cases (0 and 1) shows that, for these filters, the interfaces should ideally be correlated. However, these results must be interpreted with care, given the great sensitivity of scattering to the correlations between interfaces, a sensitivity that grows with the number of sources. This results from an unstable interference equilibrium, as shown in Figure 8.14 for PP polarization. This figure shows that the scattering level increases very rapidly with decreasing correlation. In particular, a correlation of 0.99 gives rise to an angle resolved scattering closer to the uncorrelated than to the correlated case.

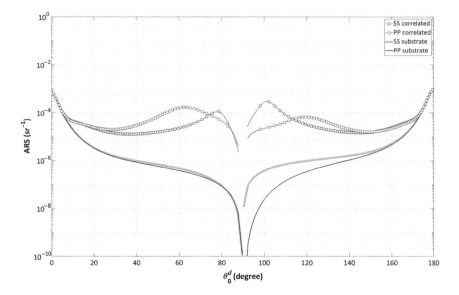

Figure 8.15 Angle resolved scattering for a Fabry–Perot filter configured as Air / HLHLH 6L HLHLH / Glass, obtained for perfectly correlated $(\alpha_{jk} = 1)$ interfaces.

Finally, Figures 8.15 and 8.16 show the scattering distributions for the same Fabry–Perot filter, but in the common case where superstrate and substrate are air and glass respectively. These results are given in the plane of incidence for polarization states SS and PP for both extreme correlation cases. The simple asymmetry due to the difference between the substrate and superstrate materials is sufficient to break the interference equilibrium and lose the ability for the back-scatter to vanish in the correlated case. Notice, however, that for correlated surfaces the backscattering of the filter is identical to that of the substrate, which makes a major difference with the uncorrelated case.

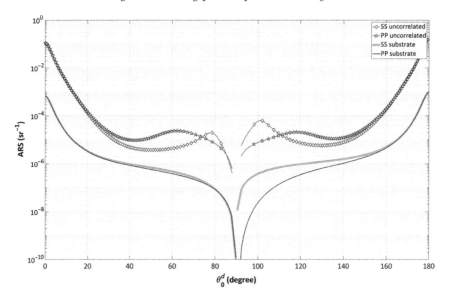

Figure 8.16 Angle Resolved Scattering for a Fabry–Perot filter configured as Air / HLHLH 6L HLHLH / Glass, obtained for perfectly uncorrelated $(\alpha_{jk} = 0)$ interfaces.

### *The Quarter-Wave Mirror*

The quarter-wave mirror is different in the sense that the scattered waves, in contrast to the previous cases, are in phase in the extreme media. Interference is therefore constructive, and the scattered flux will increase with increasing correlation: there is no antiscatter effect.

Consider a mirror defined by $(HL)^4$ H centered on $\lambda_0 = 700$ nm for normal illumination, $\lambda_0$ being also the illuminating wavelength. Figure 8.17 shows that the stationary field essentially excites the first odd interfaces on the air side. This result explains why scattering in transmission will be lower in the case of the mirror, at least at low scattering angles ($\theta^d$ close to 180 degrees); recall (see Chapter 6) that the stationary field decreases as the ratio of the refractive indices. Figures 8.18 and 8.19 show the scattering distributions for SS and PP polarizations in the plane of incidence for the two extreme correlations.

More generally, note finally that for all stacks (with identical interface spectra) scattering increases approximately with the square of roughness; this means that if scattering is modified by some ratio $\beta$ over the whole angular domain (which is equivalent to multiplying the roughness spectrum by the same value), then the roughness is modified by a factor $\sqrt{\beta}$. Albeit approximately, this last observation enables, from the scattering data of

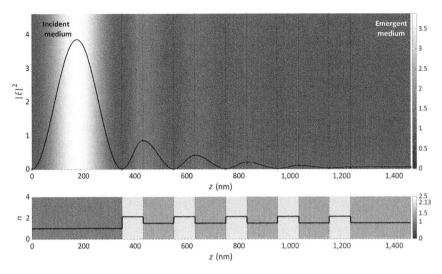

Figure 8.17 Distribution of the stationary field within the volume of an Air / $(HL)^4H$ / Glass quarter-wave mirror.

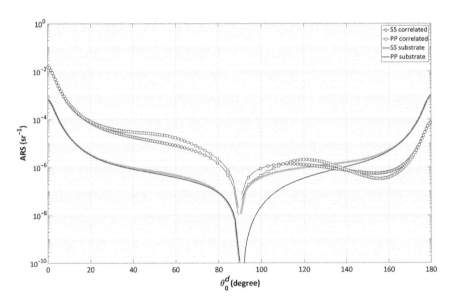

Figure 8.18 Angle resolved scattering calculated for a quarter-wave mirror in the case of perfect correlation between interfaces $(\alpha_{jk} = 1)$.

this section, the scattering level of the stacks to be quickly predicted as a function of its roughness (in the extreme cases of correlation). Note also that an attempt can be made to reduce scattering by displacing the field maxima

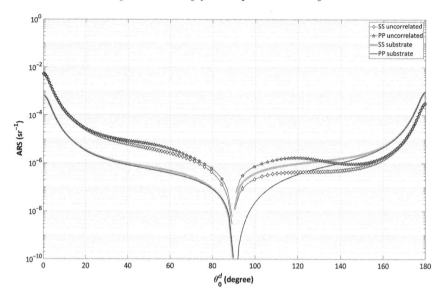

Figure 8.19 Angle resolved scattering calculated for a quarter-wave mirror in the case of no correlation between interfaces ($\alpha_{jk} = 0$).

within the volume of the mirror, though this technique has not provided convincing results.

### Sensitivity of Scattering to Cross-Correlations

A practical way of studying the sensitivity to the correlation mechanism involves using a stack constructed uniquely of half-wave layers in alternating high and low refractive indices. This is no longer about causing the retro-scattering to vanish (this would be the case for an air substrate), but ensuring that the scattering from the multilayer is identical to that from the substrate, at least in the backscatter direction. As the number of interfaces increases, the only means of achieving such an effect is by bringing destructive interference into play.

This occurs for half-wave stacks at $\lambda_0$, provided the interfaces are identical; for these so-called *absentee-layer stacks* for a wavelength of $\lambda_0$, any deviation with respect to scattering from the substrate will indicate a lack of correlation between interfaces. Hence this technique can be used to test the ability of the depositing technologies to replicate surfaces.

Note also that another way of testing the correlation is by oblique incidence deposition on the substrate. We know that under such deposition conditions, the layer also grows at another oblique incidence such that, starting from a substrate surface $h_s(\vec{r}) = h_p(\vec{r})$, the first interface can be written as $h_{p-1}(\vec{r}) = h_s(\vec{r} - \vec{a})$, where $\vec{a}$ is the lateral displacement resulting from the oblique growth of the material. The consequence is an additional phase term in the profiles Fourier transform, i.e.,

$$\widehat{h}_{p-1}(\vec{\nu}) = e^{2i\pi\vec{\nu}.\vec{a}}\,\widehat{h}_s(\vec{\nu})\,.$$

This result, obtained for a perfect replication, leads to a scattering anisotropy (a loss of the rotational symmetry otherwise characteristic of an isotropic surface illuminated by unpolarized light at normal illumination) that is easy to detect by turning the sample about its normal while measuring the scattered radiation. The phase term can also be used to alter or even

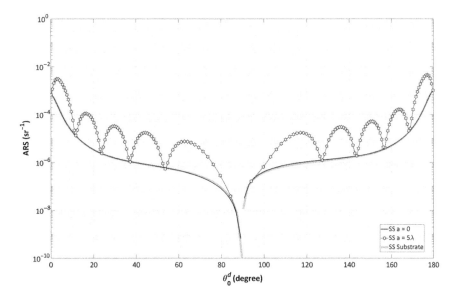

Figure 8.20 Correlated SS angular scattering from a single half-wave high-index layer on glass at $\lambda_0 = 700$ nm. The surface lateral shifts are parallel to the $x$ direction with modulus $a = 0$ and $a = 5\lambda_0$ (see text for more details).

suppress scattering in one or several directions (see Figure 8.20), by adjusting a phase match, for example. However, although this phenomenon might be of interest in some situations, its scope remains a priori limited to the extent that deposition under oblique incidence corresponds to an extremely specific situation.

To conclude, since destructive interferences of scattering are validated by experiment with multilayers, one could think of designing stacks in a way that reduces the scattering. In the case of broad-band applications, this raises a complex problem so that one is usually content with the comparison of the scattering curves from the different designs holding similar optical properties. On the other hand, for an application devoted to a single working wavelength, additional (absentee) layers may be introduced in the design so as to create extra scattering sources able to reduce scattering though the process of destructive interferences. However correlation still should be perfect with a disadvantageous sensitivity to scattering. Furthermore, this solution should be compared to that involving the former (classical) stacks deposited on superpolished samples leading to partial correlation; indeed, due to the intrinsic material microstructure, one can expect that "the lower the substrate roughness, the lower the correlation."

### Spectral Variations in Scattering

In this section dedicated to examples, we shall conclude by drawing attention to the fact that the spectral behavior of scattering also depends on the correlations between interfaces. This result is illustrated in Figure 8.21, where it can be seen that the reflection backscatter is in phase (or respectively out of phase) with the reflection coefficient for correlated (or respectively uncorrelated) interfaces. Hence this result provides an additional technique to

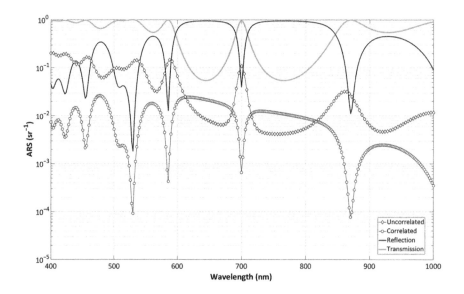

Figure 8.21 Wavelength variations of scattering for a Fabry–Perot filter at the specular scattering direction 0 degrees (black open circles, correlated interfaces – black open diamonds, uncorrelated interfaces – black line, reflectance – gray line, transmittance).

study these correlations. Note that this result is also true for total scattering (integrated over all space) to the extent that, for most stacks, scattering over small angles predominates in the integrated scattering value.

### 8.3.10 Approximate Formulae for Angle Resolved Scattering

Actually most scattering phenomena emphasized in the previous section are related to a specific property of the *specular* scattering ($\vec{v}_d = \vec{v}_i = \vec{0}$) by reflection, which is proportional to the reflectance of the stack, provided that all interfaces are fully correlated. Such property explains why specular scattering vanishes with reflectance $R$, which was the case of both the single-layer antireflective coating and the symmetrical Fabry–Perot filter (for which $R = 0$ at the design wavelength). It also explains why specular scattering is not modified by the absentee layer stack (reflection is identical to the substrate one), and why quarter-wave mirrors show no antiscattering effect (due to high reflection). In other words, destructive and constructive interferences in the correlated specular scattering by reflection are governed by the reflection factor.

### Case of a Single Surface

This is a general result which is easy to proof for a single scattering surface at normal illumination. Indeed consider the ARS formulas (8.115) and (8.116) given in Section 8.2.10, and develop them at the specular direction ($\theta_0^i = \theta_0^d = 0$). We obtain for both SS and PP polarizations in the incidence plane:

$$\text{ARS}_0^{SS}(0) = \text{ARS}_0^{PP}(0) = 16\pi^2 \left(\frac{n_0}{\lambda}\right)^4 R(0)\,\gamma_e(0)\,, \qquad (8.239)$$

where $\gamma_e$ is the effective roughness spectrum of the interface, and $R(0)$ the reflectance at 0 degree. Hence taking into account that, in the expression of the ARS curve of a single surface, the roughness spectrum shows large variations in comparison to the other scattering coefficients, the angular scattering by reflection can be approximated for normal illumination as

$$\text{ARS}_0^{SS}(\theta_0^d) = \text{ARS}_0^{PP}(\theta_0^d) = 16\pi^2 \left(\frac{n_0}{\lambda}\right)^4 R(0)\,\gamma_e(\theta_0^d)\,. \qquad (8.240)$$

The accuracy of this approximation can be evaluated in Figure 8.22, where the two ARS curves [the exact one given in the previous sections, and the approximate one defined by (8.240)] are compared. We observe a good agreement for both polarizations at low angles (below 40 degrees), but as expected this approximation fails at higher angles. Note that we did not consider the case of oblique illumination, since high polarization effects may occur in the

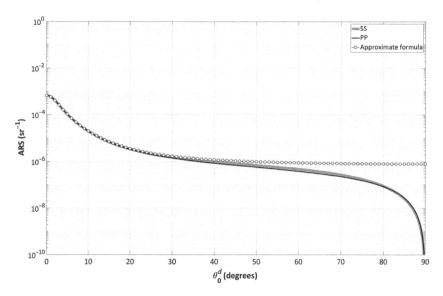

Figure 8.22 Comparison of the exact and approximate formulae for ARS ($\lambda_0 = 700$ nm, $\theta_0^i = 0$, $n_s = 1.5$, $n_0 = 1$. The roughness spectrum is that of Figure 8.6.)

angular range (see the scattering cancellation at the pseudo-Brewster angle in Figure 8.6).

### From ARS to TIS Formula

It is also useful to show how this last approximation (8.240) allows to recover the well-known Total Integrated Scattering (TIS) formula. For that the surface roughness is assumed to be isotropic, so that the spectrum is a radial function and the S or P (with S = SS+SP and P = PP+PS) polarized scattering has the symmetry of revolution at normal illumination. Under these conditions the ARS can be written with its polar ($\phi^d$) and normal ($\theta_0^d$) dependence as [see equations (8.135) and (8.136)]:

$$\mathrm{ARS}_0^{SS}(\theta_0^d, \phi^d) = \mathrm{ARS}_0^{PP}(\theta_0^d, \phi^d) = 16\pi^2 \left(\frac{n_0}{\lambda}\right)^4 R(\theta_0^i = 0)\, \gamma_e(\theta_0^d) \cos^2 \phi^d. \quad (8.241)$$

The SP and PS polarizations give similar formulas except that the $\cos^2 \phi^d$ polar dependence must be replaced by $\sin^2 \phi^d$, that is

$$\mathrm{ARS}_0^{SP}(\theta_0^d, \phi^d) = \mathrm{ARS}_0^{PS}(\theta_0^d, \phi^d) = 16\pi^2 \left(\frac{n_0}{\lambda}\right)^4 R(\theta_0^i = 0)\, \gamma_e(\theta_0^d) \sin^2 \phi^d. \quad (8.242)$$

Now let us integrate the ARS given for SS or PP polarization. With the solid angle $d\Omega = \sin \theta_0^d \, d\theta_0^d d\phi^d$, we obtain

$$\iint\limits_{\theta_0^d,\phi^d} \mathrm{ARS}_0^{SS}(\theta_0^d,\phi^d)\sin\theta_0^d\,d\theta_0^d d\phi^d = \iint\limits_{\theta_0^d,\phi^d} \mathrm{ARS}_0^{PP}(\theta_0^d,\phi^d)\sin\theta_0^d\,d\theta_0^d d\phi^d$$

$$= 16\pi^3\left(\frac{n_0}{\lambda}\right)^4 R(\theta_0^i=0)\int_{\theta_0^d}\gamma_e(\theta_0^d)\sin\theta_0^d\,d\theta_0^d\,.$$

$$(8.243)$$

It remains to compare the roughness integral to the last integral of (8.243). Following (8.127) [see Section 8.2.11] we have

$$\delta^2 = 2\pi\left(\frac{n_0}{\lambda}\right)^2\int_{\theta_0^d}\gamma(\theta_0^d)\cos\theta_0^d\sin\theta_0^d d\theta_0^d\,. \qquad (8.244)$$

Provided that the surface correlation length is enough large, the weight of roughness is predominant at low angles and this allows to write $\cos\theta_0^d \approx 1$ in (8.244). Consequently,

$$\int_{\theta_0^d}\gamma_e(\theta_0^d)\sin\theta_0^d\,d\theta_0^d \approx \frac{\delta^2}{2\pi}\left(\frac{\lambda}{n_0}\right)^2. \qquad (8.245)$$

Following (8.243) and (8.245), we obtain

$$\iint\limits_{\theta_0^d,\phi^d} \mathrm{ARS}_0^{SS}(\theta_0^d,\phi^d)\sin\theta_0^d\,d\theta_0^d d\phi^d = 8R\left(\frac{\pi n_0\delta}{\lambda}\right)^2. \qquad (8.246)$$

Eventually the addition of SS and SP polarizations yields the total scattering emitted with an unpolarized incident beam on the surface:

$$\left[\frac{\Phi_d^-}{\Phi_0^+}\right]_S = \left(\frac{4\pi n_0\delta}{\lambda}\right)^2 R\,. \qquad (8.247)$$

This last equation is that usually given for the TIS formula, and holds for both S- and P-polarized illumination at normal incidence. Such formula recalls why metallic surfaces approximately scatter 20 times more than a glass surface in the visible.

### The Case of Multilayers

The situation is different with multilayers, in the sense that the scattering coefficients may strongly vary with direction. However, a similar property holds for the specular scattering of correlated surfaces, a property which we now analyze numerically.

In Figures 8.23 and 8.24 we compare the exact angular scattering to the approximate scattering given in (8.241), except that the reflection factor is that of a multilayer.

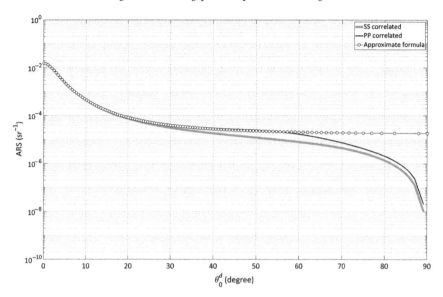

Figure 8.23 Comparison of the exact and approximate formulae for ARS of Air / (HL)$^4$H / Glass quarter-wave mirror ($\lambda_0 = 700$ nm, $\theta_0^i = 0$, perfectly correlated interfaces. The roughness spectrum is that of Figure 8.6.)

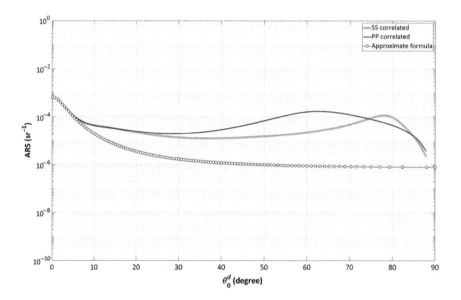

Figure 8.24 Comparison of the exact and approximate formulae for ARS of Air / HLHLH 6L HLHLH / Glass Fabry–Perot ($\lambda_0 = 700$ nm, $\theta_0^i = 0$, perfectly correlated interfaces. The roughness spectrum is that of Figure 8.6.)

Correlation between surfaces is assumed to be perfect. We observe that the approximation works at low angles (<25 degrees for the mirror, <5 degrees for the Fabry–Perot), and that strong differences occur at higher angles (see the Fabry-Perot case). Hence we conclude that these approximate formulae only work for low-angle scattering of correlated multilayers under normal illumination, and that they cannot be used for high-precision optics. Notice however that these approximations explain why the correlated specular scattering is in phase with reflection.

### 8.3.11 Effects Related to the Rear Face of the Substrate

The formulas established in Section 8.3.7 were obtained under the assumption that the substrate was considered to be a semi-infinite medium. However, in most cases this substrate is of finite thickness and we must therefore include the presence of the rear face, which is generally reflective and may also be capable of scattering.

The problem to which this corresponds is similar to the problem we dealt with in Chapter 6 in the case of the transmission and reflection properties of a stack of optical thin layers deposited on the surface of a substrate. We showed that in this case an incoherent summation of the different contributions generated by the rear face was able to correctly describe the new situation thus created.

In the case of transmission, this incoherent summation gives the same result as that we can obtain by gradually increasing the spectral bandwidth of the source; indeed, this increase is the right way to describe the cancellation of the coherent effects associated with the large thickness of the substrate. Strictly speaking, there is no evidence that these two approaches lead to the same result in the case of scattering (actually, numerical calculation shows that this is not perfectly the case for complex filters with high number of layers), but we assume that this incoherent summation can be used to obtain a realistic description of the scattering distributions. Moreover, we will consider the more general case, i.e., that for which this rear face also carries a multilayer treatment. Indeed if the filter function is particularly complex, it is often desirable to distribute it over both faces of the substrate. The model adopted can obviously be applied to simpler cases where only the front face of the substrate has been treated, or to the canonical situation where neither face of the substrate has any treatment at all.

We shall deal with this problem tackling in turn the effects associated with the rear face:

- first of all by considering only the simple effect of direct transmission of scattering from the upper face through the rear (smooth) face;
- then including the effect of all the multiple reflections that develop within the component if its rear (coated) face is assumed to be perfectly smooth (no roughness);
- and finally, by taking into account the scattering from the (rough and coated) rear face.

### Direct Transmission via the Rear Face

The situation we shall analyze now is illustrated in Figure 8.25. Light emerges from the rear face of the substrate and propagates in the semi-infinite medium whose refractive index is denoted $n_a$. In practice, it is clear that this refractive index is identical to that of the incident medium, and corresponds to air $(n_0 = n_a = 1)$.

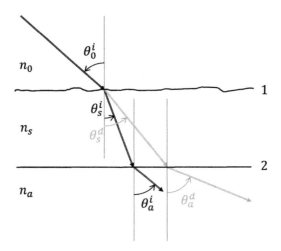

Figure 8.25 Diagram showing simple transmission from the rear face of the substrate (specular beams in black – scattered beams in gray).

The amplitude of the tangential component scattered by the front face and transmitted from the rear face is expressed as

$$\mathbb{A}_{a,d}^{T+} = t_{2,d}\mathbb{A}_{s,d}^{T+} , \qquad (8.248)$$

where $t_{2,d}$ is the transmission amplitude of the filter deposited on the rear face of the component. Furthermore, using the notation introduced in Section 8.3.7, we have

$$\vec{\mathbb{A}}_{s,d}^{T+} = \sum_{j=0}^{p_1} C_{1,j}^{+}[\widehat{h}_{1,j} \star \hat{s}_e] \vec{\mathcal{E}}_i^T , \qquad (8.249)$$

where the subscript 1 has been introduced to identify the multilayer stack concerned.

Applying equation (8.109) established in Section 8.2.9, the spectral density of the scattered flux in medium $a$ per unit solid angle is given by

$$\frac{d\Phi_{1,a}^d}{d\Omega_a} = \frac{1}{2}\left(\frac{n_a}{\lambda}\right)^2 \tilde{n}_a^d \cos\theta_a^d |t_{2,d}|^2 \left|\sum_{j=0}^{p_1} C_{1,j}^+ [\hat{h}_{1,j} \star \hat{s}_e] \vec{\mathcal{E}}_i^T\right|^2 . \tag{8.250}$$

Dividing this result by the incident flux, we deduce an expression for the scattering distribution embodied in medium $a$, i.e.,

$$\text{ARS}_{1,a} = \left(\frac{n_a}{\lambda}\right)^2 \frac{\tilde{n}_a^d}{\tilde{n}_0^i} \cos\theta_a^d |t_{2,d}|^2 \frac{1}{S} \left|\sum_{j=0}^{p_1} C_{1,j}^+ [\hat{h}_{1,j} \star \hat{s}_e]\right|^2 . \tag{8.251}$$

In arriving at the result (8.251), it would also have been possible to first calculate the expression describing the spectral density of the flux scattered per unit solid angle within the substrate (medium $s$), and then deduce the spectral density corresponding to the scattering distribution in medium $a$ after having crossed the rear face; simultaneously taking account of its energy transmission factor and any alteration in the value of the solid angle as a result of refraction, namely

$$\text{ARS}_{1,a} = \frac{1}{\Phi_0^i} \frac{d\Phi_{1,a}^d}{d\Omega_a} = \frac{1}{\Phi_0^i} T_{2,d} \frac{d\Phi_{1,s}^d}{d\Omega_s} \frac{d\Omega_s}{d\Omega_a}, \tag{8.252}$$

where

$$T_{2,d} = \frac{\Re\{\tilde{n}_a^d\}}{\Re\{\tilde{n}_s^d\}} |t_{2,d}|^2 = \frac{\tilde{n}_a^d}{\tilde{n}_s^d} |t_{2,d}|^2$$

$$n_s \sin\theta_s^d = n_a \sin\theta_a^d \quad \Rightarrow \quad \frac{d\Omega_s}{d\Omega_a} = \frac{\sin\theta_s^d d\theta_s^d d\phi_d}{\sin\theta_a^d d\theta_a^d d\phi_d} = \frac{n_a n_a \cos\theta_a^d}{n_s n_s \cos\theta_s^d} \tag{8.253}$$

$$\frac{1}{\Phi_0^i} \frac{d\Phi_{1,s}^d}{d\Omega_s} = \left(\frac{n_s}{\lambda}\right)^2 \frac{\tilde{n}_s^d}{\tilde{n}_0^i} \cos\theta_s^d \frac{1}{S} \left|\sum_{j=0}^{p_1} C_{1,j}^+ [\hat{h}_{1,j} \star \hat{s}_e]\right|^2 .$$

Combining (8.252) and (8.253) we immediately get (8.251), as expected.

### Multiple Reflections

The beams transmitted and scattered by the front face of the component are not simply transmitted by the rear face, but are also reflected by it, as illustrated in Figure 8.26.

The incident beam first encounters the front face (filter 1) at an angle

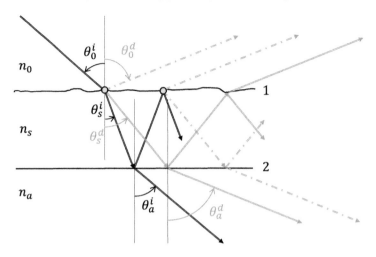

Figure 8.26 Diagram showing beams scattered by multiple reflections in a
component of finite thickness (specular beams, black continuous arrows –
beams scattered forward, gray continuous arrows – beams scattered rear-
ward, gray dash-dot arrows).

$\theta_0^i$ (left-hand gray dot on Figure 8.26), where it gives rise to a transmitted
specular beam (black arrow), to a scattered beam in the reflected half-space
(gray dash-dot arrow) and a scattered beam in the transmitted half-space
(gray arrow).

The beam scattered into the transmitted half-space at an angle $\theta_s^d$ then
encounters the rear face (Filter 2), where part of it is transmitted (we just
analyzed this situation earlier in this section) with the other part reflected
back to the front face, where it will again be partially reflected and partially
transmitted. It will contribute not only to the scattering embodied within
the transmitted half-space but also to that in the reflected half-space.

However, we must also consider the additional scattering from the upper
stack, which results from the (reversed) illumination coming after multiple
reflections at the rear face. The multiple reflections of the incident beam
within the substrate will generate a cascade of contributions to the excitation
of the currents responsible for scattering in the reflected and transmitted
half-spaces, along the lines of what we have just described.

To include these different scattering sources, we introduce the following
general notation:

$$d\Phi_{1,k}^d = \mathcal{D}_{1,jk}\Phi_j^i \, d\Omega_k \,, \tag{8.254}$$

where the subscript 1 shows that the primary scattering source here is the
filter on the front face of the component, where the subscript $k$ specifies

in which medium the scattered flux is generated ($k = 0$ or $k = s$), and where the subscript $j$ defines from which medium the illumination comes from ($j = 0$ or $j = s$).

By way of illustration, the flux scattered by filter 1 in the reflected half-space when only the incident flux impinges on this filter is written

$$\mathcal{D}_{1,00}\Phi_0^i \, d\Omega_0 = \left(\frac{n_0}{\lambda}\right)^2 \frac{\tilde{n}_0^d}{\tilde{n}_0^i} \cos\theta_0^d \frac{1}{S} \left| \sum_{j=0}^{p_1} C_{1,j}^-[\hat{h}_{1,j} \star \hat{s}_e] \right|^2 \Phi_0^i \, d\Omega_0, \quad (8.255)$$

while the flux scattered by this same filter in the transmitted half-space (still with only the incident flux) is given by

$$\mathcal{D}_{1,0s}\Phi_0^i \, d\Omega_s = \left(\frac{n_s}{\lambda}\right)^2 \frac{\tilde{n}_s^d}{\tilde{n}_0^i} \cos\theta_s^d \frac{1}{S} \left| \sum_{j=0}^{p_1} C_{1,j}^+[\hat{h}_{1,j} \star \hat{s}_e] \right|^2 \Phi_0^i \, d\Omega_s. \quad (8.256)$$

If we now include the global transmission of this flux by the rear face of the substrate, including the multiple reflections occurring within this same substrate, we obtain

$$\left[ T_{2,d} + R_{2,d}R'_{1,d}T_{2,d} + \cdots \right] \mathcal{D}_{1,0s}\Phi_0^i \, d\Omega_s \, ;$$

that is,

$$\mathcal{D}_{1,0s} \frac{T_{2,d}}{1 - R_{2,d}R'_{1,d}} \Phi_0^i \, d\Omega_s \, ,$$

where $R'_{1,d}$ is the reflection coefficient for filter 1 in retrograde propagation and $T_{2,d}$ and $R_{2,d}$ are associated with the second filter illuminated from the medium $s$. Note that the subscript $d$ indicates that the coefficient depends on the scattering angle.

Similarly, the flux resulting from this first scattering from filter 1 and detected in the upper half-space can be expressed as

$$\mathcal{D}_{1,00}\Phi_0^i \, d\Omega_0 + \mathcal{D}_{1,0s}\left[ R_{2,d}T'_{1,d} + R_{2,d}R'_{1,d}R_{2,d}T'_{1,d} + \cdots \right] \Phi_0^i \, d\Omega_s \, ,$$

or,

$$\mathcal{D}_{1,00}\Phi_0^i \, d\Omega_0 + \mathcal{D}_{1,0s}\frac{R_{2,d}T'_{1,d}}{1 - R_{2,d}R'_{1,d}}\Phi_0^i \, d\Omega_s \, .$$

However, we must also include the waves scattered from the same filter 1, both forward and backward, which result from the reverse illumination created by the rear face (the second gray dot from the left in Figure 8.26). The direct scattering generated in the upper half-space is expressed as

$$\mathcal{D}_{1,s0}T_{1,i}R_{2,i}\Phi_0^i \, d\Omega_0 \, ,$$

while, after multiple reflections, the scattering produced in the substrate leads to the following contribution:

$$\mathcal{D}_{1,ss}T_{1,i}R_{2,i}\Phi_0^i \, d\Omega_s \frac{R_{2,d}T'_{1,d}}{1 - R_{2,d}R'_{1,d}} \, .$$

Note here that though we considered multiple reflections for the scattered wave, this is not the case for the specular beam at this step. Actually the multiple cascaded reflections of the incident beam within the component will give rise to contributions similar to those just described, except that the intensity of the *exciting* flux will decrease in geometric progression with common ratio $R'_{1,i}R_{2,i}$.

Consequently, the total scattering generated in the upper half-space by filter 1, and which incorporates multiple reflections taking account of both incident beam and scattered beams, can finally be expressed as

$$d\Phi_{1,0}^d = \mathcal{D}_{1,00}\Phi_0^i \, d\Omega_0 + \mathcal{D}_{1,0s}\frac{R_{2,d}T'_{1,d}}{1 - R_{2,d}R'_{1,d}}\Phi_0^i \, d\Omega_s + \mathcal{D}_{1,s0}\frac{T_{1,i}R_{2,i}}{1 - R_{2,i}R'_{1,i}}\Phi_0^i \, d\Omega_0$$

$$+ \mathcal{D}_{1,ss}\frac{T_{1,i}R_{2,i}}{1 - R_{2,i}R'_{1,i}}\Phi_0^i \, d\Omega_s \frac{R_{2,d}T'_{1,d}}{1 - R_{2,d}R'_{1,d}} \, . \tag{8.257}$$

We can then deduce an expression for the corresponding angle resolved scattering:

$$\mathrm{ARS}_{1,0} = \frac{1}{\Phi_0^i}\frac{d\Phi_{1,0}^d}{d\Omega_0} = \left\{ \mathcal{D}_{1,00} + \frac{T_{1,i}R_{2,i}}{1 - R_{2,i}R'_{1,i}}\mathcal{D}_{1,s0} \right\}$$

$$+ \frac{R_{2,d}T'_{1,d}}{1 - R_{2,d}R'_{1,d}}\left\{ \mathcal{D}_{1,0s} + \frac{T_{1,i}R_{2,i}}{1 - R_{2,i}R'_{1,i}}\mathcal{D}_{1,ss} \right\}\frac{d\Omega_s}{d\Omega_0} \, . \tag{8.258}$$

Similarly, taking account of multiple reflections, the angle resolved scattering embodied in the lower half-space $a$ describing the scattering by filter 1 can finally be expressed as

$$\mathrm{ARS}_{1,a} = \frac{T_{2,d}}{1 - R_{2,d}R'_{1,d}}\left\{ \mathcal{D}_{1,0s} + \frac{T_{1,i}R_{2,i}}{1 - R_{2,i}R'_{1,i}}\mathcal{D}_{1,ss} \right\}\frac{d\Omega_s}{d\Omega_a} \, . \tag{8.259}$$

### *The Influence of Rear Face Roughness*

We now assume that the rear face of the substrate (which carries filter 2) also has a certain roughness. The scattering distributions embodied in the reflected and transmitted half-spaces must be completed by the new contributions associated with the white dots in Figure 8.27.

These contributions are as follows:

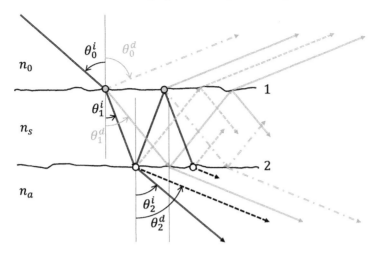

Figure 8.27 Diagram showing the beams generated by scattering from the two faces of a component [specular beams, black continuous arrows – beams scattered forward (resp. backward) by filter 1, gray continuous (resp. dash-dot) arrows – beams scattered forward (resp. backward) by filter 2, black (resp. gray) dashed arrows].

- in the upper half-space:

$$\text{ARS}_{2,0} = \mathcal{D}_{2,ss} \frac{T_{1,i}}{1 - R_{2,i}R'_{1,i}} \frac{T'_{1,d}}{1 - R_{2,d}R'_{1,d}} \frac{d\Omega_s}{d\Omega_0} ; \qquad (8.260)$$

- and in the lower half-space:

$$\text{ARS}_{2,a} = \frac{T_{1,i}}{1 - R_{2,i}R'_{1,i}} \left\{ \mathcal{D}_{2,sa} + \mathcal{D}_{2,ss} \frac{R'_{1,d}T_{2,d}}{1 - R_{2,d}R'_{1,d}} \frac{d\Omega_s}{d\Omega_a} \right\}. \qquad (8.261)$$

### Application to the Special Case of a Bare Substrate

In Section 8.2.10 we showed the particular form taken by the scattering distributions of an interface when these distributions are embodied in the plane of incidence. We shall now turn our attention to those generated by an uncoated substrate so as to better illustrate (for a simple case) the consequences of a rear face being present. As with the single interface, we need only consider the SS and PP configurations, since the other (SP and PS) coefficients are 0 in the incidence plane.

Equations (8.115) and (8.116) lead to an expression for the coefficient $\mathcal{D}_{1,00}$ for these two polarization configurations, namely

$$\mathcal{D}_{1,00}^{SS} = \left(\frac{n_0}{\lambda}\right)^2 \frac{\cos^2 \theta_0^d}{\cos \theta_0^i} \left[\frac{2k_0 \cos \theta_0^i (n_0 \cos \theta_0^i - n_s \cos \theta_s^i)}{n_0 \cos \theta_0^d + n_s \cos \theta_s^d}\right]^2 \gamma_1, \qquad (8.262)$$

$$\mathcal{D}^{PP}_{1,00} = \left(\frac{n_0}{\lambda}\right)^2 \frac{1}{\cos\theta^i_0} \left[\frac{2k_0(n_0^2 - n_s^2)}{\left(\frac{n_0}{\cos\theta^d_0} + \frac{n_s}{\cos\theta^d_s}\right)\left(\frac{n_0}{\cos\theta^i_0} + \frac{n_s}{\cos\theta^i_s}\right)}\right]^2$$

$$\times \left[1 - \frac{\sin\theta^i_0 \sin\theta^d_0}{\cos\theta^i_s \cos\theta^d_s}\right]^2 \gamma_1 . \tag{8.263}$$

We immediately deduce an expression for coefficients $\mathcal{D}_{1,ss}$ by applying a permutation of indices $s$ and $0$, then for coefficients $\mathcal{D}_{2,ss}$ replacing $\gamma_1$ by $\gamma_2$ in this last set of expressions (recall here that medium $a$ and medium $0$ are identical).

Furthermore, combining (8.95), (8.104), and (8.113), we get an expression for the coefficients $\mathcal{D}_{1,0s}$:

$$\mathcal{D}^{SS}_{1,0s} = \left(\frac{n_s}{\lambda}\right)^2 \frac{n_s \cos^2\theta^d_s}{n_0 \cos\theta^i_0} \left[\frac{2k_0 \cos\theta^i_0(n_0 \cos\theta^i_0 - n_s \cos\theta^i_s)}{n_0 \cos\theta^d_0 + n_s \cos\theta^d_s}\right]^2 \gamma_1 , \tag{8.264}$$

$$\mathcal{D}^{PP}_{1,0s} = \left(\frac{n_s}{\lambda}\right)^2 \frac{n_s}{n_0 \cos\theta^i_0} \left[\frac{2k_0(n_0^2 - n_s^2)}{\left(\frac{n_0}{\cos\theta^d_0} + \frac{n_s}{\cos\theta^d_s}\right)\left(\frac{n_0}{\cos\theta^i_0} + \frac{n_s}{\cos\theta^i_s}\right)}\right]^2$$

$$\times \left[1 + \frac{\sin\theta^i_0 \sin\theta^d_s}{\cos\theta^i_s \cos\theta^d_0}\right]^2 \gamma_1 . \tag{8.265}$$

As before, we deduce an expression for coefficients $\mathcal{D}_{1,s0}$ by permuting indices $s$ and $0$, and then that for $\mathcal{D}_{2,sa}$ by replacing $\gamma_1$ by $\gamma_2$.

To illustrate the various calculation steps we have just taken in this section, we shall consider the case of a silica substrate whose two uncoated faces are characterized by the same roughness ($\gamma_1 = \gamma_2$). The roughness spectrum common to both faces will be exactly the same as that used in Section 8.2.11 ($\delta_g = 1$ nm, $L_g = 100$ nm, $\delta_e = 1$ nm, $L_e = 2{,}000$ nm) and the illuminating wavelength will be taken to be 600 nm.

For an angle of incidence of 60 degree, the graphs combined in Figure 8.28 represent the scattering distributions embodied in the reflected ($0 \leqslant \theta^d_0 \leqslant 90$ degree) and transmitted (90 degrees $\leqslant \theta^d_0 \leqslant 180$ degrees) half-spaces for different situations summarized as follows:

- We consider only the roughness of the front face and take account of the presence of the rear face, but neglecting the weight of multiple reflections (only direct transmission is considered) [gray curve in SS and PP modes]; in this case we have simply

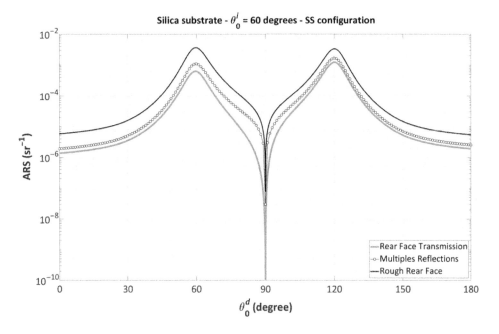

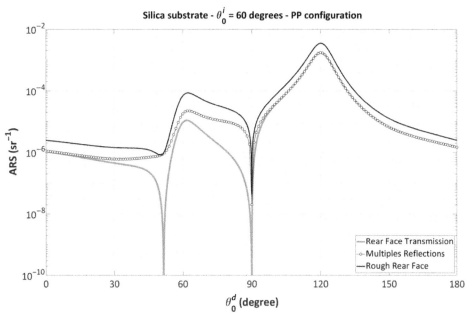

Figure 8.28 Scattering distributions (ARS) of a silica substrate of roughness of approximately 1 nm RMS, calculated in the plane of incidence
($\phi_d = 0$) for an angle of incidence $\theta_0^i$ of 60 degrees and two polarization
configurations (SS, upper graph, and PP, lower graph).

$$\text{ARS}_{1,0} = \mathcal{D}_{1,00}$$

$$\text{ARS}_{1,a} = T_{2,d}\,\mathcal{D}_{1,0s}\,\frac{d\Omega_s}{d\Omega_a} = \frac{2\tilde{n}_s\tilde{n}_a}{(\tilde{n}_s + \tilde{n}_a)^2}\,\mathcal{D}_{1,0s}\,\frac{n_s}{n_a}\,\frac{n_s}{n_a}\,\frac{\cos\theta_s^d}{\cos\theta_a^d}. \qquad (8.266)$$

Note the significant difference in behavior in the PP configuration between the reflected half-space (where we recognize the pseudo-Brewster phenomenon: see Section 8.2.11) and the transmitted half-space, which shows levels of scattering similar to those found for the SS configuration. This difference is directly related to the structure of the formulae that govern scattering in this particular configuration [see equations (8.104) in which the last term in square brackets contains a − sign for the reflected half-space and a + sign for the transmitted half-space].

- Then we include all the multiple reflections, for the incident beam as well as for the scattered beams (open circles, in SS and PP modes). This means that we sum the scattering data from the first filter, taking into account, without approximation, both direct and reverse illumination beams.

Note in this case (first for the SS configuration) the significant increase in the level of scattering for angles between 70 degrees and 100 degrees (this is directly related to the reflection coefficient on the rear face at these large scattering angles); on the other hand, note that in the PP configuration the pseudo-Brewster phenomenon disappears in the reflected half-space: this is due to the presence, in this part of the angle resolved scattering associated with the reflected half-space, of contributions normally associated with the transmitted half-space and reduced in this space by multiple reflections on the rear face. This is the reason why opaque glasses must be used to analyze in detail the pseudo-Brewster effect.

- Finally, we assume that the rear face is also rough, with a roughness root-mean-square identical to that of the front face (black curve in SS and modes). Note that the two roughness profiles are uncorrelated since they result from a polishing process (no replication effect connected with a film deposition).

As expected, the level of the various distributions increases significantly, especially in the PP configuration, though a pseudo-Brewster mechanism remains and results from the contribution of this second face to scattering in the reflected half-space.

# 9

# Planar Multilayer Microcavities

## 9.1 Introduction

Until now we have always assumed that the source of the light illuminating a multilayer stack was outside this planar structure in the half-space defined by negative values of $z$; until this point this half-space has been referred to generically as the *incident medium* or the *superstrate*.

However, this *entrant wave* condition is not in any sense obligatory, as the investigation of the light scattered by a multilayer stack developed in Chapter 8 has shown. Indeed, in that case we were able to associate *fictitious currents*, both electric and magnetic, with the roughness of each of the interfaces within the stack, and we showed how we could get from a mathematical expression for these currents to the spectral flux density per unit solid angle radiated into the two outermost media.

We now turn our attention to an altogether comparable situation, shown diagrammatically in Figure 9.1: this concerns the case in which a planar source of current – this time real (physical) rather than fictitious – is inserted into a multilayer stack. From now on, this structure will be known generically as a **multilayer planar cavity**.

For example, this planar current source might comprise

- Atoms of rare earth elements such as europium or praseodymium, implanted in the surface of one of the layers in a multidielectric stack, excited optically with an external laser source
- A thin film of electroluminescent organic material between two electrodes, of which at least one is transparent, and across which a voltage is applied

The second example is a generic description of an organic electroluminescent diode, or OLED (organic light-emitting diode), and it is around this particular case that the theoretical modeling will be constructed and illustrated,

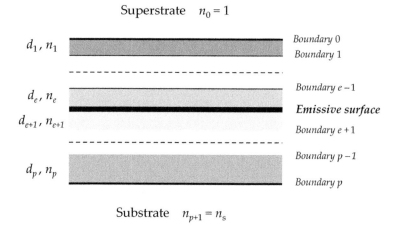

Figure 9.1 Diagrammatic representation of an emissive planar cavity.

to be developed in the sections that follow. At this stage it is important to emphasize that we assume no interaction between the source and the field created by that source.

Finally, note that if we had chosen a configuration in which excitation of the emitting elements was achieved optically using an external far-field source or a near-field source in the superstrate, we would have had to include transmission of this external wave through all the layers between the superstrate and the emissive structure (which would have meant weighting the current by the value of the exciting field at the interface where the source was located).

## 9.2  Theoretical Modeling

### *9.2.1  Current Source*

For an OLED, the emissive structure corresponds in fact not to a plane (in the mathematical sense) but to a very thin layer of luminescent material, typically a few tens of nanometers.

It would be entirely possible to describe such a *volume* emission using an appropriate formalism, namely one that models the scattering of light by a stack of layers incorporating volume heterogeneities, for example (which then play the role of fictitious currents); however, the resulting additional complexity would not really be justified by an improved or more pertinent description of the physical mechanisms underlying such an emission. Hence, we shall describe this emission as resulting from inserting into a stack a thin film of passive material (roughly 10 nm), within which is a planar current source.

This surface current source will be assumed to be located at the interface of order number $e$ and comprising the juxtaposition of several identical but independent emitters. This can therefore be described by a spatiotemporal distribution of the form

$$\vec{J}(\vec{r}, t) = \sum_n J_e(\vec{r} - \vec{r}_n, t - t_n) \vec{s}_n \quad \text{where} \quad \vec{s}_n = \begin{bmatrix} \cos \psi_n \\ \sin \psi_n \end{bmatrix}, \qquad (9.1)$$

where $\vec{r}_n$ defines the position of each of these elementary emitters, with $t_n$ and $\psi_n$ being random variables that take account of the coherence and polarization properties of the emission of such a source (see Section 5.3). Note that here we have chosen a source whose physical structure precludes any magnetic current.

To consolidate our ideas, we shall assume that variables $\vec{r}$ and $t$ are separable and that these emitters are quasi-pointwise; hence they can be represented by uniform disks whose radius $a$ is very small compared to a wavelength, i.e.,

$$J_e(x, y, t) = \text{Circ}_a(r) \, S_e(t) \qquad \text{where} \quad r = \sqrt{x^2 + y^2} \; ; \; a \ll \lambda. \qquad (9.2)$$

The positions $\vec{r}_n$ of the centers of these disks are chosen such that the plane is partially covered by nonoverlapping disks.

Recall the generic definition of the $\text{Circ}_a$ function:

$$\text{Circ}_a(r) = \begin{cases} 1 & \text{si } r \leqslant a \\ 0 & \text{si } r > a \end{cases} ; \qquad (9.3)$$

the nonoverlap condition

$$\forall n, \forall m, \quad |\vec{r}_n - \vec{r}_m| > 2a \, ; \qquad (9.4)$$

as well as an expression for the Fourier transform of this function $\text{Circ}_a$, written $\widehat{\mathcal{C}}_a$:

$$\widehat{\mathcal{C}}_a(\nu_x, \nu_y) = \pi a^2 \left[ \frac{2J_1(2\pi a \nu)}{2\pi a \nu} \right] \qquad \text{where} \quad \nu = \sqrt{\nu_x^2 + \nu_y^2}. \qquad (9.5)$$

If we now apply to this spatiotemporal current distribution the two Fourier transformations necessary for solving the Maxwell's equations that describe such a problem (see Chapters 2 and 4), we get

$$\vec{\mathcal{J}}(\vec{r}, f) = \int_t \vec{J}(\vec{r}, t) \, e^{2i\pi f t} \, dt = \sum_n S_e(f) \, e^{2i\pi f t_n} \, \vec{s}_n \, \text{Circ}_a(x - x_n, y - y_n) \, .$$

$$(9.6)$$

Then

$$\mathbb{J}(\vec{\nu}, f) = \int_{\vec{r}} \vec{J}(\vec{r}, f)\, e^{-2i\pi\vec{\nu}.\vec{r}}\, d^2\vec{r} = \widehat{\mathcal{C}}_a(\nu)\mathcal{S}_e(f) \sum_n \vec{s}_n\, e^{2i\pi f t_n}\, e^{-2i\pi\vec{\nu}.\vec{r}_n}\, . \quad (9.7)$$

We can then put this in the form

$$\mathbb{J}(\vec{\nu}, f) = \mathbb{J}_e(\nu, f) \sum_n \vec{s}_n\, e^{2i\pi f t_n}\, e^{-2i\pi\vec{\nu}.\vec{r}_n} \quad \text{where} \quad \mathbb{J}_e(\nu, f) = \widehat{\mathcal{C}}_a(\nu)\mathcal{S}_e(f)\, .$$

$$(9.8)$$

### 9.2.2 Radiated Amplitudes in the End Media

#### Taking the Structure of the Cavity into Account

We shall assume that the planar cavity here comprises $p$ elementary layers of thickness $d_j$ and refractive index $n_j$ ($j = 1, ..., p$). Taking account of the particular location of the source (within the stack, at the interface of order $e$), it will be convenient to denote by the term *superstrate* (and no longer the *incident medium*) the half-space of refractive index $n_0 = 1$ above interface $0$ and corresponding to negative values of $z$; the term *substrate* naturally continues to be used to denote the half-space of refractive index $n_s$ located below the interface $p$ and hence corresponding to positive values of $z$ greater than $z_p$.

Applying directly the theoretical results established in Section 8.3.2, the tangential components of the elementary electric fields radiated by the planar source on either side of interface $e$ are identical in the absence of any magnetic current, and given by the following equation:

$$\vec{E}_e^T(\vec{\nu}, z_e, f) = \vec{E}_{e+1}^T(\vec{\nu}, z_e, f) = -\frac{\mathbb{J}(\vec{\nu}, f)}{Y_e - Y_e'}\, , \quad (9.9)$$

where $Y_e$ and $Y_e'$ are the complex admittances either side of this interface $e$, the first being calculated from the effective refractive index of the substrate using the following recurrence equation:

$$Y_p = \tilde{n}_s \quad ; \quad Y_{j-1} = \frac{Y_j \cos\delta_j - i\tilde{n}_j \sin\delta_j}{\cos\delta_j - i(Y_j/\tilde{n}_j) \sin\delta_j} \quad j = p, \cdots, e+1\, . \quad (9.10)$$

The second is calculated from the effective refractive index of the superstrate using this same recurrence equation, though written in a different form:

$$Y_0' = -\tilde{n}_0 \quad ; \quad Y_j' = \frac{Y_{j-1}' \cos\delta_j + i\tilde{n}_j \sin\delta_j}{\cos\delta_j + i(Y_{j-1}'/\tilde{n}_j) \sin\delta_j} \quad j = 1, \cdots, e\, . \quad (9.11)$$

The tangential component of the elementary electric field radiated by the planar source within the substrate is then expressed as (see Section 8.3.6):

$$\vec{\mathbb{A}}_s^{T+}(\vec{\nu}, f) = t_{ep}\, \vec{\mathbb{E}}_{e+1}^T(\vec{\nu}, z_e, f)\,, \tag{9.12}$$

where

$$t_{ep} = \frac{1}{\prod\limits_{j=e+1}^{p} [\cos \delta_j - i(Y_j/\tilde{n}_j)\sin \delta_j]} \quad \text{if } e \leqslant p-1\,, \tag{9.13}$$

$$t_{ep} = 1 \text{ if } e = p\,.$$

Similarly, the tangential component of the electric field radiated by the planar source in the superstrate is given by (see Section 8.3.6):

$$\vec{\mathbb{A}}_0^{T-}(\vec{\nu}, f) = t'_{e0}\, \vec{\mathbb{E}}_e^T(\vec{\nu}, z_e, f)\,, \tag{9.14}$$

where

$$t'_{e0} = \frac{1}{\prod\limits_{j=1}^{e} \left[\cos \delta_j + i(Y'_{j-1}/\tilde{n}_j)\sin \delta_j\right]} \quad \text{if } e \geqslant 1\,, \tag{9.15}$$

$$t'_{e0} = 1 \text{ if } e = 0\,.$$

### Taking the Polarization State into Account

The tangential components radiated into the end media are defined by the following general equations:

$$\vec{\mathbb{A}}_s^{T+}(\vec{\nu}, f) = -t_{ep}\frac{\vec{\mathbb{J}}(\vec{\nu}, f)}{Y_e - Y'_e} \quad ; \quad \vec{\mathbb{A}}_0^{T-}(\vec{\nu}, f) = -t'_{e0}\frac{\vec{\mathbb{J}}(\vec{\nu}, f)}{Y_e - Y'_e}\,, \tag{9.16}$$

where

$$\vec{\mathbb{J}}(\vec{\nu}, f) = \mathbb{J}_e(\nu, f) \sum_n (\cos \psi_n \hat{x} + \sin \psi_n \hat{y})\, e^{2i\pi f t_n}\, e^{-2i\pi \vec{\nu}\cdot\vec{r}_n} = \mathbb{J}_x \hat{x} + \mathbb{J}_y \hat{y}\,. \tag{9.17}$$

To calculate the quantities $t_{ep}$, $t'_{e0}$, $Y_e$, and $Y'_e$ in (9.16), the polarization state, TE or TM, of the elementary waves emitted by the planar source needs to be known. To this end, Figure 9.2 shows diagrammatically the orientation of the various constituent vectors, where $\phi$ is the same polar angle as that introduced in Chapter 8 when the scattered fields were being calculated.

For the elementary wave emitted toward the substrate we get

$$\vec{\mathbb{A}}_{s,\text{TE}}^{T+}(\vec{\nu}, f) = t_{ep}\frac{\mathbb{J}_x \sin \phi - \mathbb{J}_y \cos \phi}{Y_e - Y'_e}\,,$$

$$\vec{\mathbb{A}}_{s,\text{TM}}^{T+}(\vec{\nu}, f) = -t_{ep}\frac{\mathbb{J}_x \cos \phi + \mathbb{J}_y \sin \phi}{Y_e - Y'_e}\,, \tag{9.18}$$

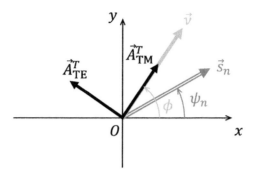

Figure 9.2 Diagram showing the orientation of the various vectors used in modeling emission from a multilayer planar cavity [$\vec{s}_n$ is a unit vector associated with the elementary current source; $\vec{\nu}$ is the tangential component of the wave vector for the elementary wave radiated by the planar source; $\vec{A}_{\text{TE}}^T$ (respectively $\vec{A}_{\text{TM}}^T$) is the tangential component of the amplitude of the radiated wave of polarization TE (respectively TM)].

and similarly for that emitted toward the superstrate:

$$\vec{A}_{0,\text{TE}}^{T-}(\vec{\nu}, f) = t_{e0}' \frac{\mathbb{J}_x \sin \phi - \mathbb{J}_y \cos \phi}{Y_e - Y_e'} ,$$

$$\vec{A}_{0,\text{TM}}^{T-}(\vec{\nu}, f) = -t_{e0}' \frac{\mathbb{J}_x \cos \phi + \mathbb{J}_y \sin \phi}{Y_e - Y_e'} .$$

(9.19)

### 9.2.3 Radiated Intensities in the End Media

We start by considering the radiated intensity in the substrate. For each polarization state this is given by the following equation (see Section 8.2.9):

$$I_s = \frac{d\Phi_s}{d\Omega_s} = \frac{1}{2} \left(\frac{n_s}{\lambda}\right)^2 \Re\{\tilde{n}_s\} \cos \theta_s \, |\vec{A}_s^{T+}(\vec{\nu}, f)|^2 ,$$

(9.20)

where the elementary amplitude is defined by equations (9.18). For TE polarization, the quantity $|\vec{A}_s^{T+}(\vec{\nu}, f)|^2$ is therefore written as

$$|\vec{A}_{s,\text{TE}}^{T+}(\vec{\nu}, f)|^2 = \frac{|t_{ep}|^2}{|Y_e - Y_e'|^2} \left\{|\mathbb{J}_x|^2 \sin^2 \phi + |\mathbb{J}_y|^2 \cos^2 \phi + 2\Re[\mathbb{J}_x \mathbb{J}_y^*] \sin \phi \cos \phi\right\},$$

where

$$\mathbb{J}_x = \mathbb{J}_e(\nu, f) \sum_n \cos \psi_n \, e^{2i\pi f t_n} \, e^{-2i\pi \vec{\nu}.\vec{r}_n} ,$$

$$\mathbb{J}_y = \mathbb{J}_e(\nu, f) \sum_n \sin \psi_n \, e^{2i\pi f t_n} \, e^{-2i\pi \vec{\nu}.\vec{r}_n} .$$

We now need to calculate the three quantities $|\mathbb{J}_x|^2$, $|\mathbb{J}_y|^2$ et $\Re[\mathbb{J}_x\mathbb{J}_y^*]$. For the first of these we get

$$|\mathbb{J}_x|^2 = |\mathbb{J}_e(\nu, f)|^2 \left| \sum_n \cos\psi_n\, e^{2i\pi f t_n}\, e^{-2i\pi\vec{\nu}.\vec{r}_n} \right|^2 = |\mathbb{J}_e(\nu, f)|^2 \sum_n \cos^2\psi_n$$

$$+ |\mathbb{J}_e(\nu, f)|^2 \sum_n \sum_{m\neq n} \cos\psi_n \cos\psi_m\, e^{2i\pi f(t_n - t_m)}\, e^{-2i\pi\vec{\nu}.(\vec{r}_n - \vec{r}_m)}. \quad (9.21)$$

Since the emitter characteristics are distributed randomly, the second term of (9.21) goes to 0, while the sum of the cosine squared terms in the first is equal to $N/2$. Consequently,

$$|\mathbb{J}_x|^2 = \frac{N}{2} |\mathbb{J}_e(\nu, f)|^2 = \frac{N}{2} |\mathcal{S}_e(f)|^2 (\pi a^2)^2 \left[ \frac{2J_1(2\pi a\nu)}{2\pi a\nu} \right]^2, \quad (9.22)$$

where $N$ is the number of elementary emitters.

When calculating the far-field intensities radiated toward one of the end media, the spatial frequency $\sigma = 2\pi\nu$ is bounded by the wave vector $2\pi n/\lambda$ associated with this medium, so the argument $x = 2\pi a\nu$ of the function $2J_1(x)/x$ appearing in equation (9.22) is always much smaller than 1, since $a \ll \lambda$. We then deduce that

$$|\mathbb{J}_x|^2 \approx \frac{N}{2}(\pi a^2)^2 |\mathcal{S}_e(f)|^2 = \frac{N}{2}|\mathbb{J}_e(f)|^2. \quad (9.23)$$

Using an entirely similar process, it can be shown that

$$|\mathbb{J}_y|^2 = \frac{N}{2}|\mathbb{J}_e(f)|^2 \quad \text{and} \quad \Re[\mathbb{J}_x\mathbb{J}_y^*] = 0. \quad (9.24)$$

Substituting expressions (9.23) and (9.24) into (9.25), we get

$$I_s(\theta_s, \lambda) = \frac{1}{4}\left(\frac{n_s}{\lambda}\right)^2 \Re\{\tilde{n}_s\} \cos\theta_s\, |t_{ep}|^2 \frac{N|\mathbb{J}_e(\lambda)|^2}{|Y_e - Y_e'|^2}, \quad (9.25)$$

where, in the expression for the spectral dependence of the elementary current, we have replaced frequency $f$ by wavelength $\lambda = c/f$ so as to create a variable more appropriate to our needs. On the other hand, we have omitted to indicate the polarization state, since the expression obtained for TM is identical (wee keep in mind that the coefficients $\tilde{n}_s$, $t_{ep}$, $Y_e$, and $Y_e'$ are polarization dependent).

Similarly, it can be shown that for each of the polarization states, the radiated intensity in the superstrate is described by

$$I_0(\theta_0, \lambda) = \frac{1}{4}\left(\frac{n_0}{\lambda}\right)^2 \Re\{\tilde{n}_0\} \cos\theta_0\, |t_{e0}'|^2 \frac{N|\mathbb{J}_e(\lambda)|^2}{|Y_e - Y_e'|^2}. \quad (9.26)$$

### 9.2.4 Radiation Patterns and Emission Spectra

Now that we have general formulas for calculating the variation in emitted intensity as a function of angle $\theta$ (we speak of radiation patterns) or wavelength $\lambda$ (we speak of emission spectra), we need to specify the intrinsic spectral properties of this emissive surface. However, this would require having as many results as there are possible sources, which would make any general formulation of conclusions somewhat problematic.

Consequently, we prefer to use the concept of an **ideal** source emitting equally over all wavelengths, and then analyze how inserting this source into a particular cavity will affect its emission properties. In order to make a simple comparison of these various results, it is convenient to use a **reference configuration** that corresponds to inserting this ideal source into a thin layer of passive material deposited onto the surface of a glass substrate. Hence, we shall begin our analyses with this reference configuration.

## 9.3 Examples of Structures

The radiation patterns of the various structures we shall be analyzing in this section will be shown at a wavelength of $\lambda_0 = 600$ nm, which also corresponds to the design wavelength of thin layers denoted by H or L. Emission spectra will be analyzed in the window between 400 and 800 nm.

### 9.3.1 Reference Configuration

This reference configuration, shown diagrammatically in Figure 9.3, corresponds to the *minimum* structure realizable in practice, in which a thin film of emissive material of thickness $d_e$ and refractive index $n_e$, assumed nondispersive ($n_e = 1.7$), is deposited on the surface of a substrate of N-BK7, refractive index $n_s$ ($n_s = 1.516$ at 600 nm).

#### Semi-infinite Substrate

Assume initially that the substrate is semi-infinite. The different stages of calculation are detailed as follows:

- Calculation of admittance $Y_e$:

$$Y_2 = \tilde{n}_s \quad ; \quad Y_e = \frac{\tilde{n}_s \cos \delta_e - i\tilde{n}_e \sin \delta_e}{\cos \delta_e - i(\tilde{n}_s/\tilde{n}_e) \sin \delta_e} \qquad \text{avec} \quad \delta_e = \alpha_e \frac{d_e}{2}. \qquad (9.27)$$

- Calculation of admittance $Y_e'$:

$$Y_0' = -\tilde{n}_0 \quad ; \quad Y_e' = \frac{-\tilde{n}_0 \cos \delta_e + i\tilde{n}_e \sin \delta_e}{\cos \delta_e - i(\tilde{n}_0/\tilde{n}_e) \sin \delta_e}. \qquad (9.28)$$

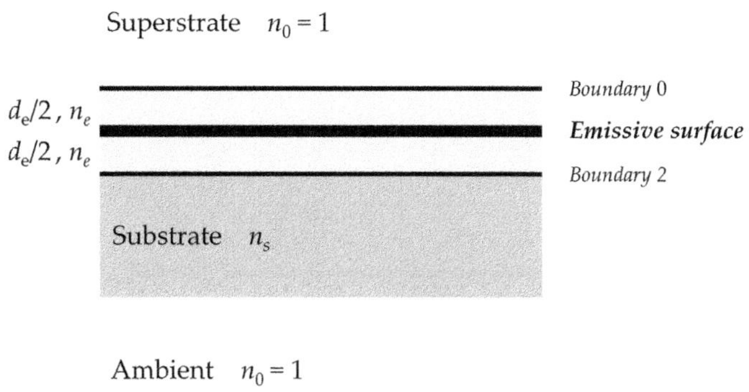

Superstrate   $n_0 = 1$

$d_e/2, n_e$   Boundary 0
$d_e/2, n_e$   Emissive surface
   Boundary 2

Substrate   $n_s$

Ambient   $n_0 = 1$

Figure 9.3 Diagrammatic representation of the reference configuration.

- Calculation of transmission $t_{ep}$:

$$t_{ep} = \frac{1}{\cos \delta_e - i(\tilde{n}_s/\tilde{n}_e) \sin \delta_e} .$$ (9.29)

- Calculation of transmission $t'_{e0}$:

$$t'_{e0} = \frac{1}{\cos \delta_e - i(\tilde{n}_0/\tilde{n}_e) \sin \delta_e} .$$ (9.30)

To conclude the calculation, it is helpful to make use of the very small thickness of the passive layer that contains the emissive surface, since this will simplify the analytical form of the formulas, especially after normalization. To first order in $\delta_e$ we get

$$Y_e \approx \tilde{n}_s + i\left\{\frac{\tilde{n}_s^2}{\tilde{n}_e} - \tilde{n}_e\right\}\delta_e \quad ; \quad Y'_e \approx -\tilde{n}_0 - i\left\{\frac{\tilde{n}_0^2}{\tilde{n}_e} - \tilde{n}_e\right\}\delta_e ,$$
$$t_{ep} \approx 1 + i\frac{\tilde{n}_s}{\tilde{n}_e}\delta_e \quad ; \quad t'_{e0} \approx 1 + i\frac{\tilde{n}_0}{\tilde{n}_e} .$$ (9.31)

and hence, still to first order in $\delta_e$:

$$\frac{|t_{ep}|^2}{|Y_e - Y'_e|^2} \approx \frac{|t'_{e0}|^2}{|Y_e - Y'_e|^2} \approx \frac{1}{|\tilde{n}_s + \tilde{n}_0|^2} .$$ (9.32)

It remains only to apply equations (9.26) and (9.25) to obtain the intensities $I_0(\theta_0, \lambda)$ and $I_s(\theta_s, \lambda)$ emitted toward the superstrate and substrate respectively.

### Substrate of Finite Thickness

The presence of the rear face of the N-BK7 substrate means that we must consider how this additional interface is traversed, but also the multiple reflections that it creates, as shown in Figure 9.4. Indeed, the calculation is entirely similar to that used for the scattering case in Section 8.3.11.

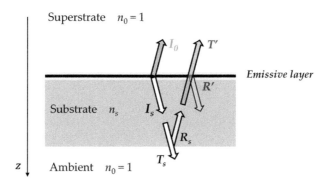

Figure 9.4 Diagrammatic representation of the reference configuration, including the multiple reflections generated by the rear face of the substrate.

The emitted intensity in the direction $z > 0$, with the end medium being now identical to that of the superstrate, therefore has the general expression

$$\frac{d\Phi_+}{d\Omega_0} = I_s \left\{ T_s + R_s R' T_s + ... \right\} \frac{d\Omega_s}{d\Omega_0} = I_s \frac{T_s}{1 - R_s R'} \frac{d\Omega_s}{d\Omega_0}, \tag{9.33}$$

where $R'$ is the reflection coefficient of the multilayer stack in the retrograde configuration, and $T_s$ (respectively $R_s$) is the energy transmission coefficient (respectively reflection) of the rear face of the substrate, i.e.,

$$T_s = \frac{\Re[\tilde{n}_0]}{\Re[\tilde{n}_s]} \left| \frac{2\tilde{n}_s}{\tilde{n}_s + \tilde{n}_0} \right|^2 \quad \text{et} \quad R_s = \left| \frac{\tilde{n}_s - \tilde{n}_0}{\tilde{n}_s + \tilde{n}_0} \right|^2 . \tag{9.34}$$

Furthermore,

$$n_0 \sin \theta_0 = n_s \sin \theta_s \quad \Rightarrow \quad \frac{d\Omega_s}{d\Omega_0} = \frac{\sin \theta_s d\theta_s d\phi_s}{\sin \theta_0 d\theta_0 d\phi_0} = \frac{n_0^2 \cos \theta_0}{n_s^2 \cos \theta_s} . \tag{9.35}$$

Consequently, we get

$$\frac{d\Phi^+}{d\Omega_0}(\theta_0, \lambda) = \frac{n_0^2 \cos \theta_0}{n_s^2 \cos \theta_s} \frac{T_s}{1 - R_s R'} I_s(\theta_s, \lambda) . \tag{9.36}$$

Combining (9.25) and (9.36), we get the following general expression:

$$\frac{d\Phi^+}{d\Omega_0}(\theta_0, \lambda) = \frac{1}{4} \left( \frac{n_0}{\lambda} \right)^2 \cos \theta_0 \, \Re[\tilde{n}_s] \frac{T_s}{1 - R_s R'} |t_{ep}|^2 \frac{N |\mathcal{J}_e(\lambda)|^2}{|Y_e - Y'_e|^2} . \tag{9.37}$$

In the particular case of the reference configuration it can easily be shown that $R' \approx R_s$ to first order in $\delta_e$. Also using equation (9.32), we get

$$\frac{d\Phi_{\text{ref}}^+}{d\Omega_0}(\theta_0, \lambda) = \frac{1}{4}\left(\frac{n_0}{\lambda}\right)^2 \cos\theta_0 \, \frac{\Re[\tilde{n}_s]}{1 + R_s} \frac{N|\mathbb{J}_e(\lambda)|^2}{|\tilde{n}_s + \tilde{n}_0|^2}. \qquad (9.38)$$

Similarly, for the intensity emitted in the direction $z < 0$:

$$\frac{d\Phi^-}{d\Omega_0}(\theta_0, \lambda) = I_0(\theta_0, \lambda) + \frac{n_0^2 \cos\theta_0}{n_s^2 \cos\theta_s} \frac{R_s T'}{1 - R_s R'} I_s(\theta_s, \lambda). \qquad (9.39)$$

This time combining (9.25), (9.26), and (9.39), we obtain the following general expression:

$$\frac{d\Phi^-}{d\Omega_0}(\theta_0, \lambda) = \frac{1}{4}\left(\frac{n_0}{\lambda}\right)^2 \cos\theta_0 \left\{ \Re[\tilde{n}_0]\,|t'_{e0}|^2 + \Re[\tilde{n}_s]\,|t_{ep}|^2 \frac{R_s T'}{1 - R_s R'} \right\} \frac{N|\mathbb{J}_e(\lambda)|^2}{|Y_e - Y'_e|^2},$$

and for the reference configuration,

$$\frac{d\Phi_{\text{ref}}^-}{d\Omega_0}(\theta_0, \lambda) = \frac{1}{4}\left(\frac{n_0}{\lambda}\right)^2 \cos\theta_0 \left\{ \Re[\tilde{n}_0] + \Re[\tilde{n}_s] \frac{R_s}{1 + R_s} \right\} \frac{N|\mathbb{J}_e(\lambda)|^2}{|\tilde{n}_s + \tilde{n}_0|^2}. \qquad (9.40)$$

In what follows we shall describe the performance of a multilayer planar cavity using the following variables:

- Normalized monochromatic radiation patterns, represented in polar coordinates at one wavelength $\lambda_0$ for each polarization state, and defined as the ratio of the intensity emitted at each direction $\theta_0$ by the multilayer cavity being considered, to the intensity along the axis ($\theta_0 = 0$) at the same wavelength $\lambda_0$ emitted by the reference configuration, i.e.,

$$\mathcal{R}_{\lambda_0}^\pm(\theta_0) = \frac{\dfrac{d\Phi^\pm}{d\Omega_0}(\theta_0, \lambda_0)}{\dfrac{d\Phi_{\text{ref}}^\pm}{d\Omega_0}(0, \lambda_0)}. \qquad (9.41)$$

- Normalized emission spectra on the axis, defined as the ratio of the spectral dependence of the flux emitted along the axis by the multilayer cavity being considered to that emitted on the axis by the reference configuration, i.e.,

$$\mathcal{S}_0^\pm(\lambda) = \frac{\dfrac{d\Phi^\pm}{d\Omega_0}(0, \lambda)}{\dfrac{d\Phi_{\text{ref}}^\pm}{d\Omega_0}(0, \lambda)}. \qquad (9.42)$$

For the two polarization states, Figure 9.5 shows the normalized monochromatic radiation patterns corresponding to this reference configuration, while Figure 9.6 shows the normalized emission spectra on the axis.

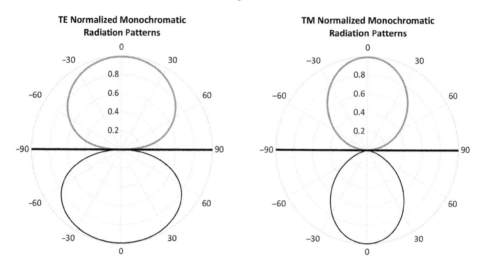

Figure 9.5 Normalized monochromatic radiation patterns ($\lambda_0 = 600$ nm) for the reference configuration (gray in the superstrate, black in the substrate).

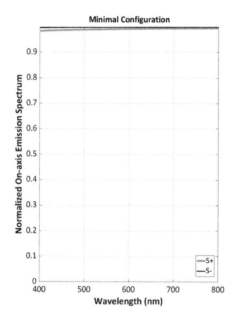

Figure 9.6 Reference configuration. Normalized emission spectra on the axis.

These patterns were calculated by normalizing the data obtained numerically for the reference configuration by those obtained analytically for the same case using equations (9.38) and (9.40). For the normalized emission spectra along the axis, note the very slight difference between the results given and the expected value of 100%; this difference is due simply to the limit of validity of the first-order approximation in $\delta_e$ that was used to carry out the analytical calculations. Likewise, the overall efficiency $\gamma$ actually comes out at 99.3% (rather than 100%), which is sufficiently precise for the validity of the comparison we shall now make.

### 9.3.2 All-Dielectric Quarter Wave Mirror

We start by considering a structure comprising an alternating stack of quarter wave H and L layers; we shall use the same terminology as was introduced in Chapter 7 when synthesizing giant field enhancements (Section 7.8.2), namely

$$Q_1 = (\mathrm{LH})^N S \quad Q_2 = (\mathrm{HL})^N S \quad Q_3 = \mathrm{H}(\mathrm{LH})^N S \quad Q_4 = \mathrm{L}(\mathrm{HL})^N S. \quad (9.43)$$

This leads to an expression with which we can calculate the admittance of the upper interface of these various stacks in the order $Q_1, ..., Q_4$, i.e.,

$$Y_0 = \begin{cases} \tilde{n}_s/\beta^{2N} \\ \tilde{n}_s\beta^{2N} \\ \tilde{n}_H^2\beta^{2N}/\tilde{n}_s \\ \tilde{n}_L^2/\beta^{2N}\tilde{n}_s \end{cases}, \quad (9.44)$$

where $\beta$ is the ratio $\tilde{n}_H/\tilde{n}_L$ of the effective refractive indices.

Now suppose that the planar source is located at the upper interface of this stack. Under the assumption of a semi-infinite substrate, the intensity radiated toward the superstrate is given by equation (9.26) and is therefore proportional to the quantity

$$|t'_{e0}|^2 \frac{N|\mathbb{J}_e|^2}{|Y_0 - Y'_0|^2}, \quad \text{i.e., here} \quad \frac{N|\mathbb{J}_e|^2}{|Y_0 + \tilde{n}_0|^2}. \quad (9.45)$$

To maximize the emitted intensity in the direction $\theta_0 = 0$, the quantity $|Y_0 + \tilde{n}_0|$ must be minimized which, on the axis, amounts to minimizing the quantity $Y_0$ alone; hence we need to select a structure corresponding to configurations $Q_4$ or $Q_1$. To illustrate the behavior obtained, we shall choose configuration $Q_4$ using the same number of quarter-wave layer pairs as in Section 7.8.4, i.e., $N = 6$. The graphs demonstrating the angular and spectral characteristics of this quarter-wave mirror configuration are shown in Figures 9.7 and 9.8.

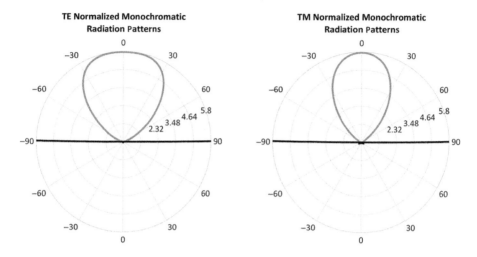

Figure 9.7 Normalized monochromatic radiation patterns ($\lambda_0 = 600$ nm) for a quarter-wave configuration of type $Q_4$ ($N = 6$) when the emissive surface is on the first interface of the stack (gray in the superstrate, black in the substrate).

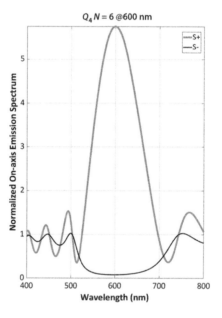

Figure 9.8 Quarter-wave configuration of type $Q_4$ ($N = 6$) when the emissive surface is on the first interface of the stack. Normalized emission spectra along the axis.

Note that the emitted intensity on the axis in the direction of the super-strate (see Figure 9.7) is nearly six times greater than that emitted in the same direction by the reference configuration.

Note also that this enhancement of emission along the axis obtained us-ing a $Q_4$ type configuration could be replaced by an inhibition of emission along the axis by choosing a type $Q_2$ or $Q_3$ configuration. In this case, the admittance $Y_0$ would tend to infinity with the number $N$ of quarter-wave layer pairs, while it tends to 0 for configurations $Q_1$ and $Q_4$.

As a consequence, the electric field is close to 0 at the top interface for configurations $Q_2$ and $Q_3$ since we have

$$Y_0 \to \infty \ \text{for} \ N \to \infty \quad \Rightarrow \quad r \to -1 \quad \Rightarrow \quad 1 + r \to 0. \tag{9.46}$$

It is not the case for configurations $Q_4$ and $Q_1$, for which we have

$$Y_0 \to 0 \ \text{for} \ N \to \infty \quad \Rightarrow \quad r \to 1 \quad \Rightarrow \quad 1 + r \to 2. \tag{9.47}$$

Eventually we notice that the $Q_4$ or $Q_1$ mirror-induced enhancement of the on-axis emission is somewhat limited in comparison to that of the reference configuration; indeed this emission is governed by the $1/|Y_0 + \tilde{n}_0|^2$ factor, equal to $1/(n_s + n_0)^2$ for the reference configuration at $\theta_0 = 0$ and close to $1/n_0^2$ for a $Q_1$ or $Q_4$ configuration. Hence there is an upper bound of $(n_s/n_0 + 1)^2 \approx 6.25$ for this enhancement factor, due to the fact that this enhancement factor is connected with the field at this top surface, and not with the reflectance of the mirror, which is much higher than that of the reference configuration by a factor close to 25.

Consider now that the source is located on the upper face of the substrate. In this case, the quantity to be maximized is proportional to

$$|t_{ep}|^2 \frac{N|\mathbb{J}_e|^2}{|Y_e - Y_e'|^2} \quad \text{soit ici} \quad \frac{N|\mathbb{J}_e|^2}{|\tilde{n}_s - Y_p'|^2}. \tag{9.48}$$

For the stacks of type $Q_i$ that we shall consider, the admittance $Y_p'$ is expressed as

$$Y_p' = \begin{cases} -\,\tilde{n}_0 \beta^{2N} \\ -\,\tilde{n}_0/\beta^{2N} \\ -\,\tilde{n}_H^2 \beta^{2N}/\tilde{n}_0 \\ -\,\tilde{n}_L^2/\beta^{2N}\tilde{n}_0 \end{cases}, \tag{9.49}$$

the optimum configuration here being type $Q_2$. The graphs illustrating the angular and spectral characteristics of this alternative quarter wave config-uration are shown in Figures 9.9 and 9.10. Note that the gain in emission on the axis in the direction of the substrate is smaller (roughly 2.8).

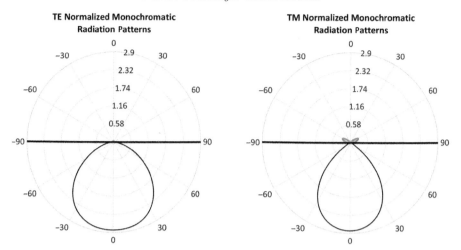

**Figure 9.9** Normalized monochromatic radiation patterns ($\lambda_0 = 600$ nm) of the quarter-wave mirror configuration of type $Q_2$ ($N = 6$) when the emissive surface is located at the last interface in the stack (gray in the superstrate, black in the substrate).

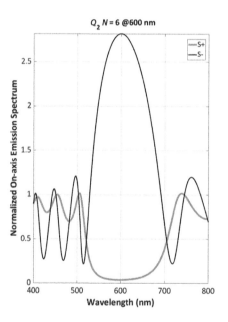

**Figure 9.10** Quarter-wave configuration of type $Q_2$ ($N = 6$) when the emissive surface is located at the last interface in the stack. Normalized emission spectra along the axis.

Clearly, the approach just described is not limited to maximizing emission along the axis. To obtain a favorable emission in the spatial directions that correspond, for example, to an angle $\theta_0$ of 60 degrees, it is again sufficient to locate the planar source at the upper interface of a quarter wave stack of type $Q_4$, but in which the layers are matched for this particular incidence ($\delta = \alpha d = \lambda_0/4$, see Chapters 6 and 7).

Under these conditions, the effective indices for layers of high and low index at a wavelength of 600 nm are

- TE polarization:

$$\tilde{n}_H = \sqrt{n_H^2 - n_0^2 \sin^2 \theta_0} = 1.9667 \quad \text{and} \quad \tilde{n}_L = \sqrt{n_L^2 - n_0^2 \sin^2 \theta_0} = 1.2152.$$

- TM polarization:

$$\tilde{n}_H = \frac{n_H^2}{\sqrt{n_H^2 - n_0^2 \sin^2 \theta_0}} = 2.3480 \quad \text{and} \quad \tilde{n}_L = \frac{n_L^2}{\sqrt{n_L^2 - n_0^2 \sin^2 \theta_0}} = 1.8324.$$

Hence coefficient $\beta$ takes different values depending on the polarization state of the light (1.618 in TE, 1.281 in TM), resulting in different behaviors as a function of the angle of incidence $\theta_0$, as shown in Figure 9.11.

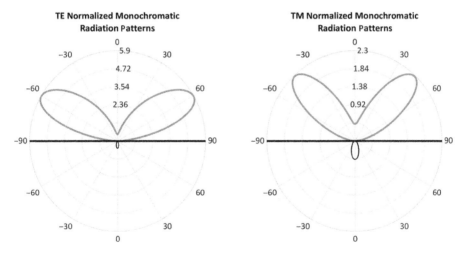

Figure 9.11 Normalized monochromatic radiation patterns ($\lambda_0 = 600$ nm) for the configuration comprising a planar source at the upper interface of a quarter wave stack of type $Q_4$ with $N = 6$, matched for an incidence of 60 degrees (gray in the superstrate, black in the substrate).

Note that emission is effectively maximum at 60 degrees for TE polarization and that the normalized intensity obtained in this particular direction is similar to that obtained on the axis for the initial configuration.

Note also that the radiation diagram for the microcavity in TM polarization shows a preferential direction of emission (47 degrees), which is different from that obtained in TE polarization, and is characterized by a smaller normalized maximum intensity (2.3 rather than 5.9). To understand the origins of these differences in behavior, it is sufficient to consider the angular dependence of the constituent terms in the complex quantity $Y_0 + \tilde{n}_0$, as shown in Figure 9.12 for the two TE and TM polarization states.

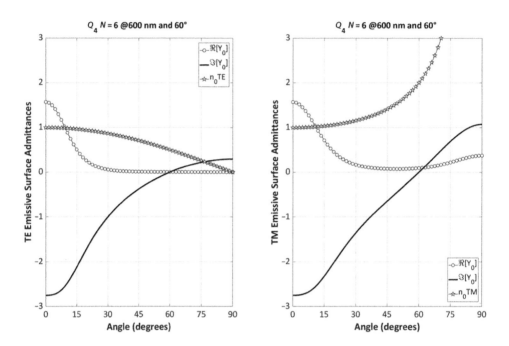

Figure 9.12 Angular dependence of the real and imaginary parts of the admittances characteristic of the emissive surface (left, TE; right, TM) for a planar source at the upper interface of a quarter wave stack of type $Q_4$ with $N = 6$, matched for an incidence of 60 degrees (circles, $\Re[Y_0]$; black line, $\Im[Y_0]$; pentagrams, $\tilde{n}_0$).

In TE polarization, we observe the almost perfect cancellation of the admittance $Y_0$ for an angle of 60 degrees; this gives a minimum value for this particular angle to the real quantity $Y_0 + \tilde{n}_0$ while in TM, only the imaginary part of $Y_0$ vanishes at 60 degrees, the real part combining with a high value (around 2) of effective index: the minimum of the quantity $|Y_0 + \tilde{n}_0|$ therefore shifts toward a lower angle, close to 45 degrees.

### 9.3.3 All-Dielectric Fabry–Perot

We shall now turn our attention to the case in which the planar source is in the middle of the *spacer* of a Fabry–Perot cavity. The stacks above and below the interface $e$ therefore both consist of quarter-wave mirrors of type $Q_i$ with $i = 1, ..., 4$. Among all the possible combinations, only eight generate a half-wave spacer by the juxtaposition of the quarter wave mirrors that comprise it; these are configurations $Q_1$ ES $Q_2$, $Q_1$ ES $Q_3$, $Q_2$ ES $Q_1$, $Q_2$ ES $Q_4$, $Q_3$ ES $Q_2$, $Q_3$ ES $Q_3$, $Q_4$ ES $Q_1$ and $Q_4$ ES $Q_4$, where ES denotes the emissive surface.

Assume first of all that the substrate is air, so we do not have to take account of the contribution of multiple reflections generated by the rear face of the N-BK7 substrate. To optimize emission along the axis of this microcavity, we need to maximize the quantities

$$\frac{|t_{ep}|^2}{|Y_e - Y_e'|^2} \quad \text{and/or} \quad \frac{|t_{e0}'|^2}{|Y_e - Y_e'|^2}.$$

An exhaustive analysis of the characteristics of the various structures we have just listed shows the optimum configurations equate to $Q_2$ ES $Q_4$, then $Q_2$ ES $Q_1$; the first of these demonstrates characteristics that are very slightly better than the second.

The structure of this optimal Fabry–Perot microcavity is therefore described by the following formula:

$$\text{Air} \ / \ (\text{HL})^N \ \text{ES L} \ (\text{HL})^N \ / \ \text{N-BK7} \ .$$

Actually this corresponds to a structure in which the field is maximum at the middle of the 2L spacer layer where the source is located. The normalized intensities that it emits on the axis toward the substrate and superstrate respectively are defined by the following equations (the multiple reflections generated by the rear face of the substrate are again taken into account):

$$
\begin{aligned}
S_+ &= \frac{4n_0 n_s n_L^2}{(n_0 n_s + n_L^2)^2} \frac{T_s}{1 - R_s R'} \beta^{2N} \,, \\
S_- &= \frac{4n_0 n_s^2}{(n_0 n_s + n_L^2)^2} \left\{ n_0 + \frac{n_L^2}{n_s} \frac{R_s T'}{1 - R_s R'} \right\} \beta^{2N} \,,
\end{aligned}
\tag{9.50}
$$

i.e., more than 80 times the intensity emitted on the axis by the reference configuration!

The graphs illustrating the angular and spectral characteristics of this Fabry–Perot configuration are shown in Figures 9.13 and 9.14. In particular, note the narrowness of the spectrum emitted on the axis by this structure

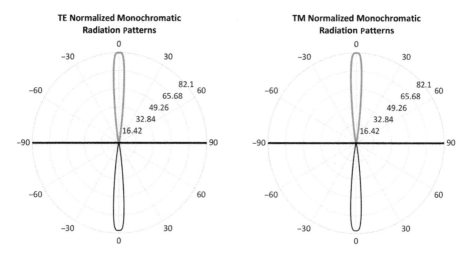

Figure 9.13 Normalized monochromatic radiation patterns ($\lambda_0 = 600$ nm) for the configuration comprising a planar source in the middle of the Fabry–Perot spacer formed by superposing two quarter wave stacks of types $Q_2$ and $Q_4$ with $N = 6$ (gray in the superstrate, black in the substrate).

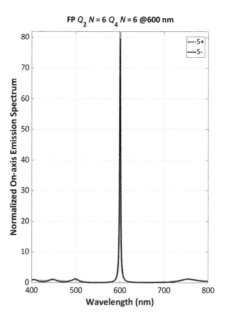

Figure 9.14 Fabry–Perot configuration formed by superposing two quarter wave stacks of types $Q_2$ and $Q_4$ with $N = 6$, when the emissive surface is in the middle of the spacer, Normalized emission spectra along the axis.

(2.4 nm total width at mid-height) as well as the directional nature of the emission lobe (roughly 12 degrees of total divergence).

To conclude this section, it is important to emphasize that obtaining such levels of performance requires the emissive source to be situated in the middle of the spacer in the case of a structure of type $Q_2$ ES $Q_4$, and that replacing this structure by a configuration of type $Q_1$ ES $Q_3$ almost totally inhibits emission along the axis (0.0098 toward the substrate and 0.0050 towards the superstrate, in normalized values). This is again connected with the relative locations of the source and field maxima. We shall return to the underlying mechanisms later in this chapter.

### 9.3.4 OLED

#### Description of the Structure

We now turn our attention to an emissive structure corresponding to the typical operational layout of an OLED, as shown in Figure 9.15.

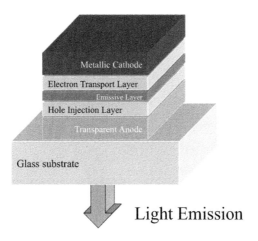

Figure 9.15 Diagrammatic representation of the structure of an OLED.

From top to bottom, this comprises

- An opaque metal cathode forming a mirror (e.g., silver 150 nm thick)
- An electron transport layer (typical thickness 50 nm, refractive index = 1.7448 at 600 nm)
- An emissive layer 20 nm thick, whose (nondispersive) refractive index = 1.7
- A hole injection layer (typical thickness 50 nm, refractive index = 1.7915 at 600 nm)

- A transparent anode, comprising, e.g., a 100-nm layer of indium tin oxide (ITO)
- A glass substrate, e.g., N-BK7

The performance of this microcavity is shown in Figures 9.16 and 9.17. It corresponds to a very wide band emission directed entirely toward the substrate, with a radiation diagram close to that of a Lambertian source.

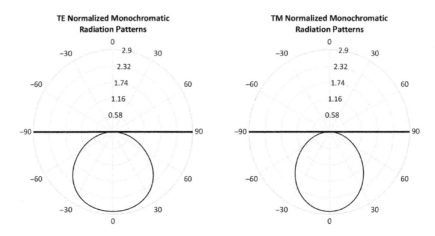

Figure 9.16 Normalized monochromatic radiation patterns for an OLED of standard structure (gray in the superstrate, black in the substrate).

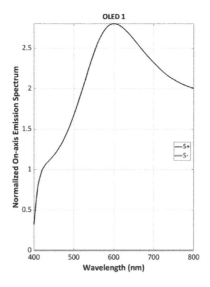

Figure 9.17 Standard structure OLED. Normalized emission spectra along the axis.

To improve the directivity and spectral narrowness of the emission, it is sufficient to boost the reflection coefficient of the ITO layer by adding a thin layer of silver (e.g., 30 nm thick) just above it. The performance of the thus-modified emissive structure is shown in Figures 9.18 and 9.19.

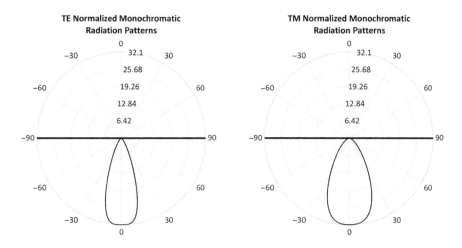

Figure 9.18 Normalized monochromatic radiation patterns of an OLED incorporating a thin layer of silver (30 nm) in the lower part (gray in the superstrate, black in the substrate).

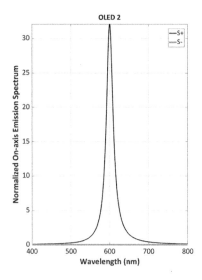

Figure 9.19 OLED incorporating a thin layer of silver (30 nm) in the lower part. Normalized emission spectra along the axis.

It is clearly of good quality (spectral width in the order of 20 nm, normalized emission on the axis corresponding to 30 times more than that of the reference configuration), in particular bearing in mind the great simplicity of the structure, close to that of a Fabry–Perot with metallic mirrors.

Such results give rise to two questions: first, what causes such good emission efficiency? Second, is the configuration as shown optimal, in the sense that we established in Section 9.3.3 for a Fabry–Perot quarter wave mirror cavity. We shall address this question in the next section.

### *Optimization of a Metal-Dielectric Structure*

Just as we did for a Fabry–Perot quarter wave mirror structure, we can consider the OLED structure shown in Figure 9.15 as the superposition of two distinct multilayer stacks (see Figure 9.20), namely

- An upper stack, denoted US, in which emissive layer plays the role of semi-infinite substrate
- A lower stack, denoted LS, in which this same material plays the role of superstrate

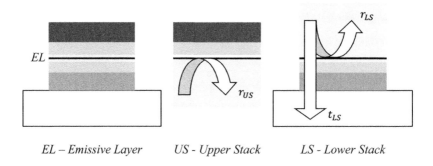

EL – Emissive Layer    US - Upper Stack    LS - Lower Stack

Figure 9.20 Decomposition of the structure of an OLED into two substacks.

Given that the light emitted by an OLED is directed solely toward the substrate, optimizing its operation amounts to maximizing the quantity

$$\frac{|t_{ep}|^2}{|Y_e - Y'_e|^2} \quad \text{where} \quad t_{ep} = t_{\text{LS}}.$$

The reflection coefficients of the two substacks we introduced are defined by

$$r_{\text{US}} = \frac{-\tilde{n}_{\text{EL}} - Y'_e}{-\tilde{n}_{\text{EL}} + Y'_e} \quad ; \quad r_{\text{LS}} = \frac{\tilde{n}_{\text{EL}} - Y_e}{\tilde{n}_{\text{EL}} + Y_e}. \tag{9.51}$$

From these expressions we can express the admittances $Y_e$ and $Y'_e$, i.e.,

$$Y_e = \tilde{n}_{EL} \frac{1 - r_{LS}}{1 + r_{LS}} \quad ; \quad Y'_e = -\tilde{n}_{EL} \frac{1 - r_{US}}{1 + r_{US}}$$

$$\Rightarrow \quad Y_e - Y'_e = 2\tilde{n}_{EL} \frac{1 - r_{LS} r_{US}}{(1 + r_{LS})(1 + r_{US})} . \tag{9.52}$$

Consequently, the intensity emitted toward the substrate, here assumed to be semi-infinite, can be put in the form

$$I_s(\theta_s, \lambda) = \frac{1}{4} \left(\frac{n_s}{\lambda}\right)^2 \Re\{\tilde{n}_s\} \cos\theta_s \, |t_{ep}|^2 \frac{N |\mathbb{J}_e(\lambda)|^2}{|Y_e - Y'_e|^2} , \tag{9.53}$$

and, only taking account of the passage across the rear face of this substrate (and not of the multiple reflections that it generates),

$$\frac{d\Phi_+}{d\Omega_0}(\theta_0, \lambda) = \frac{n_0^2 \cos\theta_0 \, \Re[\tilde{n}_0]}{n_s^2 \cos\theta_s \, \Re[\tilde{n}_s]} \left| \frac{2\tilde{n}_s}{\tilde{n}_s + \tilde{n}_0} \right|^2 I_s(\theta_s, \lambda)$$

$$= \frac{N}{16} \left(\frac{n_0}{\lambda}\right)^2 \Re\{\tilde{n}_0\} \cos\theta_0 \left| \frac{2\tilde{n}_s t_{ep} \mathbb{J}_e(\lambda)}{(\tilde{n}_s + \tilde{n}_0)\tilde{n}_{EL}} \right|^2 \left| \frac{(1 + r_{LS})(1 + r_{US})}{1 - r_{LS} r_{US}} \right|^2 . \tag{9.54}$$

Let the last term in (9.54) be $\mathcal{K}$, and put

$$r_{LS} = \sqrt{R_{LS}} \, e^{i\phi_{LS}} \quad \text{et} \quad r_{US} = \sqrt{R_{US}} \, e^{i\phi_{US}} . \tag{9.55}$$

Hence

$$\mathcal{K}(\phi_{LS}, \phi_{US}) = \frac{(1 + R_{LS} + 2\sqrt{R_{LS}} \cos\phi_{LS})(1 + R_{US} + 2\sqrt{R_{US}} \cos\phi_{US})}{1 + R_{LS} R_{US} - 2\sqrt{R_{LS} R_{US}} \cos(\phi_{LS} + \phi_{US})} . \tag{9.56}$$

This coefficient $\mathcal{K}$ thus comprises three factors, of the same structure and involving cosine terms. To maximize the emission along the axis of a microcavity, we need to maximize the values taken by the first two factors (in the numerator) and to minimize the value taken by the third (in the denominator). Taking account of the sign difference affecting the third term, this comes down to making the arguments of all the cosines 0, i.e.,

$$\phi_{LS} = 0 \, [2\pi] \quad ; \quad \phi_{US} = 0 \, [2\pi] \tag{9.57}$$

$$\phi_{LS} + \phi_{US} = 0 \, [2\pi] . \tag{9.58}$$

This shows that the intensity emitted along the axis by a microcavity is maximum when two of these three resonance conditions are satisfied simultaneously. Indeed, this is sufficient to guarantee that the third condition is also satisfied. Note that condition (9.58) is particularly critical, since it is in the denominator. Conversely, the two conditions (9.57) are more tolerant, since they are in the numerator.

In the case of the last OLED structure we described in Section 9.3.4, the reflection coefficients $R_{US}$ and $R_{LS}$ are 98% and 80% respectively at a wavelength of 600 nm.

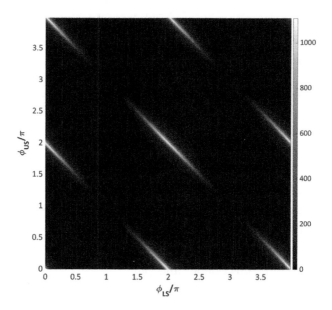

Figure 9.21 False color representation of the variations in the coefficient $\mathcal{K}(\phi_{LS}, \phi_{US})$ when phase shifts $\phi_{LS}$ and $\phi_{US}$ vary independently between 0 and $4\pi$.

Figure 9.21 uses gray levels to show the variation of the coefficient $\mathcal{K}$, which is associated with the coefficients $R_{US}$ and $R_{LS}$ when phase shifts $\phi_{LS}$ and $\phi_{US}$ vary independently between 0 and $4\pi$. Note that $\phi_{LS}$ and $\phi_{US}$ can deviate by 0.1 radians from the target value of $2\pi$ without affecting the value of coefficient $\mathcal{K}$ too drastically provided their sum remains equal to $2\pi$. This explains the needle-shaped zones where the coefficient $\mathcal{K}$ is greater than 800 (light gray) and their orientation parallel to the diagonal of equation $y = -x$.

Still with this same OLED structure incorporating a thin coating of silver on the lower part, the phase shifts $\phi_{LS}$ and $\phi_{US}$ are $-0.0718$ and $0.0745$ radians respectively, showing that condition (9.58) is effectively fulfilled, while conditions (9.57) are satisfied only approximately.

To improve the behavior of this structure, we might consider increasing the thickness of electron transport and hole injection layers until the phase shifts $\phi_{LS}$ and $\phi_{US}$ are strictly equal to $2\pi$. This is realized for thicknesses of 225.95 nm for the electron transport layer and 218.28 nm for the hole injection layer respectively.

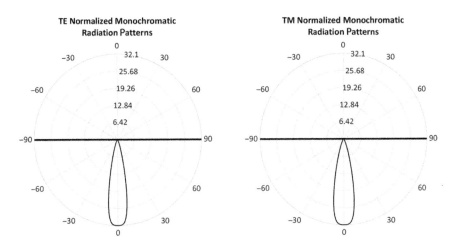

Figure 9.22 Normalized monochromatic radiation patterns for an OLED incorporating a thin layer of silver (30 nm) in the lower stack; the electron transport and hole injection layer thicknesses are 225.95 nm and 218.29 nm respectively (gray in the superstrate, black in the substrate).

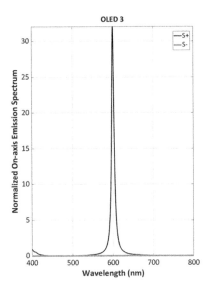

Figure 9.23 OLED incorporating a thin layer of silver (30 nm) in the lower stack; the electron transport and hole injection layer thicknesses are 225.95 nm and 218.29 nm respectively. Normalized emission spectra along the axis.

However, this modification clearly brings about (see Figures 9.22 and 9.23) a reduction in the width of the emission line on the axis (7 nm rather than 20 nm) and a reduction in the divergence of this emission.

To identify the causes of such behavior, we need to analyze the manner in which the emitted intensity on the axis varies with wavelength. This intensity is given by equation (9.54) where we put $\theta_0 = 0$, i.e.,

$$\frac{d\Phi_+}{d\Omega_0}(0, \lambda) = \frac{N}{16}\left(\frac{n_0}{\lambda}\right)^2 \frac{4n_0 n_s^2}{(n_s + n_0)^2 n_{EL}^2} |t_{ep}(0, \lambda)|^2 |\mathbb{J}_e|^2 \mathcal{K}(\phi_{LS}, \phi_{US}). \quad (9.59)$$

The maximum value of this intensity is obtained for the wavelength $\lambda_0$, with the two resonance conditions assumed satisfied for this particular wavelength. Consequently,

$$\left[\frac{d\Phi_+}{d\Omega_0}\right]_{max} = \frac{N}{16}\left(\frac{n_0}{\lambda_0}\right)^2 \frac{4n_0 n_s^2}{(n_s + n_0)^2 n_{EL}^2} |t_{ep}(0, \lambda_0)|^2 |\mathbb{J}_e|^2 \frac{(1 + \sqrt{R_{LS}})^2 (1 + \sqrt{R_{US}})^2}{(1 - \sqrt{R_{LS} R_{US}})^2}.$$

To a first approximation in the neighborhood of the central wavelength $\lambda_0$, we can put the expression for the intensity of emission on the axis in the form

$$\frac{d\Phi_+}{d\Omega_0}(0, \lambda) \approx \left[\frac{d\Phi_+}{d\Omega_0}\right]_{max} \frac{\left\{1 - \frac{4\sqrt{R_{LS}}}{(1+\sqrt{R_{LS}})^2}\sin^2\frac{\phi_{LS}}{2}\right\}\left\{1 - \frac{4\sqrt{R_{US}}}{(1+\sqrt{R_{US}})^2}\sin^2\frac{\phi_{US}}{2}\right\}}{\left\{1 + \frac{4\sqrt{R_{LS} R_{US}}}{(1-\sqrt{R_{LS} R_{US}})^2}\sin^2\frac{\phi_{LS}+\phi_{US}}{2}\right\}}.$$

where only the phases are assumed to vary with wavelength to any significant extent. At the central wavelength $\lambda_0$, the resonance conditions (9.57) and (9.58) are fulfilled; after expanding the phase terms to first order in $\delta\lambda$, we get

$$\frac{\frac{d\Phi_+}{d\Omega_0}(0, \lambda_0 + \delta\lambda)}{\left[\frac{d\Phi_+}{d\Omega_0}\right]_{max}} \approx \frac{\left\{1 - \frac{4\sqrt{R_{LS}}}{(1+\sqrt{R_{LS}})^2}\left[\frac{\partial\phi_{LS}}{\partial\lambda}\frac{\delta\lambda}{2}\right]^2\right\}\left\{1 - \frac{4\sqrt{R_{US}}}{(1+\sqrt{R_{US}})^2}\left[\frac{\partial\phi_{US}}{\partial\lambda}\frac{\delta\lambda}{2}\right]^2\right\}}{\left\{1 + \frac{4\sqrt{R_{LS} R_{US}}}{(1-\sqrt{R_{LS} R_{US}})^2}\left[\frac{\partial(\phi_{LS}+\phi_{US})}{\partial\lambda}\frac{\delta\lambda}{2}\right]^2\right\}}.$$

$$(9.60)$$

The minus ($-$) sign in the denominator of the third term makes its effect significantly greater than the other two. By way of illustration, using the example of the OLED cavity with a thin silver layer we have already considered, we have

$$R_{LS} = 0.80 \text{ and } R_{US} = 0.98 \quad \Rightarrow \quad \begin{cases} \frac{4\sqrt{R_{LS}}}{(1 + \sqrt{R_{LS}})^2} \approx \frac{4\sqrt{R_{US}}}{(1 + \sqrt{R_{US}})^2} \approx 1 \\ \frac{4\sqrt{R_{LS} R_{US}}}{(1 - \sqrt{R_{LS} R_{US}})^2} = 270 \end{cases}. \quad (9.61)$$

Consequently (9.60) can be put into the following simplified form:

$$\frac{d\Phi_+}{d\Omega_0}(0, \lambda_0 + \delta\lambda) \approx \frac{\left[\frac{d\Phi_+}{d\Omega_0}\right]_{\max}}{1 + \frac{4\sqrt{R_{\mathrm{LS}}R_{\mathrm{US}}}}{(1-\sqrt{R_{\mathrm{LS}}R_{\mathrm{US}}})^2}\left[\frac{\partial(\phi_{\mathrm{LS}}+\phi_{\mathrm{US}})}{\partial\lambda}\frac{\delta\lambda}{2}\right]^2} . \tag{9.62}$$

The half-width at half-maximum $\delta\lambda$ of the emission line on the axis is then given by

$$\frac{4\sqrt{R_{\mathrm{LS}}R_{\mathrm{US}}}}{(1-\sqrt{R_{\mathrm{LS}}R_{\mathrm{US}}})^2}\left[\frac{\partial(\phi_{\mathrm{LS}}+\phi_{\mathrm{US}})}{\partial\lambda}\frac{\delta\lambda}{2}\right]^2 = 1, \tag{9.63}$$

so, the corresponding total width $\Delta\lambda$ is defined by

$$\Delta\lambda = 2\delta\lambda = \frac{2}{\left|\frac{\partial(\phi_{\mathrm{LS}}+\phi_{\mathrm{US}})}{\partial\lambda}\right|_{\lambda_0}}\frac{1-R}{\sqrt{R}} \quad \text{where} \quad R = \sqrt{R_{\mathrm{LS}}R_{\mathrm{US}}}. \tag{9.64}$$

This shows that the width of the emission line on the axis for a cavity is inversely proportional to the sum of the phase dispersion of the constituent substacks. This also shows the advantage of using metallic mirrors when we want to broaden the operating spectral band, given that the spectral phase variations in reflection for these metallic mirrors are known to be slow (compared with all-dielectric mirrors).

Along the same lines, the overall emission efficiency of this microcavity will be proportional to the quantity $M$ defined by

$$M = |t_{ep}(0, \lambda_0)|^2 \frac{(1+\sqrt{R_{\mathrm{LS}}})^2(1+\sqrt{R_{\mathrm{US}}})^2}{(1-\sqrt{R_{\mathrm{LS}}R_{\mathrm{US}}})^2}\Delta\lambda, \tag{9.65}$$

and

$$M = 2\frac{(1+\sqrt{R_{\mathrm{LS}}})^2(1+\sqrt{R_{\mathrm{US}}})^2}{(1-R)\sqrt{R}}\frac{|t_{ep}(0, \lambda_0)|^2}{\left|\frac{\partial(\phi_{\mathrm{LS}}+\phi_{\mathrm{US}})}{\partial\lambda}\right|_{\lambda_0}} . \tag{9.66}$$

Table 9.1 Main characteristics of the three different emissive planar microcavity structures

| Structure | $R_{\mathrm{US}}$ | $R_{\mathrm{LS}}$ | $R$ | $|t_{ep}(0, \lambda_0)|^2$ | $\partial\phi/\partial\lambda$ | $M$ | $\Delta\lambda$ (nm) | $\gamma$ |
|---|---|---|---|---|---|---|---|---|
| FP $Q_2Q_4$ | 0.967 | 0.952 | 0.959 | 0.012 | $-0.0349$ | 268 | 2.4 | 1.05 |
| OLED 2 | 0.980 | 0.803 | 0.887 | 0.052 | $-0.0117$ | 1193 | 20.6 | 1.19 |
| OLED 3 | 0.980 | 0.802 | 0.887 | 0.052 | $-0.0351$ | 397 | 6.9 | 0.38 |

These results are illustrated in Table 9.1 where the main characteristics of the 3 cavities we have studied are displayed together: the Fabry–Perot cavity of type $Q_2Q_4$, the OLED incorporating a thin layer of silver on its lower stack (OLED 2) and that of the same structure, but where the thicknesses of the electrically functional layers have been increased (OLED 3).

The spectral line widths $\Delta\lambda$ in this table are those that result from application of equation (9.64) and are in perfect agreement with the numerical modeling results shown graphically in the previous sections.

### 9.3.5 Partial Conclusion

Our investigation of the performance of the several types of emissive planar microcavities carried out in Section 8.3 highlighted the following two key points:

- The proper functioning of a structure presupposes that **two resonance conditions** are satisfied; one shows that the overall phase shift corresponding to a round trip in the cavity must be a multiple of $2\pi$ (*strong* condition), while the other stipulates that the phase shift accumulated over a round trip between the emissive surface and any one of the two constituent substacks must also be a multiple of $2\pi$ (*weak* condition).
- The width of the emission line on the axis of a microcavity varies in inverse proportion to the total **spectral dispersion of phase** accumulated during a round trip in this cavity.

In a more general way, it is surely useful to stress on the fact that the optimization of the microcavities intensity patterns involves both intensity and phase terms [see equations (9.54) and (9.56)], which is different from the synthesis of optical filters used in free space.

## 9.4 Energy Balance

We have just seen that, for a given surface current, the structure and position of the emissive surface in this cavity influence both the radiation pattern and the overall emission efficiency. This result, obtained under the assumption that the source and the field do not interact, shows that the emission properties are not intrinsic to the source, but are influenced by the environment in which the source is placed. It is important in this context to consider the energy balance: this will be the purpose of this section.

In Chapter 2, Section 2.10, we saw that the energy balance in the harmonic regime could be written in the form

$$F = \Phi + A. \tag{9.67}$$

In this expression $\Phi$ is the Poynting vector flux radiated by the source through a surface $\Sigma$ delimiting a volume $\Omega$ surrounding the cavity; $A$ is the absorption in this same volume and $F$ is the optical power supplied by the source.

In Chapter 4, equation (4.52) we also saw that the flux density per unit spatial frequency per unit polar angle took the form

$$\frac{d^2\Phi_j}{d\nu d\phi} = \frac{1}{2\omega\tilde{\mu}_j}\Re[\alpha_j(\nu)]\,\nu\,|\vec{\mathbb{A}}_j(\vec{\nu})|^2 \quad j = 0, s. \tag{9.68}$$

This expression is given for nonmagnetic media, with $j = s$ and $\vec{\mathbb{A}}_j(\vec{\nu}) = \vec{\mathbb{A}}_s^+(\vec{\nu})$ for the flux emitted into the substrate (here assumed semi-infinite) and $j = 0$, and $\vec{\mathbb{A}}_j(\vec{\nu}) = \vec{\mathbb{A}}_0^-(\vec{\nu})$ for the flux emitted into the superstrate.

In a similar manner, it now remains to establish an expression giving the optical spectral power density supplied by the source. In the monochromatic regime and in the absence of any magnetic source, and as shown in Chapter 2, equation (2.79), the total power is written in the following form:

$$F = -\int_\Omega \frac{1}{2}\Re[\vec{\mathcal{J}}.\vec{\mathcal{E}}^*]dV. \tag{9.69}$$

Since $\vec{\mathcal{J}}$ is a surface current and located at ordinate $z = z_e$, it takes the form of a Dirac distribution centered on $z_e$, i.e.,

$$\vec{\mathcal{J}}(\vec{r}, z) = \vec{\mathcal{J}}_e(\vec{r})\,\delta(z - z_e); \tag{9.70}$$

hence

$$F = -\frac{1}{2}\Re\left[\int_{\vec{r}} \vec{\mathcal{J}}_e(\vec{r}).\vec{\mathcal{E}}^*(\vec{r}, z_e)\right] d^2\vec{r}. \tag{9.71}$$

Note at this stage that since the surface current is necessarily tangential (see Chapter 2, Section 2.12), the field components involved in the integral (9.71) are also tangential. Applying Parseval's theorem to this integral, we get

$$F = -\frac{1}{2}\Re\left[\int_{\vec{\nu}} \vec{\mathbb{J}}_e(\vec{\nu}).\vec{\mathbb{E}}^*(\vec{\nu}, z_e)\right] d^2\vec{\nu}. \tag{9.72}$$

As is customary, we put

$$\vec{\nu} = \frac{n\sin\theta}{\lambda}\begin{Bmatrix} \cos\phi \\ \sin\phi \end{Bmatrix} = \nu\begin{Bmatrix} \cos\phi \\ \sin\phi \end{Bmatrix}, \tag{9.73}$$

which gives the power density per unit spatial frequency per unit polar angle, i.e.,

$$\frac{d^2 F}{d\nu d\phi} = -\frac{\nu}{2} \Re[\vec{\mathbb{J}}_e(\vec{\nu}) . \vec{\mathbb{E}}^*(\vec{\nu}, z_e)] . \tag{9.74}$$

So, using equation (9.9):

$$\frac{d^2 F}{d\nu d\phi} = \frac{\nu}{2} |\vec{\mathbb{J}}_e(\vec{\nu})|^2 \Re\left[\frac{1}{Y_e - Y_e'}\right] . \tag{9.75}$$

Finally, the energy balance given in (9.67) expressed in terms of spectral densities, takes the following form:

$$\frac{d^2 F}{d\nu d\phi} = \frac{d^2 \Phi_0}{d\nu d\phi} + \frac{d^2 \Phi_s}{d\nu d\phi} + \frac{d^2 A}{d\nu d\phi} , \tag{9.76}$$

so, in the absence of absorption and in accordance with equation (9.68),

$$\frac{\nu}{2} |\vec{\mathbb{J}}_e(\vec{\nu})|^2 \Re\left[\frac{1}{Y_e - Y_e'}\right] = \frac{1}{2\omega\tilde{\mu}_0} \Re[\alpha_0(\nu)] \, \nu \, |\vec{\mathbb{A}}_0^-(\vec{\nu})|^2$$

$$+ \frac{1}{2\omega\tilde{\mu}_s} \Re[\alpha_s(\nu)] \, \nu \, |\vec{\mathbb{A}}_s^+(\vec{\nu})|^2 . \tag{9.77}$$

In this expression, the term on the left is related to the power emitted by the source, while those on the right relate to the flux radiated into the two end media. Note that all the terms appearing in (9.77) depend on the structure of the cavity, so the energy balance is properly respected, even though the power emitted by the current depends on the formula of the multilayer stack.

To be convinced that this energy balance is indeed respected, the following analytical approach can also be used:

- Observe first of all that the power spectral density expressed as a function of current in (9.75) can also be expressed as a function of the field. Indeed, taking account of the boundary conditions in the presence of a single electric current [see Chapter 8, equation (8.150)],

$$\vec{\mathbb{E}}_{e+1}^T(\vec{\nu}, z_e) = \vec{\mathbb{E}}_e^T(\vec{\nu}, z_e) \;\Rightarrow\; \vec{z} \wedge [\vec{\mathbb{H}}_{e+1}^T(\vec{\nu}, z_e) - \vec{\mathbb{H}}_e^T(\vec{\nu}, z_e)] = \vec{\mathbb{J}}_e(\vec{\nu}) , \tag{9.78}$$

and of the definition of the associated complex admittances at interface $e$, i.e.,

$$\vec{\mathbb{H}}_e^T(\vec{\nu}, z_e) = Y_e' \, [\vec{z} \wedge \vec{\mathbb{E}}_e^T(\vec{\nu}, z_e)] \;;\; \vec{\mathbb{H}}_{e+1}^T(\vec{\nu}, z_e) = Y_e \, [\vec{z} \wedge \vec{\mathbb{E}}_{e+1}^T(\vec{\nu}, z_e)] , \tag{9.79}$$

we have

$$\vec{\mathbb{J}}_e(\vec{\nu}) = (Y_e' - Y_e) \, \vec{\mathbb{E}}_e^T(\vec{\nu}, z_e) \;\Rightarrow\; \vec{\mathbb{J}}_e(\vec{\nu}) . \vec{\mathbb{E}}_e^T(\vec{\nu}, z_e) = (Y_e' - Y_e) \, |\vec{\mathbb{E}}_e^T(\vec{\nu}, z_e)|^2 , \tag{9.80}$$

and hence

$$\frac{d^2F}{d\nu d\phi} = \frac{\nu}{2} |\vec{\mathbb{E}}_e^T (\vec{\nu}, z_e)|^2 \, \mathfrak{R}[Y_e - Y_e'] \,. \tag{9.81}$$

- We then use the result established in Chapter 6, equation (6.137) to calculate the flux of the Poynting vector radiated in the neighborhood of, and on either side of, an optical interface, namely

$$\Phi_e = \frac{\nu}{2} \, \mathfrak{R}[Y_e] \, |\vec{\mathbb{E}}_e^T (\vec{\nu}, z_e)|^2 \quad ; \quad \Phi_e' = -\frac{\nu}{2} \, \mathfrak{R}[Y_e'] \, |\vec{\mathbb{E}}_e^T (\vec{\nu}, z_e)|^2 \,. \tag{9.82}$$

Note in these two expressions the spatial frequency $\nu$, which does not appear in equation (6.137); this results from the fact that we are here considering a spectral density in 3D geometry (and not in 2D). At this step we observe, by comparison of (9.81) and (9.82), that the total power emitted by the source at surface $e$ is equal to the sum of the two fluxes emitted from each part of this surface.

- To go further we consider the case of a nonabsorbing cavity, so that the two fluxes in (9.82) are recovered in the end media, i.e.,

$$\begin{aligned} \frac{\nu}{2} \, \mathfrak{R}[Y_e] \, |\vec{\mathbb{E}}_e^T (\vec{\nu}, z_e)|^2 &= \frac{\nu}{2} \, \mathfrak{R}[\tilde{n}_s] \, |\vec{\mathbb{A}}_s^{T+} (\vec{\nu})|^2 \\ -\frac{\nu}{2} \, \mathfrak{R}[Y_e'] \, |\vec{\mathbb{E}}_e^T (\vec{\nu}, z_e)|^2 &= \frac{\nu}{2} \, \mathfrak{R}[\tilde{n}_0] \, |\vec{\mathbb{A}}_0^{T-} (\vec{\nu})|^2 \,. \end{aligned} \tag{9.83}$$

- Finally, we combine equations (9.81), (9.82), and (9.83) to get the desired expression:

$$\begin{aligned} \frac{d^2F}{d\nu d\phi} &= \frac{\nu}{2} \, \mathfrak{R}[Y_e] \, |\vec{\mathbb{E}}_e^T (\vec{\nu}, z_e)|^2 - \frac{\nu}{2} \, \mathfrak{R}[Y_e'] \, |\vec{\mathbb{E}}_e^T (\vec{\nu}, z_e)|^2 \\ &= \frac{\nu}{2} \, \mathfrak{R}[\tilde{n}_s] \, |\vec{\mathbb{A}}_s^{T+} (\vec{\nu})|^2 + \frac{\nu}{2} \, \mathfrak{R}[\tilde{n}_0] \, |\vec{\mathbb{A}}_0^{T-} (\vec{\nu})|^2 \,. \end{aligned} \tag{9.84}$$

This expression is equivalent to equation (9.77), since the tangential components are related to the corresponding total field by the expression below [see Chapter 6, equations (6.132) and (6.133)]:

$$\tilde{n}_s |\vec{\mathbb{A}}_s^{T+}|^2 = \frac{\alpha_s}{\omega \tilde{\mu}_s} |\vec{\mathbb{A}}_s^+|^2 \quad \text{et} \quad \tilde{n}_0 |\vec{\mathbb{A}}_0^{T-}|^2 = \frac{\alpha_0}{\omega \tilde{\mu}_0} |\vec{\mathbb{A}}_0^-|^2 \,. \tag{9.85}$$

In summary, we have verified analytically that in the absence of absorption, the power emitted by the source is equal to the flux of the Poynting vector in the end media, with each of the two terms (power and flux) being related to the structure of the cavity and the position of the source within the multilayer stack.

## 9.5 Trapped Light

For greater precision in the energy balance just established, we now turn our attention not to the light radiated in the far field by the cavity in the two end media, but to the light that might be trapped in this cavity and that will propagate within the multilayer in the form of guided modes, with or without attenuation.

To this end we consider the modal window introduced in Chapter 7 and defined by

$$\max(n_0, n_s) < \bar{\nu} < \max(n_j) \quad j = 1, ..., p, \tag{9.86}$$

where $\bar{\nu}$ is the normalized spatial frequency ($\bar{\nu} = \lambda\nu$).

It can be seen immediately that any incautious calculation of the guided flux using the same expressions as those used in Section 9.4 [e.g., (9.68)] would lead to a value of 0 since, in the modal window defined by (9.86), we saw (see Chapter 7) that in the absence of absorption in the end media,

$$\Re[\alpha_0] = \Re[\alpha_s] = 0. \tag{9.87}$$

Hence any treatment of trapped light requires a particular procedure, to be set out in detail in the text that follows.

### 9.5.1 *General Considerations*

Consider once more equation (9.75), which expresses the spectral density of the power provided by the source, and let $f_F(\nu, \phi)$ be the associated complex quantity, i.e.,

$$f_F(\nu, \phi) = \frac{\nu}{2}|\vec{\mathbb{J}}_e(\vec{\nu})|^2 \frac{1}{Y_e - Y'_e}. \tag{9.88}$$

Under these conditions, the power spectral density per unit polar angle is written as

$$\frac{dF}{d\phi} = \Re\left[\int_0^\infty f_F(\nu, \phi)\, d\nu\right] = \Re\left[\int_0^\infty \frac{\nu}{2}\frac{|\vec{\mathbb{J}}_e(\vec{\nu})|^2}{Y_e(\nu) - Y'_e(\nu)}\, d\nu\right], \tag{9.89}$$

where the frequency dependence of the complex admittances on either side of the emissive surface appears explicitly.

The calculation of this integral presupposes that there might be poles in the expression for function $f_F$, these poles being identified with the zeroes of the difference $Y_e(\nu) - Y'_e(\nu)$. Let $\nu_c$ be one of these zeroes. For this particular value of the spatial frequency, the absence of any discontinuity at the interface $e$ allows the difference of the admittances at the interface $e-1$

to be calculated by applying the same recurrence relation (see Chapter 6, Section 6.3.7) for the two quantities involved, i.e.,

$$Y_{e-1}(\nu_c) = \frac{Y_e(\nu_c)\cos\delta_e - i\tilde{n}_e\sin\delta_e}{\cos\delta_e - i[Y_e(\nu_c)/\tilde{n}_e]\sin\delta_e},$$

$$Y'_{e-1}(\nu_c) = \frac{Y'_e(\nu_c)\cos\delta_e - i\tilde{n}_e\sin\delta_e}{\cos\delta_e - i[Y'_e(\nu_c)/\tilde{n}_e]\sin\delta_e}.$$

(9.90)

To improve the clarity of these expressions, we have made no explicit mention of the dependence on $\nu_c$ of the effective refractive index $\tilde{n}_e$ and phase shift $\delta_e$. A direct comparison of these two relations shows that the equality between $Y_e(\nu_c)$ and $Y'_e(\nu_c)$ automatically leads to equality between $Y_{e-1}(\nu_c)$ and $Y'_{e-1}(\nu_c)$. Hence, layer by layer, we get to interface 0, for which we have

$$Y_0(\nu_c) = Y'_0(\nu_c) \quad \text{and also} \quad Y_0(\nu_c) + \tilde{n}_0(\nu_c) = 0. \tag{9.91}$$

Equation (9.91) shows (see Chapter 7) that frequency $\nu_c$ is identified with one of the guided modes $(\nu_{g,0})$ of the multilayer structure of which the planar cavity is composed.

This procedure also highlights the fact that the notion of pole (or guided mode) involves the entire volume of the component: using the same approach as that used to get up to the upper interface of the stack, we could descend to the lower interface of this same stack and hence show that, for each interface $j$, the difference between the admittances taken on either side of the interfaces was 0; and hence by extension that this difference was everywhere 0 within the thickness of the stack regardless of the direction of the illumination (coming from the substrate or the superstrate).

As pointed out in Chapter 7, these poles cannot be real in the frequential window associated with free space, defined by

$$0 < \bar{\nu} < \max(n_0, n_s). \tag{9.92}$$

Consequently, over this frequency interval, the integral in equation (9.89) can be calculated without any special precautions. However, note that if the substrate is of finite thickness and of index greater than that of the superstrate (as is the case for all the planar cavities studied in Section 9.3), the first integral must be split into two distinct contributions, as shown in the following:

$$\frac{dF}{d\phi}\{0 \leqslant \nu \leqslant n_s/\lambda\} = \Re\left[\int_0^{n_0/\lambda} f_F(\nu,\phi)\,d\nu\right] + \Re\left[\int_{n_0/\lambda}^{n_s/\lambda} f_F(\nu,\phi)\,d\nu\right]. \tag{9.93}$$

The first integral describes the power density provided by the source and radiated in free space into the two end media, while the second describes the power density provided by the same source, but that remains trapped in the substrate by total internal reflection.

Now and concerning the integral in the modal frequency window defined in (9.86), we know that real poles may exist in this window. Hence the situation is different and two cases must then be considered (the associated comments are illustrated by the figures that follow in this section):

- If the multilayer is absorbing, then these poles are complex and peaks will appear in the frequency dependence of the power spectral density; the smaller the imaginary part of the frequency of the pole, the narrower the width of the peaks. Integration of the function $f_F$ in the neighborhood of these peaks will therefore give the amount of trapped energy that propagates in modal form in the cavity.
- If the multilayer is transparent, the situation requires a more careful analysis. We saw in Chapter 7, that in the absence of absorption, all the admittances were imaginary in the modal window; this amounts to saying that, restricted to this single frequency domain, the integral (9.89) would be 0 since the quantity $Y_e - Y'_e$ would always be imaginary there. However, this result is no longer true if the functions being integrated contain singularities. Indeed, we showed in Chapter 7 that when we decrease the imaginary part of the refractive index ($\kappa$) that is responsible for thin films absorption in the stack, the widths of the peaks appearing in the spectral dependence of the power density decrease, while the amplitudes of these peaks increase; in fact, this sequence of functions behaves as if it was converging to a Dirac distribution.

Finally, this problem can be treated rigorously by making use of the residue theorem (the function $f_F$ satisfies the requisite conditions of analyticity), leading to the following equation:

$$
\frac{dF}{d\phi}\{\max(n_0, n_s) \leqslant \nu \leqslant \max(n_j)\} = \Re \left[ \int_{\max(n_0,n_s)/\lambda}^{\max(n_j)/\lambda} f_F(\nu, \phi)\, d\nu \right]
$$

$$
= \Re \left[ i\pi \sum_q \mathrm{Res}(f_F; \nu_q) \right], \quad (9.94)
$$

where $\nu_q$ is the frequency of the pole of order $q$. Note that the coefficient in front of the sum of residues is $i\pi$ (not $2i\pi$) since the poles concerned are located on the real axis, which is here the boundary of the integration path used in the complex plane.

Two possible approaches are therefore available for treating the problem of light trapped in a transparent multilayer cavity. The first consists of using equation (9.94) and hence requires an analytical extension of the function $f_F$ (the real frequency $\nu$ is extended to a complex variable frequency) whose residues are extracted. The second (for a more immediate numerical implementation) consists of reproducing the method used in Chapter 7, Section 7.7 to study the relationships between resonances and guided modes. This involved assigning an imaginary index $\kappa$, of small but arbitrary value (e.g. between $10^{-5}$ and $10^{-7}$), to all the layers in the stack and, in accordance with equation (9.89), integrating the function $f_F$ along the real axis, around each of the peaks appearing in the power spectral density. Provided the stability of the result with respect to any modification in the value of $\kappa$ can be verified, the result of this integration in the neighborhood of the characteristic frequency of each mode of the cavity will yield the amount of energy transported by it ; since the values of $\kappa$ are small, the energy trapped in the transparent cavities can be calculated with great precision.

For completeness, note that in the case of a nonabsorbent multilayer, no power is emitted at frequencies outside the modal window, i.e., those defined by

$$\bar{\nu} > \max(n_j) \quad j = 1, ..., p. \tag{9.95}$$

Indeed we showed in Chapter 7 that in this domain of frequencies there were no real poles, while admittances $Y_e$ and $Y'_e$ are imaginary. However, this result is no longer true when absorption is present, and restricting the integration to this frequency domain then gives the amount of light emitted (and absorbed). As already highlighted in Chapters 4 and 7 this illustrates the fact that the high frequencies, often described as evanescent, can nevertheless contribute to the energy balance via the dissipative component.

We conclude this section by emphasizing the fact that introducing the power spectral density (rather than the spectral flux density) is essential for any proper treatment of the problem of trapped light, at least for absorbing media. With the energy balance in the modal window given by

$$F_m = \Phi_m + A_m, \tag{9.96}$$

then indeed $F_m = \Phi_m$ for a transparent cavity, so in this case it makes no difference whether we work with power or flux.

Conversely, for an absorbent cavity we have $F_m = A_m$ and hence $\Phi_m = 0$, since the energy transported by the modes is progressively (and totally) absorbed over the course of its propagation within the structure, with the surface serving as a reference for the balance of energy being rejected into infinity in the directions of the plane $xOy$. In this case, the formulae involving the flux spectral density allow no use to be made of the quantity of trapped light.

### 9.5.2 Examples of Implementation

#### Reference Configuration

This reference configuration was introduced in Section 9.3.1 and corresponds to the minimum possible structure that can be created in practice: this simply consists of a thin (10 nm) layer of emissive material, of refractive index 1.7 and deposited onto the surface of an N-BK7 substrate of finite thickness. As explained in Section 9.5.1, we have assigned a small valued imaginary index $\kappa$ (here $\kappa = 10^{-6}$) to this emissive material.

For both light polarization states and at a wavelength of 600 nm, Figure 9.24 shows the variation of the source power spectral density per unit frequency and polar angle as a function of the normalized spatial frequency $\bar{\nu}$. Three frequential windows can be distinguished: the window in which the power is radiated into free space ($0 \leqslant \bar{\nu} \leqslant 1$, white circles on the graph), the window in which the power from the source is trapped by total internal reflection in the substrate ($1 < \bar{\nu} < 1.5163$, gray line on the graph) and finally, the window in which this power is dissipated ($1.5163 < \bar{\nu}$, black line on the graph). For this minimal structure there are no guided modes so that, in the modal window and beyond, the power density is 0 when there is no absorption. As in this case where absorption is weak ($\kappa = 10^{-6}$), the power density falls asymptotically to 0 in the frequency window $\bar{\nu} > 1.5163$.

On the basis of these graphs, the power density per unit polar angle, and hence variation with wavelength, can be obtained by integrating over each of these frequency windows. This is shown in arbitrary units on the left of Figure 9.25 and as a percentage of the total flux (the sum of radiative and trapped contributions for both polarization states) on the right of the same Figure. Note the significant amount of flux trapped by total reflection in the substrate, particularly in TE polarization (roughly 45% of the total). More generally, we observe that the amount of light trapped in the substrate is of the same order of magnitude as the total flux emitted into the end media. This fact cannot be altered by depositing an antireflection coating on the rear face of the substrate, since the total reflection condition depends only on the refractive index of the emergent medium (see Chapter 7). Finally, for this reference configuration, the contribution of the guided flux is asymptotically zero (when $\kappa$ tends to 0), since the structure has modes in neither TE nor TM.

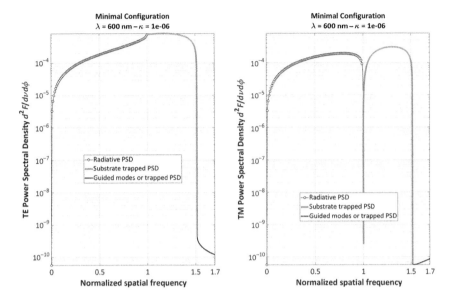

Figure 9.24 Reference configuration ($\kappa = 10^{-6}$). Variation of the power spectral density of the source as a function of the normalized spatial frequency (left, TE; right, TM; see text for more details).

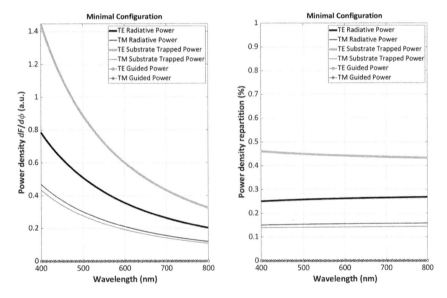

Figure 9.25 Reference configuration. Wavelength dependence of the source power density per unit polar angle in various frequency windows (left: in arbitrary units; right: percentage of the total power from the source; see text for more details).

### Fabry–Perot

Now consider the Fabry–Perot structure studied in Section 9.3.3: this comprises a stack of two quarter-wave mirrors of type $Q_2$ and $Q_4$, with the emissive planar source located in the middle of the *spacer*, thus ensuring that the resonance conditions introduced in Section 9.3.4 are met. As in the case of the reference configuration, we can plot the variation of the power spectral density of this source as a function of the normalized spatial frequency (see Figure 9.26).

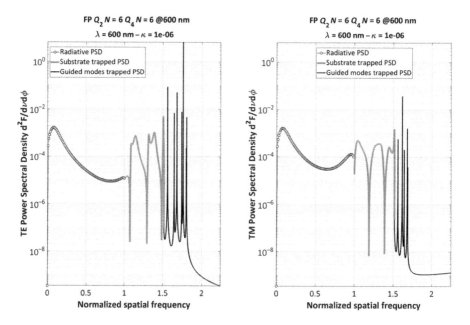

Figure 9.26 Fabry–Perot Air / $(\text{HL})^6$ ES $\text{L}(\text{HL})^6$ / N-BK7 / Air. Variation of the power spectral density of this source as a function of the normalized spatial frequency (left, TE; right, TM; see text for more details).

This time, the modal window extends from the index of the substrate (N-BK7) at the wavelength being considered (600 nm) to that of the high-index layer ($\text{Ta}_2\text{O}_5$) at the same wavelength, i.e., $1.5163 < \bar{\nu} < 2.1489$. As expected in the presence of low absorption ($\kappa = 10^{-6}$), there are peaks representing the amount of light transported by the guided modes of the multilayer structure.

Clearly, the normalized spatial frequencies of these peaks depend on the wavelength, as can be seen in Figures 9.27 and 9.28:

- On the one hand, Figure 9.27 shows, for the two polarization states, the variation with wavelength of the normalized spatial frequencies of the guided modes in the multilayer structure being considered; note that certain modes gradually vanish as the wavelength increases: in fact, the normalized spatial frequency tends toward the lower bound of the modal window, this being the substrate index at that wavelength (black curve on the graph); beyond this value the guided wave condition is no longer satisfied.

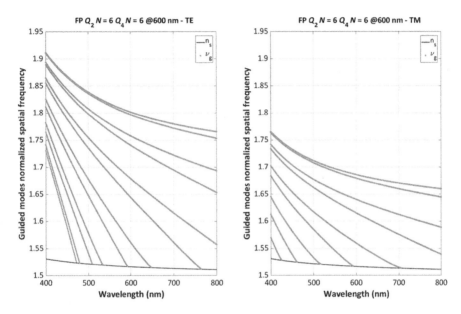

Figure 9.27 Fabry–Perot Air / (HL)$^6$ ES L(HL)$^6$ / N-BK7 / Air. Variation with wavelength of the normalized spatial frequencies of the guided modes of the multilayer structure (left, TE; right, TM). The black curve shows the spectral dependence of the index of the N-BK7 substrate.

- On the other hand, the function $f_F(\bar{\nu}, \lambda)$ is represented in gray levels in Figure 9.28. Recall at this point that the structures that appear in the frequency window between zero frequency and the substrate index are scarcely affected by any alteration in the value of $\kappa$, to the extent that this value remains small (typically less than $10^{-5}$), while the amplitudes of the peaks in the modal window vary in inverse proportion to the value of this same parameter $\kappa$.

As explained in Section 9.5.1, integrating the function $f_F(\nu, \phi)$ in the neighborhood of the characteristic frequency of one of the structures guided modes, and using a sequence of decreasing values for parameter $\kappa$, allows us to obtain a sequence of values that converge toward the quantity of energy transported by this mode. An example of the result obtained for TE

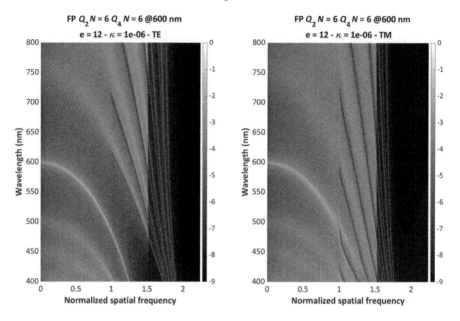

Figure 9.28 Fabry–Perot Air / (HL)$^6$ ES L(HL)$^6$ / N-BK7 / Air. Gray-level representation of the function $f_F(\bar{\nu}, \lambda)$ [log units: left, TE; right, TM].

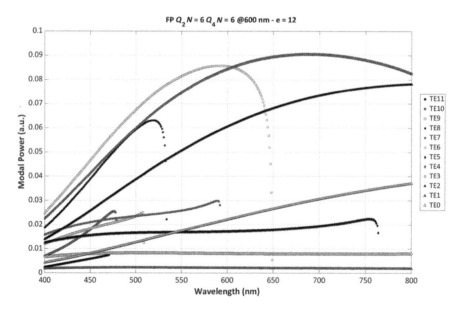

Figure 9.29 Fabry–Perot Air / (HL)$^6$ ES L(HL)$^6$ / N-BK7 / Air Variation in the power transported by each mode as a function of wavelength (TE polarization).

polarization is shown in Figure 9.29. The 12 modes appearing on the graph to the left of Figure 9.27 are set out in increasing order of their normalized spatial frequencies.

The power transported by each of the first 7 modes vanishes when the normalized spatial frequency of one of these modes becomes equal to the index of the substrate at the corresponding wavelength (e.g., see curves TE 8 or TE 6 in Figure 9.29).

Now that this result is available for both TE and TM polarization, we can provide a detailed distribution for the energy balance (see Figure 9.30), showing the following contributions:

1. Power radiated into free space (TE, TM, and total)
2. Power provided by the source and trapped by total internal reflection in the substrate (TE and TM)
3. Power provided by the source and transported by the guided modes of the structure (TE and TM)
4. Trapped power, which refers to the sum (TE and TM) of items 2 and 3

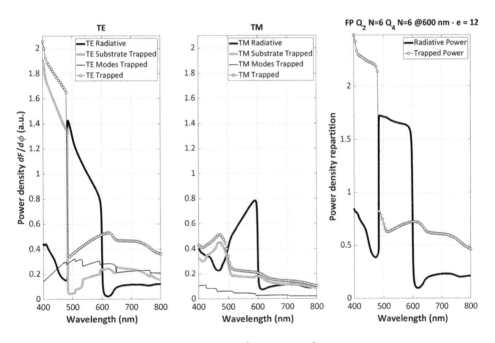

Figure 9.30 Fabry–Perot Air / (HL)$^6$ ES L(HL)$^6$ / N-BK7 / Air. Spectral dependence of energy balance contributions [left, TE – center, TM – right, overall (TE + TM), see text].

Note that the **trapped power**, in both TE and TM, reveals a relatively gentle dependence as a function of wavelength, while the variation in power transported by the guided modes of the multilayer (item 2) shows a more discontinuous structure, on account of the abrupt disappearance of a mode each time its normalized spatial frequency becomes the same as the value of the substrate index at that frequency. This highlights the fact that, when the overall power transported by the guided modes shows an abrupt fall, this is compensated by an increase in the same value of power trapped by total internal reflection within the substrate.

The graph on the right shows a comparison between the spectral dependence of the total radiated power in free space (TE + TM) and that of the total trapped power, as previously defined. Note that the radiative part is broadly dominant between 500 and 600 nm, while it is the trapped contribution that dominates outside this spectral domain.

In extending the concepts established in Section 9.3.4, we might analyze the consequences of failing to meet one of the resonance conditions [(9.57) and (9.58)] that govern the effectiveness of emission along the axis of a multilayer planar microcavity. So as to allow easy comparison between these two situations, we have chosen to keep the multilayer stack just as it is (the strong condition is therefore still satisfied) and simply displace the position of the source by locating it not in the middle of the spacer ($e = 12$), but above it ($e = 11$). Hence the weak conditions are no longer met.

Figure 9.31 shows the new distribution of power from the source in this modified configuration.

We first observe that, for TE polarization, the power transported by the guided modes is, over almost the entire spectral domain, greater than that trapped by total internal reflection within the substrate; this was not the case when the source was located in the middle of the spacer of the Fabry–Perot cavity. Moreover, and more generally, we observe that:

- The power radiated in free space by this new structure is very significantly reduced, especially in the spectral band (500–600 nm) where previously it dominated; this phenomenon is particularly marked for TE polarization.
- As in the case $e = 12$, we find contrary power variations, on the one hand trapped by total internal reflection within the substrate, and on the other hand transported by the guided modes of the multilayer structure, when the normalized spatial frequency of one of the guided modes becomes the same as the substrate's refractive index at that wavelength.
- The trapped light is now broadly dominant at all wavelengths in the energy balance of this microcavity.

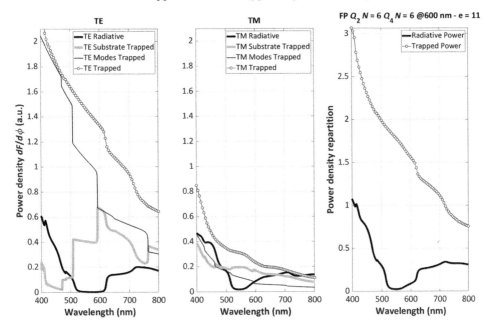

Figure 9.31 Fabry–Perot Air / (HL)⁵ H ES 2L (HL)⁶ / N-BK7 / Air. Spectral dependence of energy balance contributions [left, TE – center, TM – right, overall (TE + TM), see text].

If necessary, these results confirm that the position of the source in the spacer of a Fabry–Perot cavity crucially governs the effects of enhancement or inhibition on emission.

## 9.6 Application to Trapped Light Scattering

### 9.6.1 Introduction

It goes without saying that the results established in the previous section can be extended to the general case where each stack interface is the support for electric $(\vec{J_j})$ and magnetic $(\vec{M_j})$ sources. In Chapter 8 we encountered a case that corresponded precisely to such a description, except for the fact that the sources were *fictitious*: this involved the scattering of light by a multilayer stack. The parallel inclusion of the results established in Chapter 8 with those just established in this chapter will enable us to tackle the problem of scattered light trapped in a multilayer structure, a problem not addressed thus far. We showed in Chapter 8 how to calculate the intensity of the light scattered by the roughness of the interfaces in a stack, but we are limited to calculating the flux radiated in the far field. To complete this energy balance, we now need to include this additional component, henceforth described as *trapped light scattering.*

This process of trapped light scattering is similar to that of coupling by a grating, although coupling here is caused by the roughness of the disordered structure at the interfaces. As with planar microcavities, this trapped light scattering is transported by the electromagnetic modes of the multilayer waveguide and becomes progressively attenuated as it propagates. In cases in which the waveguide is short and absorption is weak, this light can strike the side faces of the planar guide and create new diffraction and scattering phenomena on exiting the guide. However, bearing in mind the usual dimensions of optical interference filters (in the order of a centimeter) and of the values of the imaginary parts of the refractive indices of thin film materials (greater than $10^{-6}$, roughly), it is more likely that the scattered trapped light will be absorbed and will consequently contribute to the absorption component (A) of the energy balance. We then speak of **absorption induced by interface roughness**. This situation is illustrated in Figure 9.32, which shows the difference between the *classical* absorption related to the integral of the modulus squared of the field under the illuminated spot, and the absorption *induced by the roughness of the interfaces* resulting from the attenuation of the modes, and generally involves a volume that extends beyond the illuminated spot.

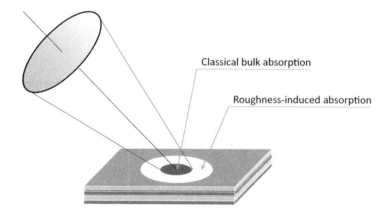

Classical bulk absorption

Roughness-induced absorption

Figure 9.32 *Classical* absorption (dark gray spot) and absorption *induced by interface roughness* (white spot). See text for more detail.

In this context it is important to know, especially for very low loss (e.g., less than a few tens of ppm) interference filters, whether this trapped component is of the same order of magnitude as the classical component, normally measured in the far field. This question is far from trivial since, if trapped light scattering predominates over classical absorption, then any effort to reduce interface roughness should be prioritized, concentrating on polishing techniques (which condition the roughness of the substrate) and reducing the number of localized defects generated by the deposition process, rather than optimizing the stoichiometry of the deposited materials or improving the purity of the targets that are evaporated or sputtered in the filter fabrication chambers.

Finally, it is important to make a distinction between two types of trapped light scattering (see Figure 9.33), namely *coherent* and *incoherent*. The *co-*

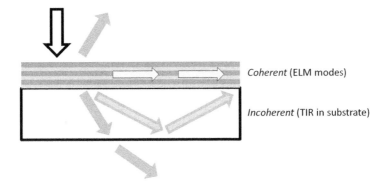

Figure 9.33 *Coherent* (electromagnetic modes) and *incoherent* (total internal reflection in substrate) trapped light [see text for more details].

*herent* component requires a knowledge of the electromagnetic modes of the multilayer structure. As with the microcavities, the first thing to do is to express the power spectral density associated with all the scattering sources within the multilayer, a density whose poles (characteristic of the modes of propagation) govern the trapped light; then to integrate this density function either in the complex plane (using the method of residues) or over the real axis, making the coefficient $\kappa$ tend to 0, this coefficient defining the absorption of the multilayer (see Section 9.5).

The *incoherent* component involves the quantity of light trapped by total internal reflection (TIR) within the substrate (see Figure 9.33). In this case, and taking account of the great thickness of the substrate in relation to the illuminating wavelength, the calculation will be more direct and it does not matter whether we integrate the flux spectral density $f_\Phi$ or the power spectral density $f_F$ in the corresponding frequency window, i.e.,

$$\frac{dF_{\mathrm{TIR}}}{d\phi} = \Re \left[ \int_{n_0/\lambda}^{n_s/\lambda} f_F(\nu, \phi) \, d\nu \right]. \tag{9.97}$$

Note that this light trapped in the substrate is often quite considerable, and by illuminating the side faces of the substrate, can then create an additional parasitic light: hence at this level of accuracy, it is important to polish the side faces of the substrates to minimize this contribution.

### 9.6.2 Power Spectral Density

To calculate the power emitted by a distribution of electric and magnetic surface sources, we proceed as in Section 9.2 provided the currents are written as

$$\vec{\mathcal{J}}(\vec{r}, z) = \sum_{j=0}^{p} \vec{\mathcal{J}}_j(\vec{r})\, \delta(z - z_j) \quad ; \quad \vec{\mathcal{M}}(\vec{r}, z) = \sum_{j=0}^{p} \vec{\mathcal{M}}_j(\vec{r})\, \delta(z - z_j), \quad (9.98)$$

where $\delta$ is the Dirac distribution and the $z_j$ denote the positions of the interfaces in the multilayer stack. In the case of surface light scattering, these currents are given by equations (8.21) established in Chapter 8.

The power associated with this scattering is obtained by expanding the divergence of the Poynting vector $\vec{P}$, and including the presence of a magnetic current (see Chapter 2, Sections 2.10 and 2.12), i.e.,

$$\mathrm{div}\,\vec{P} = \frac{1}{2}\left\{ (\vec{\mathcal{M}}^{*}.\vec{\mathcal{H}} - \vec{\mathcal{J}}.\vec{\mathcal{E}}^{*}) + i\omega\tilde{\epsilon}|\vec{\mathcal{E}}|^2 - i\omega\tilde{\mu}^{*}|\vec{\mathcal{H}}|^2 \right\}. \quad (9.99)$$

Hence, we get

$$F = \frac{1}{2}\Re\left[ \int_{\Omega} (\vec{\mathcal{M}}^{*}.\vec{\mathcal{H}} - \vec{\mathcal{J}}.\vec{\mathcal{E}}^{*})\, d\Omega \right], \quad (9.100)$$

and, using equations (9.98),

$$F = \frac{1}{2}\Re\left[ \int_{\vec{r},z} \left\{ \vec{\mathcal{H}}(\vec{r}, z).\sum_{j=0}^{p} \vec{\mathcal{M}}_j^{*}(\vec{r})\, \delta(z - z_j) - \vec{\mathcal{E}}^{*}(\vec{r}, z).\sum_{j=0}^{p} \vec{\mathcal{J}}_j(\vec{r})\, \delta(z - z_j) \right\} \right].$$

So, using the properties of the Dirac distribution,

$$F = \frac{1}{2}\Re\left[ \sum_{j=0}^{p} \int_{\vec{r}} \vec{\mathcal{M}}_j^{*}(\vec{r}).\vec{\mathcal{H}}(\vec{r}, z_j)\, d^2\vec{r} - \sum_{j=0}^{p} \int_{\vec{r}} \vec{\mathcal{J}}_j(\vec{r}).\vec{\mathcal{E}}^{*}(\vec{r}, z_j)\, d^2\vec{r} \right]. \quad (9.101)$$

Applying Parseval's theorem then leads to

$$F = \frac{1}{2}\Re\left[ \sum_{j=0}^{p} \int_{\vec{\nu}} \vec{\mathbb{M}}_j^{*}(\vec{\nu}).\vec{\mathbb{H}}(\vec{\nu}, z_j)\, d^2\vec{\nu} - \sum_{j=0}^{p} \int_{\vec{\nu}} \vec{\mathbb{J}}_j(\vec{\nu}).\vec{\mathbb{E}}^{*}(\vec{\nu}, z_j)\, d^2\vec{\nu} \right], \quad (9.102)$$

and, for the power spectral density per unit spatial frequency and polar angle,

$$\frac{d^2 F}{d\nu d\phi} = \frac{\nu}{2}\Re\left[\sum_{j=0}^{p}\vec{\mathbb{M}}_j^*(\vec{\nu}).\vec{\mathbb{H}}(\vec{\nu}, z_j) - \sum_{j=0}^{p}\vec{\mathbb{J}}_j(\vec{\nu}).\vec{\mathbb{E}}^*(\vec{\nu}, z_j)\right]. \qquad (9.103)$$

### 9.6.3 Choice of Tangential Component

Recall once again that, since the currents are tangential (see Chapter 2, Section 2.12), the fields $\vec{\mathbb{E}}$ and $\vec{\mathbb{H}}$ in equation (9.103) are also tangential. However, since these tangential fields are both discontinuous in the presence of electric and magnetic currents, equation (9.103) is ambiguous, since it fails to say on which side of interface $j$ field $\vec{\mathbb{E}}$ or field $\vec{\mathbb{H}}$ should be chosen.

To resolve this issue, we need to reconsider equation (9.99) in a distributions sense. In the case of currents on the single interface $j$, this equation can be written as

$$\mathrm{div}\vec{P} = \frac{1}{2}\left\{\left[\vec{\mathcal{M}}_j^*(\vec{r}).\vec{\mathcal{H}}(\vec{r}, z_j) - \vec{\mathcal{J}}_j(\vec{r}).\vec{\mathcal{E}}^*(\vec{r}, z_j)\right]\delta(z - z_j)\right.$$
$$\left. + i w\tilde{e}|\vec{\mathcal{E}}(\vec{r}, z)|^2 - i w\tilde{\mu}^*|\vec{\mathcal{H}}(\vec{r}, z)|^2\right\}. \qquad (9.104)$$

However, we also know that, in the sense of distributions, the divergence can be expanded as (see Chapter 2, Section 2.12)

$$\mathrm{div}\vec{P} = (\mathrm{div}\vec{P}) + [\vec{z}.\Delta\vec{P}_j]\delta(z - z_j), \qquad (9.105)$$

where $(\mathrm{div}\vec{P})$ is the usual expression (regular in the sense of functions) and $\vec{z}.\Delta\vec{P}_j$ is the jump in the Poynting vector on crossing the plane $z = z_j$, for a normal oriented in the direction $z > 0$.

Consequently, identifying the singular parts of (9.104) and (9.105) leads to

$$\vec{z}.\Delta\vec{P}_j = \frac{1}{2}\left[\vec{\mathcal{M}}_j^*(\vec{r}).\vec{\mathcal{H}}(\vec{r}, z_j) - \vec{\mathcal{J}}_j(\vec{r}).\vec{\mathcal{E}}^*(\vec{r}, z_j)\right] \text{ with } \Delta\vec{P}_j = \vec{P}_j - \vec{P}_j', \qquad (9.106)$$

and where for the vector $\vec{P}_j$ we have used the same convention as for the complex admittance when this admittance is discontinuous on crossing an interface. Again, note that only the tangential field components are involved in equation (9.106).

Now express this vector $\Delta\vec{P}_j$ using the boundary conditions between tangential field components at interface $j$. These are written as

$$\vec{z} \wedge [\vec{\mathcal{E}}_{j+1}^T(\vec{r}, z_j) - \vec{\mathcal{E}}_j^T(\vec{r}, z_j)] = \vec{\mathcal{M}}_j(\vec{r}),$$
$$\vec{z} \wedge [\vec{\mathcal{H}}_{j+1}^T(\vec{r}, z_j) - \vec{\mathcal{H}}_j^T(\vec{r}, z_j)] = \vec{\mathcal{J}}_j(\vec{r}), \qquad (9.107)$$

and, using the properties of the vector triple product,

$$
\begin{aligned}
\vec{\mathcal{E}}_{j+1}^T(\vec{r}, z_j) - \vec{\mathcal{E}}_j^T(\vec{r}, z_j) &= -\vec{z} \wedge \vec{\mathcal{M}}_j(\vec{r}) , \\
\vec{\mathcal{H}}_{j+1}^T(\vec{r}, z_j) - \vec{\mathcal{H}}_j^T(\vec{r}, z_j) &= -\vec{z} \wedge \vec{\mathcal{J}}_j(\vec{r}) .
\end{aligned}
\tag{9.108}
$$

On either side of interface $j$ the Poynting vector is expressed as

$$
\begin{aligned}
\vec{P}_j &= \frac{1}{2} \vec{\mathcal{E}}_{j+1}^{T*}(\vec{r}, z_j) \wedge \vec{\mathcal{H}}_{j+1}^T(\vec{r}, z_j) , \\
\vec{P}_j' &= \frac{1}{2} \vec{\mathcal{E}}_j^{T*}(\vec{r}, z_j) \wedge \vec{\mathcal{H}}_j^T(\vec{r}, z_j) .
\end{aligned}
\tag{9.109}
$$

and, using equations (9.108),

$$
\begin{aligned}
\vec{P}_j &= \frac{1}{2} \left[ \vec{\mathcal{E}}_j^{T*}(\vec{r}, z_j) - \vec{z} \wedge \vec{\mathcal{M}}_j^*(\vec{r}) \right] \wedge \vec{\mathcal{H}}_{j+1}^T(\vec{r}, z_j) , \\
\vec{P}_j' &= \frac{1}{2} \vec{\mathcal{E}}_j^{T*}(\vec{r}, z_j) \wedge \left[ \vec{\mathcal{H}}_{j+1}^T(\vec{r}, z_j) + \vec{z} \wedge \vec{\mathcal{J}}_j(\vec{r}) \right] .
\end{aligned}
\tag{9.110}
$$

From this can be deduced a new formulation for the quantity $\vec{z}.\Delta\vec{P}_j$:

$$
\vec{z}.\Delta\vec{P}_j = \vec{z}.\vec{P}_j - \vec{z}.\vec{P}_j' = \frac{1}{2} \left[ \vec{\mathcal{M}}_j^*(\vec{r}).\vec{\mathcal{H}}_{j+1}^T(\vec{r}, z_j) - \vec{\mathcal{J}}_j(\vec{r}).\vec{\mathcal{E}}_j^{T*}(\vec{r}, z_j) \right] .
\tag{9.111}
$$

This expression tells us which fields to choose in order to express power, initially in equation (9.101), and then, after a Fourier transformation, in equation (9.103), i.e.,

$$
\frac{d^2 F}{d\nu d\phi} = \frac{\nu}{2} \Re \left[ \sum_{j=0}^{p} \vec{\mathbb{M}}_j^*(\vec{\nu}).\vec{\mathbb{H}}_{j+1}^T(\vec{\nu}, z_j) - \sum_{j=0}^{p} \vec{\mathbb{J}}_j(\vec{\nu}).\vec{\mathbb{E}}_j^{T*}(\vec{\nu}, z_j) \right] .
\tag{9.112}
$$

Hence, we observe that the fields $\vec{\mathbb{E}}(\vec{\nu}, z_j)$ and $\vec{\mathbb{H}}(\vec{\nu}, z_j)$ must not be taken on the same side of interface $j$.

For completeness, note that other formulations for $\vec{z}.\Delta\vec{P}_j$ can be used, since similar expansions also lead to

$$
\vec{z}.\Delta\vec{P}_j = \frac{1}{2} \left[ \vec{\mathcal{M}}_j^*(\vec{r}).\vec{\mathcal{H}}_j^T(\vec{r}, z_j) - \vec{\mathcal{J}}_j(\vec{r}).\vec{\mathcal{E}}_{j+1}^{T*}(\vec{r}, z_j) \right] ,
\tag{9.113}
$$

or to

$$
\vec{z}.\Delta\vec{P}_j = \frac{1}{2} \left[ \vec{\mathcal{M}}_j^*(\vec{r}).\vec{\mathcal{H}}_{j,a}^T(\vec{r}, z_j) - \vec{\mathcal{J}}_j(\vec{r}).\vec{\mathcal{E}}_{j,a}^{T*}(\vec{r}, z_j) \right] ,
\tag{9.114}
$$

with the average fields

$$
\vec{\mathcal{H}}_{j,a}^T(\vec{r}, z_j) = \frac{1}{2} \left\{ \vec{\mathcal{H}}_{j+1}^T(\vec{r}, z_j) + \vec{\mathcal{H}}_j^T(\vec{r}, z_j) \right\} ,
\tag{9.115}
$$

and

$$
\vec{\mathcal{E}}_{j,a}^T(\vec{r}, z_j) = \frac{1}{2} \left\{ \vec{\mathcal{E}}_{j+1}^T(\vec{r}, z_j) + \vec{\mathcal{E}}_j^T(\vec{r}, z_j) \right\} .
\tag{9.116}
$$

### 9.6.4 Expression for Power Density as a Function of Currents

In equation (9.112), it is important to point out that, following on from what we encountered in Section 8.3.6 when calculating the emerging scattered waves, each field at interface $j$ results from the summation of those generated by fictitious currents on the various interfaces ($k = 1, ..., p$) of the stack. From now on we shall use the index $\Sigma$ for the fields resulting from this summation to avoid confusing them with those radiated by a single interface. The expression for the power spectral density given in (9.112) then becomes

$$f_F(\nu) = \frac{\nu}{2} \Re \left[ \sum_{j=0}^{p} \vec{\mathbb{M}}_j^*(\vec{\nu}).\vec{\mathbb{H}}_{\Sigma,j+1}^T(\vec{\nu}, z_j) - \sum_{j=0}^{p} \vec{\mathbb{J}}_j(\vec{\nu}).\vec{\mathbb{E}}_{\Sigma,j}^{T*}(\vec{\nu}, z_j) \right], \quad (9.117)$$

with

$$\vec{\mathbb{H}}_{\Sigma,j+1}^T(\vec{\nu}, z_j) = \sum_{k=0}^{p} \vec{\mathbb{H}}_{k+1}^T(\vec{\nu}, z_j) \quad \text{et} \quad \vec{\mathbb{E}}_{\Sigma,j}^T(\vec{\nu}, z_j) = \sum_{k=0}^{p} \vec{\mathbb{E}}_k^T(\vec{\nu}, z_j), \quad (9.118)$$

where we note that, in each of the last two summations, the field at interface $k$ is associated with ordinate $z_j$. Hence this involves the value at interface $j$ of the field created by interface $k$.

It is important to point out here that the fields created by the electric and magnetic currents located at interface $k$ are discontinuous at the crossing of this interface $k$, as defined by the following boundary conditions:

$$\begin{aligned} \vec{z} \wedge [\vec{\mathbb{E}}_{k+1}^T(\vec{r}, z_k) - \vec{\mathbb{E}}_k^T(\vec{r}, z_k)] &= \vec{\mathbb{M}}_k(\vec{r}), \\ \vec{z} \wedge [\vec{\mathbb{H}}_{k+1}^T(\vec{r}, z_k) - \vec{\mathbb{H}}_k^T(\vec{r}, z_k)] &= \vec{\mathbb{J}}_k(\vec{r}). \end{aligned} \quad (9.119)$$

but conversely, the values at interface $j$ of the fields created by the currents located at interface $k$ are continuous at the crossing of this interface $j$, when $j$ is different from $k$. As a result, this discontinuity requires some precautions to be taken in calculating the transmission factors between interfaces $k$ and $j$. Hence there are three distinct cases we need to consider, namely $k < j$, $k = j$ and $k > j$.

For the electric field we have

$$\begin{aligned} \vec{\mathbb{E}}_{\Sigma,j}^T(\vec{\nu}, z_j) &= \sum_{k=0}^{j-1} \vec{\mathbb{E}}_k^T(\vec{\nu}, z_j) + \vec{\mathbb{E}}_j^T(\vec{\nu}, z_j) + \sum_{k=j+1}^{p} \vec{\mathbb{E}}_k^T(\vec{\nu}, z_j) \\ &= \sum_{k=0}^{j-1} t_{kj} \vec{\mathbb{E}}_{k+1}^T(\vec{\nu}, z_k) + \vec{\mathbb{E}}_j^T(\vec{\nu}, z_j) + \sum_{k=j+1}^{p} t'_{kj} \vec{\mathbb{E}}_k^T(\vec{\nu}, z_k). \end{aligned} \quad (9.120)$$

Similarly, for the magnetic field we get

$$\vec{\mathbb{H}}^T_{\Sigma,j+1}(\vec{\nu},z_j) = \vec{z} \wedge Y_j\,\vec{\mathbb{E}}^T_{\Sigma,j+1}(\vec{\nu},z_j)$$

$$= Y_j\vec{z} \wedge \left\{ \sum_{k=0}^{j-1} \vec{\mathbb{E}}^T_{k+1}(\vec{\nu},z_j) + \vec{\mathbb{E}}^T_{j+1}(\vec{\nu},z_j) + \sum_{k=j+1}^{p} \vec{\mathbb{E}}^T_{k+1}(\vec{\nu},z_j) \right\}$$

$$= Y_j\vec{z} \wedge \left\{ \sum_{k=0}^{j-1} t_{kj}\vec{\mathbb{E}}^T_{k+1}(\vec{\nu},z_k) + \vec{\mathbb{E}}^T_{j+1}(\vec{\nu},z_j) + \sum_{k=j+1}^{p} t'_{kj}\vec{\mathbb{E}}^T_{k}(\vec{\nu},z_k) \right\},$$

where, in the two cases (see Chapter 8, Sections 8.3.2 and 8.3.6),

$$\vec{\mathbb{E}}^T_k(\vec{\nu},z_k) = -\frac{\vec{\mathbb{J}}_k(\vec{\nu}) - Y_k\,\vec{z} \wedge \vec{\mathbb{M}}_k(\vec{\nu})}{Y_k - Y'_k},$$

$$\vec{\mathbb{E}}^T_{k+1}(\vec{\nu},z_k) = -\frac{\vec{\mathbb{J}}_k(\vec{\nu}) - Y'_k\,\vec{z} \wedge \vec{\mathbb{M}}_k(\vec{\nu})}{Y_k - Y'_k}. \tag{9.121}$$

$$t'_{kj} = \frac{1}{\prod_{m=k+1}^{j} \left[\cos\delta_m + i(Y'_{m-1}/\tilde{n}_m)\sin\delta_m\right]},$$

$$t_{kj} = \frac{1}{\prod_{m=j}^{k} \left[\cos\delta_m - i(Y_m/\tilde{n}_m)\sin\delta_m\right]}. \tag{9.122}$$

Combining ((9.117) and (9.118), we can write

$$f_F(\nu) = \frac{\nu}{2}\Re\left[\sum_{j=0}^{p}\vec{\mathbb{M}}^*_j(\vec{\nu}).\sum_{k=0}^{p}\vec{\mathbb{H}}^T_{k+1}(\vec{\nu},z_j) - \sum_{j=0}^{p}\vec{\mathbb{J}}_j(\vec{\nu}).\sum_{k=0}^{p}\vec{\mathbb{E}}^{T*}_k(\vec{\nu},z_j)\right]. \tag{9.123}$$

First isolate the terms for which $k = j$ within these sum products, and let $f_0(\nu)$ be the result of the summation over $j$ of just these terms, i.e.,

$$f_0(\nu) = \frac{\nu}{2}\Re\left[\sum_{j=0}^{p}\vec{\mathbb{M}}^*_j(\vec{\nu}).\left\{Y_j\,\vec{z}\wedge\vec{\mathbb{E}}^T_{j+1}(\vec{\nu},z_j)\right\} - \sum_{j=0}^{p}\vec{\mathbb{J}}_j(\vec{\nu}).\vec{\mathbb{E}}^{T*}_j(\vec{\nu},z_j)\right]. \tag{9.124}$$

Substituting the expressions for the fields given in (9.121) into this equation, we get

$$f_0(\nu) = \frac{\nu}{2}\Re\left[\sum_{j=0}^{p}\left\{\frac{|\vec{\mathbb{J}}_j|^2}{Y^*_j - Y'^*_j} - \frac{Y_jY'_j}{Y_j - Y'_j}|\vec{\mathbb{M}}_j|^2 \right.\right.$$

$$\left.\left. -\frac{Y_j}{Y_j - Y'_j}(\vec{z}\wedge\vec{\mathbb{J}}_j).\vec{\mathbb{M}}^*_j - \frac{Y^*_j}{Y^*_j - Y'^*_j}\vec{\mathbb{J}}_j.(\vec{z}\wedge\vec{\mathbb{M}}^*_j)\right\}\right], \tag{9.125}$$

where, in the interests of clarity, we have not mentioned the dependence of the currents on $\vec{\nu}$.

Now consider the quantity $(\vec{z} \wedge \vec{\mathbb{J}}_j).\vec{\mathbb{M}}_j^*$. Since the currents are exclusively tangential, we have

$$(\vec{z} \wedge \vec{\mathbb{J}}_j).\vec{\mathbb{M}}_j^* = \mathbb{J}_{j,x}\mathbb{M}_{j,y}^* - \mathbb{J}_{j,y}\mathbb{M}_{j,x}^* = \vec{z}.(\vec{\mathbb{J}}_j \wedge \vec{\mathbb{M}}_j^*)\,, \qquad (9.126)$$

and similarly

$$\vec{\mathbb{J}}_j.(\vec{z} \wedge \vec{\mathbb{M}}_j^*). = -\mathbb{J}_{j,x}\mathbb{M}_{j,y}^* + \mathbb{J}_{j,y}\mathbb{M}_{j,x}^* = -\vec{z}.(\vec{\mathbb{J}}_j \wedge \vec{\mathbb{M}}_j^*)\,. \qquad (9.127)$$

We then put

$$a_j = -\frac{Y_j Y_j'}{Y_j - Y_j'} \quad ; \quad b_j = \frac{1}{Y_j^* - Y_j'^*} \quad ; \quad c_j = -2i\Im\left[\frac{Y_j}{Y_j - Y_j'}\right]. \qquad (9.128)$$

This allows us to write equation (9.125) in the following condensed form:

$$f_0(\nu) = \frac{\nu}{2}\Re\left[\sum_{j=0}^p \left\{a_j|\vec{\mathbb{M}}_j|^2 + b_j|\vec{\mathbb{J}}_j|^2 + c_j\vec{z}.(\vec{\mathbb{J}}_j \wedge \vec{\mathbb{M}}_j^*)\right\}\right]. \qquad (9.129)$$

In this last equation, note the presence of terms related to the power provided by magnetic currents, then by electric currents, and finally by interactions between electric and magnetic currents: these latter terms become zero if the currents are not correlated. The coefficients $a_j$, $b_j$, and $c_j$ depend on the structure of the multilayer stack and govern the emission of these various sources.

It now remains to include all those terms associated with the sum products in equation (9.123) that do not correspond to the condition $k = j$. Separate the contributions corresponding to $k > j$ from those corresponding to $k < j$, so we can write

$$f(\nu) = f_0(\nu) + f_1(\nu) \text{ where } f_1(\nu) = \frac{\nu}{2}\Re\left[\sum_{j=0}^p\left\{\sum_{k=0}^{j-1} g_{jk} + \sum_{k=j+1}^p g_{jk}'\right\}\right], \qquad (9.130)$$

and

$$
\begin{aligned}
g_{jk} &= Y_j t_{kj}\,\vec{\mathbb{M}}_j^*.[\vec{z} \wedge \vec{\mathbb{E}}_{k+1}^T(\vec{\nu}, z_k)] - t_{kj}^*\,\vec{\mathbb{J}}_j.\vec{\mathbb{E}}_{k+1}^{T*}(\vec{\nu}, z_k)\,, \\
g_{jk}' &= Y_j t_{kj}'\,\vec{\mathbb{M}}_j^*.[\vec{z} \wedge \vec{\mathbb{E}}_k^T(\vec{\nu}, z_k)] - t_{jk}'^*\,\vec{\mathbb{J}}_j.\vec{\mathbb{E}}_k^{T*}(\vec{\nu}, z_k)\,.
\end{aligned}
\qquad (9.131)
$$

Using equations (9.121), we get

$$
g_{jk} = \frac{t^*_{kj}}{Y^*_k - Y'^*_k} \vec{J}_j . \vec{J}^*_k - \frac{t_{kj} Y_j Y'_k}{Y_k - Y'_k} \vec{M}^*_j . \vec{M}_k
$$

$$
- \frac{t_{kj} Y_j}{Y_k - Y'_k} \vec{z} . (\vec{J}_k \wedge \vec{M}^*_j) + \frac{t^*_{kj} Y'^*_k}{Y^*_k - Y'^*_k} \vec{z} . (\vec{J}_j \wedge \vec{M}^*_k); \qquad (9.132)
$$

$$
g'_{jk} = \frac{t'^*_{kj}}{Y^*_k - Y'^*_k} \vec{J}_j . \vec{J}^*_k - \frac{t'_{kj} Y_j Y_k}{Y_k - Y'_k} \vec{M}^*_j . \vec{M}_k
$$

$$
- \frac{t'_{kj} Y_j}{Y_k - Y'_k} \vec{z} . (\vec{J}_k \wedge \vec{M}^*_j) + \frac{t'^*_{kj} Y^*_k}{Y^*_k - Y'^*_k} \vec{z} . (\vec{J}_j \wedge \vec{M}^*_k). \qquad (9.133)
$$

Using notation analogous to that introduced in (9.128), put

$$
\alpha_{jk} = -\frac{t_{kj} Y_j Y'_k}{Y_k - Y'_k} \quad ; \quad \alpha'_{jk} = -\frac{t'_{kj} Y_j Y_k}{Y_k - Y'_k}, \qquad (9.134)
$$

$$
\beta_{jk} = \frac{t^*_{kj}}{Y^*_k - Y'^*_k} \quad ; \quad \beta'_{kj} = \frac{t'^*_{kj}}{Y^*_k - Y'^*_k}, \qquad (9.135)
$$

and finally

$$
\gamma_{jk} = -\frac{t_{kj} Y_j}{Y_k - Y'_k} \quad ; \quad \delta_{jk} = \frac{t^*_{kj} Y'^*_k}{Y^*_k - Y'^*_k}, \qquad (9.136)
$$

$$
\gamma'_{jk} = -\frac{t'_{kj} Y_j}{Y_k - Y'_k} \quad ; \quad \delta'_{jk} = \frac{t'^*_{kj} Y^*_k}{Y^*_k - Y'^*_k}. \qquad (9.137)
$$

This allows us to write equations (9.132) and (9.133) in the following more condensed form:

$$
g_{jk} = \alpha_{jk} \vec{M}^*_j . \vec{M}_k + \beta_{jk} \vec{J}_j . \vec{J}^*_k + \gamma_{jk} \vec{z} . (\vec{J}_k \wedge \vec{M}^*_j) + \delta_{kj} \vec{z} . (\vec{J}_j \wedge \vec{M}^*_k), \qquad (9.138)
$$

$$
g'_{jk} = \alpha'_{jk} \vec{M}^*_j . \vec{M}_k + \beta'_{jk} \vec{J}_j . \vec{J}^*_k + \gamma'_{jk} \vec{z} . (\vec{J}_k \wedge \vec{M}^*_j) + \delta'_{jk} \vec{z} . (\vec{J}_j \wedge \vec{M}^*_k). \qquad (9.139)
$$

Apart from the $'$ exponent, the two expressions are now similar, and we observe that, appearing in all these quantities is a coefficient in $1/(Y_k - Y'_k)$ or in $1/(Y^*_k - Y'^*_k)$. As a result, the situation is similar overall to the situation encountered in Section 9.5 for a single electric current, namely that it is the poles of the reflection coefficient that govern the amount of trapped light.

Finally, note that in the case of uncorrelated currents, i.e., scattering by uncorrelated surfaces (see Chapter 8, Section 8.3.8), the component $f_1(\nu)$ becomes 0.

### 9.6.5 Example of Implementation

We now turn our attention to applying the results obtained in the previous section to a very simple case, namely that of a single 8H layer at 633 nm with uncorrelated interfaces. It is sufficient to calculate the component $f_0(\nu)$ [the component $f_1(\nu)$ goes to 0] and plot the result obtained for the case in which this single layer has a low value imaginary index ($\kappa = 10^{-5}$). This is shown in Figure 9.34 for an illumination angle of 60 degrees.

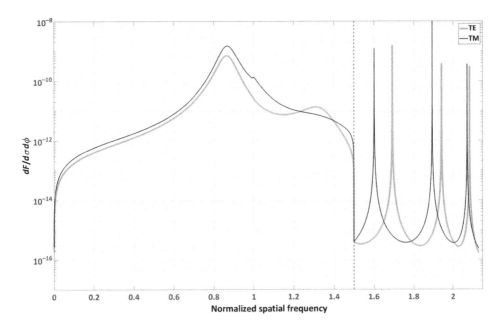

Figure 9.34 Variation of the power spectral density scattered by a single layer as a function of the normalized spatial frequency (gray, TE; black, TM).

We observe three modes per polarization in the modal window, where the power density is maximum. The energy carried by these modes approaches a few percents of the radiative scattering merging in free space in the extreme media. This percentage is much less than that previously calculated for the microcavities, and this is due to the high decrease of the roughness spectrum before the modal window is reached [see Chapter 8, equation (8.133)]. Hence we keep in mind that this trapped scattering will increase with illumination incidence and with the inverse correlation length of the surface. It should also be analyzed in detail for complex stacks.

# Conclusion

Over the past 30 years, the science of electromagnetic optics has clearly made extraordinary progress, drawing on new concepts such as metamaterials, abnormal refractive indices, and slow light; new constructions such as metasurfaces, photonic crystal fibers, nano-antennas, and nano-wires; and new calculation and modeling methods; this is also due to the continuing boom in nanotechnologies. New avenues have surfaced for controlling light with greater and greater precision, structuring matter over one or more dimensions quasi-periodically or in the form of gradients, sometimes even using isolated nano-objects. These developments have been led by communities using vocabularies or types of formalism that are sometimes very different. However, within this cauldron of activity, the notion of **interference** is omnipresent, enhancing or diminishing an electromagnetic field, but also shaping or confining it, spatially, temporally, or spectrally.

For this reason, the **interference filter** plays a central role in this field of ideas, since the simplicity of its planar structure allows the desired optical function to be designed, in particular through the idea of *admittance engineering* (which can be found in other fields of physics). Put another way, the interference multilayer is a response to the limited number of materials available, since this provides the user with an artificial dispersion, spatial and spectral, having many degrees of freedom, nearly all of them controllable. The primary aim of this book is to highlight this remarkable characteristic, and we hope we have achieved this by means of a few judiciously chosen examples.

However, we cannot claim this overview to be exhaustive, and it would be unfortunate if our presentation (reduced to theoretical aspects) left the impression that the area of optical filtering is a closed discipline, in all senses of the word, i.e., without any conceptual interaction with other fields or that excluded the possibility of any new developments.

In the field of **filters design** we might imagine that the next innovations would be related to the control of phase dispersion, albeit constrained by causality's conditions. We might also imagine developing an optical function solely on the basis of its specification for the poles and zeros of its reflection function, or make better use of the homographic nature of the recurrence relation between complex admittances.

The study of **giant field enhancements** in the **total reflection** regime has also highlighted the potential advantages of being able to shape the value of the stationary field in the very structure of the multilayer so as to locally enhance or inhibit the field's value, and hence respond to extremely varied functional requirements. There would also be considerable benefit in adapting to the domain of **microcavities** the approach presented in this book for giant field enhancements, and hence allow the creation of intense planar sources with a controlled emission diagram. Broadening the field enhancement spectral window constitutes yet another challenge, joining the problem of controlling phase dispersion.

As for **light scattering**, this is currently witnessing a strategic development with regard to the incredible levels of performance required today for a number of key components or systems, such as the mirrors for detecting gravitational waves or pixel-based detectors for space instruments. While designing the component, we would expect the specular and scattering properties to be considered jointly, or otherwise try to inhibit scattering in certain spectral windows, provided we know how to model, measure, and control the correlation laws between interfaces. In the very low loss domain, where each ppm must be vigorously pursued, the trapped light balance also becomes essential and must complement that of free space.

To conclude in a very general sense (and assuming it is even possible), it would be valuable if we could associate with each multilayer filter a 2D or 3D structured component that replicated its optical function. Such a result would greatly advance our understanding of the filtering process.

The future for optical thin films is bright!

# Selected Books

*To complete, compare, and go further*

1. Vladimir M. Agranovich and Douglas L. Mills, *Surface Polaritons: Electromagnetic Waves at Surfaces and Interfaces*, Elsevier Science (1982)
2. Roger Petit, *Ondes électromagnétiques en radioélectricité et en optique*, Masson (1989)
3. Alfred Thelen, *Design of Optical Interference Coatings*, McGraw-Hill (1989)
4. Michael R. Jacobson, *Selected Papers on Design of Optical Coatings*, SPIE Press (1990)
5. Carl E. Pearson, *Handbook of Applied Mathematics: Selected Results and Methods*, 2nd ed., Van Nostrand Reinhold (1990)
6. Roger Petit, *L'outil Mathématique*, 3rd ed., Masson (1991)
7. Frederick W. Byron and Robert W. Fuller, *Mathematics of Classical & Quantum Physics*, Dover Publications (1992)
8. Weng Cho Chew, *Waves and Fields in Inhomogeneous Media*, Wiley-IEEE Press (1995)
9. François Flory, *Thin Films for Optical Systems*, CRC Press (1995)
10. Rolf E. Hummel and Karl H. Guenther, *Handbook of Optical Properties: Thin Films for Optical Coatings*, Volume I, CRC Press (1995)
11. Leonard Mandel and Emil Wolf, *Optical Coherence and Quantum Optics*, Cambridge University Press (1995)
12. Sh. A. Furman and Alexander V. Tikhonravov, *Basics of Optics of Multilayer Systems*, World Scientific Publishing (1996)
13. Laurent Schwartz, *Méthodes mathématiques pour les sciences physiques*, Hermann (1997)
14. John D. Jackson, *Classical Electrodynamics*, 3rd ed., John Wiley & Sons (1998)
15. Jean M. Bennett and Lars Mattsson, *Introduction to Surface Roughness and Scattering*, 2nd ed., Optical Society of America (1999)
16. Max Born and Emil Wolf, *Principles of Optics*, 7th ed., Cambridge University Press (1999)
17. Joseph W. Goodman, *Statistical Optics*, Wiley-Blackwell (2000)

18. François Roddier, *Distributions et Transformation de Fourier: à l'usage des physiciens et des ingénieurs*, McGraw-Hill (2000)

19. Ronald R. Wiley, *Practical Design and Production of Optical Thin Films*, 2nd ed., CRC Press (2002)

20. Norbert Kaiser and Hans K. Pulker, *Optical Interference Coatings*, Springer Verlag (2003)

21. Philip W. Baumeister, *Optical Coating Technology*, SPIE Press (2004)

22. Joseph W. Goodman, *Introduction to Fourier Optics*, 3rd ed., W. H. Freeman (2005)

23. Pochi Yeh, *Optical Waves in Layered Media*, 2nd ed., Wiley-Blackwell (2005)

24. Emil Wolf, *Introduction to the Theory of Coherence and Polarization of Light*, Cambridge University Press (2007)

25. Angel Alastuey, Marc Magro, and Pierre Pujol, *Physique et outils mathématiques: méthodes et exemples*, EDP Sciences (2009)

26. John F. James, *A Student's Guide to Fourier Transforms: With Applications in Physics and Engineering*, 2nd ed., Cambridge University Press (2011)

27. George B. Arfken, Hans J. Weber, and Frank E. Harris, *Mathematical Methods for Physicists: A Comprehensive Guide*, 7th ed., Academic Press (2012)

28. Gregory Harry, Timothy P. Bodiya, and Riccardo DeSalvo, *Optical Coatings and Thermal Noise in Precision Measurement*, Cambridge University Press (2012)

29. Cheng-Chung Lee, Kai Wu, and Tzu-Ling Ni, *Optical Admittance Loci Monitoring for Thin Film Deposition: Optical Monitoring for Thin Film Coatings*, Lambert Academic Publishing (2012)

30. John C. Stover, *Optical Scattering: Measurements and Analysis*, 3rd ed., SPIE Press (2012)

31. Michel Lequime and Claude Amra, *De l'Optique électromagnétique à l'Interférométrie: Concepts et Illustrations*, EDP Sciences (2013)

32. Angela Piegari and François Flory, *Optical Thin Films and Coatings: From Materials to Applications*, Woodhead Publishing (2013)

33. Olaf Stenzel, *The Physics of Thin Film Optical Spectra: An Introduction*, 2nd ed., Springer Science+Business Media (2015)

34. Eugene Hecht, *Optics*, 5th ed., Pearson (2016)

35. Olaf Stenzel, *Optical Coatings: Materials Aspects in Theory and Practice*, Springer Science+Business Media (2016)

36. H. Angus Macleod, *Thin Film Optical Filters*, 5th ed., CRC Press (2018)

37. Olaf Stenzel and Miloslav Ohlidal, *Optical Characterization of Thin Solid Films*, Springer Science+Business Media (2018)
38. James E. Harvey, *Understanding Surface Scatter Phenomena: A Linear Systems Formulation*, SPIE Press (2019)
39. Bahaa E. A. Saleh and Malvin C. Teich, *Fundamentals of Photonics*, 3rd ed., Wiley (2019)

# Index